International Conference on Modern Practice in Stress and Vibration Analysis (MPSVA 2018)

Journal of Physics: Conference Series Volume 1106

Cambridge, United Kingdom
2 – 4 July 2018

ISBN: 978-1-5108-7517-3
ISSN: 1742-6588

Printed from e-media with permission by:

Curran Associates, Inc.
57 Morehouse Lane
Red Hook, NY 12571

Some format issues inherent in the e-media version may also appear in this print version.

Printed by Curran Associates, Inc. (2019)

For permission requests, please contact the Institute of Physics at the address below.

Institute of Physics
Dirac House, Temple Back
Bristol BS1 6BE UK

Phone: 44 1 17 929 7481
Fax: 44 1 17 920 0979

techtracking@iop.org

Additional copies of this publication are available from:

Curran Associates, Inc.
57 Morehouse Lane
Red Hook, NY 12571 USA
Phone: 845-758-0400
Fax: 845-758-2633
Email: curran@proceedings.com
Web: www.proceedings.com

Table of contents

Volume 1106

Modern Practice in Stress and Vibration Analysis (MPSVA) 2018

2–4 July 2018, Clare College, Cambridge, United Kingdom

Accepted papers received: 4 October 2018
Published online: 5 November 2018

Computer Techniques

Railway

Fluid/Structural Interaction

Dynamics theory

Experimental techniques

Friction

Manufacturing

Composites

Damping

Crack growth

Damage

IOP Conf. Series: Journal of Physics: Conf. Series **1106** (2018) 011001 doi:10.1088/1742-6596/1106/1/011001

Foreword

The present MPSVA Conference is the 9th in an international series starting in Liverpool in 1992. The conference is well-established for the presentation of research findings in stress analysis, fracture mechanics, elasticity and vibration dynamics with the focus on the state-of-the-art in theoretical, numerical and experimental methodology.

The conference has been organised by the Applied Mechanics Group of the UK's Institute of Physics with the aim to bring together physicists, mathematicians, materials scientists and engineers from academia and industry, and to provide a platform for the presentation of creative and novel research findings, and providing guidance for future research directions.

The MPSVA 2018 Conference was organised at Clare College, Cambridge, the second oldest of the university's colleges, and during the 3 days, 55 papers were presented, creating a forum for discussions and cross fertilisation of ideas across a diverse range of engineering disciplines.

IOP Conf. Series: Journal of Physics: Conf. Series **1106** (2018) 011001 doi:10.1088/1742-6596/1106/1/011001

The Organising Committee would like to thank the authors and the 4 invited speakers for their contribution, and the members of the scientific committee for their effort in reviewing the conference papers.

Finally, we are also grateful to the IoP Conference Department and Clare College staff who provided professional support for the conference organisation.

Cristinel Mares
Brunel University, London
October 2018

Organising Committee

Scientific Committee

IOP Conf. Series: Journal of Physics: Conf. Series **1106** (2018) 011002 doi:10.1088/1742-6596/1106/1/011002

Key note speakers

Janice Barton

Steve Beeby

Keith Worden

Robin Langley

Peer review statement

All papers published in this volume of *Journal of Physics: Conference Series* have been peer reviewed through processes administered by the proceedings Editors. Reviews were conducted by expert referees to the professional and scientific standards expected of a proceedings journal published by IOP Publishing.

IOP Conf. Series: Journal of Physics: Conf. Series **1106** (2018) 012001 doi:10.1088/1742-6596/1106/1/012001

A principled multiresolution approach for signal decomposition

Iason Iakovidis, Elizabeth J. Cross and Keith Worden

Dynamics Research Group, Department of Mechanical Engineering, University of Sheffield , Mappin Street, Sheffield S1 3JD, UK

E-mail: iiakovidis1@sheffield.ac.uk

Abstract. One of the common issues with the Structural Health Monitoring (SHM) of civil infrastructure, is that signals measured from structures are nonstationary. The nonstationarity is the result of uncontrolled variations in the environmental and operational variations (EOVs) in the structure, which result from the fact that monitoring can only be carried out *in situ*. Unfortunately, the EOVs prove to be a problem for damage detection, as they can mask the onset of damage, itself a nonstationary process. The aim of the current paper is to demonstrate a principled, but simple, decomposition method, which can separate signals into stationary and nonstationary components; the stationary component can then be used as a feature for SHM. The method is based on a wavelet/multiresolution analysis of the signal of interest, followed by a sequence of wavelet level-by-level tests for stationary using the ADF (Augmented Dickey-Fuller) test statistic. The decomposition is illustrated using data from the famous SHM campaign on the Z24 Bridge. The paper should be of interest, not just to engineers, but to econometricians, or anyone concerned with nonstationary signal processing.

1. Introduction

In the context of Structural Health Monitoring (SHM), signals obtained by continuous monitoring exhibit variability, both in the shorter and longer terms, which can be associated with the impact of Environmental and Operational Variations (EOVs). This is often manifested in the form of signal nonstationarity and has been observed to mask the nonstationarity of that signals that can be a signature of damage; something that makes damage detection difficult [1]. Within SHM, it is common practice to employ methods such as Statistical Process Control (SPC) charts [2] for damage detection purposes, which require the signal to be stationary. Thus, it is desirable to quantify and project out any nonstationarity introduced into SHM signals by confounding influences such as EOVs.

An important conclusion from the SHM literature is that the impact of EOVs on SHM signals can occur on widely disparate time scales; their quantification and elimination is not a straightforward matter. For such purposes, refined means originated from the field of signal processing have developed throughout the years and are used within SHM. Of particular interest here is the concept of Multiresolution Analysis (MRA) [3], which has been employed in SHM in order to decompose a given SHM signal into band-limited frequency components and evaluate the damage sensitivity of each one [4]. The latter reference also considered the method of *cointegration* [5, 6], as a means of combining multiple nonstationary signals into a stationary residual which could employed as an SHM feature. An important precursor to cointegration is to establish which of the signals or interest are actually nonstationary; this is usually accomplished by using a statistical hypothesis test. The most common test for stationarity of this sort is the Augmented Dickey-Fuller test, which is a type of t-test; the ADF t-statistic is computed for a signal and then compared to a critical value.

In [7], the current authors proposed a simple interpretation of the ADF statistic, based on dimensional analysis. A byproduct of the analysis was to show that there existed a critical frequency f_c for given sample parameters, such that any signal components containing only frequencies below f_c would be judged nonstationary by the ADF test. This result is used here to motivate the definition of a simple, but principled, decomposition method for signals, which can resolve them into their stationary and nonstationary components.

The layout of this paper is as follows; firstly, a brief description of the analysis methods including the MRA, the ADF test and autocorrelation functions is given. The second section provides an illustration of the method on the Z24 bridge data. Finally, brief conclusions are presented.

2. A Signal Decomposition Method based on Wavelet Analysis and the ADF Test

Based on the analysis presented in [7], it can be assumed that one can decompose a signal into its stationary and nonstationary components by separating out the components with frequency content above and below a critical frequency f_c, established using the ADF test criterion. However, in the previous analysis the existence of the critical frequency was only demonstrated for the case of a harmonic signal corrupted by noise; in the general case, the critical frequency would require a more complicated analysis. It will be shown in this section that the decomposition idea can be implemented in the context of wavelet analysis, without an *a priori* calculation of f_c. This will prove important in the context of SHM.

The issue in SHM, is that there will generally be nonstationary (periodic) components in time series at more than one frequency. If one considers the example the Z24 Bridge, Figure 1 shows the fluctuations in natural frequency associated with seasonal variations in temperature over a period of one year. However, other environmental and operational variations (EOVs) generate nonstationarity at other time scales e.g. day-night variations in temperature and traffic loading will create nonstationarity with a daily period; patterns of traffic loading will induce weekly variations as a result of the different loads during the working week. SHM signals will thus require the more general decomposition method alluded to earlier. As mentioned in the introduction, it is necessary to remove all the confounding influences caused by benign EOVs, so that any potential signatures of damage can be detected.

Previous work on confounding influences and cointegration has exploited multiresolution analysis (MRA) and wavelet analysis [3], as a means of decomposing time series into frequency bands so that the EOVs at different timescales were exposed [4]. In particular, a scale-by-scale analysis performed in [4] considered the effects of using cointegration at different time scales. It was shown that the most nonstationary time scales, associated with lower-frequency signal components, manifested greater damage sensitivity than the higher-frequency ones, due to the fact that the onset of damage leads to a stronger reassertion of nonstationarity than in the original signal and correspondingly to a greater effect on the cointegration residual.

The *level decomposition* used in [4], will be used here to motivate the required decomposition into stationary and nonstationary components. In fact, by making use of autocorrelation functions, it will prove quite simple to decompose signals – in a principled manner – into four main components: (*i*) a mean, (*ii*) a stationary part, (*iii*) a non-stationary part and (*iv*) a noise component. In essence, the main idea will be to obtain the MRA signal levels and then assess them using the ADF test to identify those that can be judged as nonstationary and those as stationary. In this way, a *critical level* can be found and used to decompose the original signal into a nonstationary and stationary part. This critical level (denoted L^*) is based on the

IOP Conf. Series: Journal of Physics: Conf. Series **1106** (2018) 012001 doi:10.1088/1742-6596/1106/1/012001

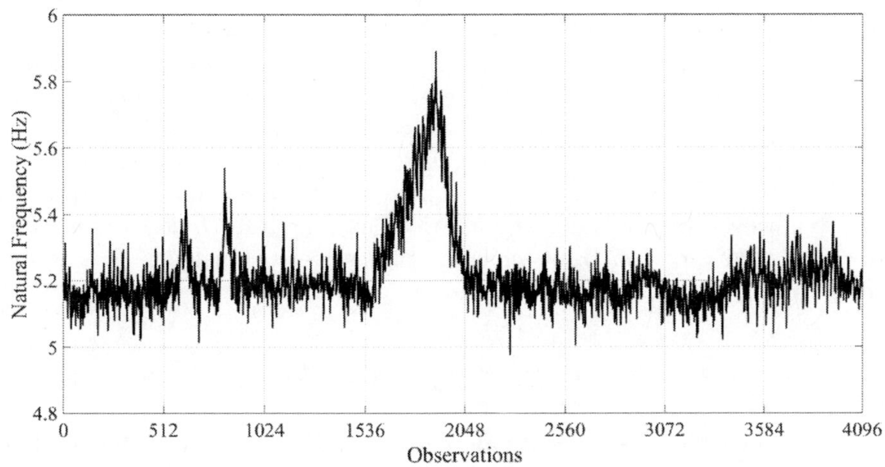

Figure 1. Second natural frequency of Z24 Bridge over a single year of monitoring.

ADF statistic associated with the 99.73% confidence interval, which corresponds to a critical t-statistic of −3.045. The second step in the method proposed is use *autocorrelation functions* (ACFs) in order to test the MRA levels and identify those that are *delta-correlated*, and can therefore be considered to be noise. In order to explain the decomposition, it will be necessary to summarise a little wavelet analysis.

2.1. The Orthogonal Wavelet Transform and MRA

As is now well known, the wavelet transform is a linear transform of the function $x(t)$,

$$x_k^m = \int_{-\infty}^{\infty} x(t)\psi_{m,k}(t)dt \tag{1}$$

into a translation (k) and scale (m) domain [3]. Unlike the Fourier transform, the expansion basis is in terms of functions localised in time, and this allows the wavelet transform to represent nonstationary signals, among its other benefits. If the set of scale and translation parameters are reduced to a discrete set, one obtains the *discrete wavelet transform.* A common choice for the discrete parameters is the dyadic framework in which, $a_j = 2^j$ and $b_{j,k} = k/2^j$. In this framework, the expansion basis functions are defined by,

$$\psi_{m,k}(t) = 2^{m/2}\psi(2^m t - k), \qquad (m, k \in Z) \tag{2}$$

where ϕ is termed a *mother wavelet.* The mother wavelet must satisfy certain properties e.g. it must be localised in the variable t, and this can be ensured by using functions ψ with compact support for example. The conditions on the mother wavelet allow many families of wavelet functions, but the most important families are those which have additional orthogonality properties,

$$< \psi_{m,k}, \psi_{n,l} >= \delta_{mn}\delta_{kl} \tag{3}$$

where $<,>$ is the standard inner product,

$$< h, g >= \int_{-\infty}^{\infty} h^*(t)g(t)dt \tag{4}$$

where the $*$ denotes complex conjugation. The term δ_{mn} represents the *Kronecker Delta function*, which is equal to unity when $m = n$, while zero for $m \neq n$.

With the extra condition on the wavelet functions, the transform in equation (1) becomes the *orthogonal wavelet transform* (OWT). The great advantage of the OWT is that it allows the definition of extremely efficient algorithms for computation, for example Mallat's pyramidal algorithm, which can calculate the expansion coefficients in $O(N)$ time, where N is the number of sample points in the time series of interest. There are various families of orthogonal wavelets that can be used; however, by far the most commonly used are the *Daubechies* families [3]. There do not exist closed forms for the Daubechies mother wavelets; instead, they are computed in terms of scaling functions ϕ defined by recursion relations [8],

$$\phi(t) = \sqrt{2} \sum_{k=0}^{2r-1} c_k \phi(2t - k) \tag{5}$$

where different sets of coefficients c_k yield different families. The order of the family is controlled by the integer r; mother wavelets of higher order are smoother than those of lower order (i.e. more differentiable), but the low-order families have more compact support. The lowest-order mother wavelet in the Daubechies family is the Haar wavelet which essentially yields a square-wave expansion basis.

Once the wavelet coefficients have been computed, one can construct a decomposition in terms of *wavelet levels*,

$$x_m(t) = \sum_k x_k^m \psi_{m,k}(t) \tag{6}$$

where the sum is over all translation parameters at a common scale (which is roughly 1/frequency), and thus provides a decomposition in terms of frequency band-limited time series. The different levels potentially capture different low-frequency components corresponding to EOVs, so the level decomposition is used here as the basis of a new decomposition into stationary and nonstationary parts.

The idea behind the decomposition is quite simple. Having established the level decomposition, one simply applies the ADF test at each level in order to establish if a given level time series is stationary or not. It is clear that one level will capture the critical frequency f_c, so that level – denoted by the index L^* – has a frequency band covering the transition from stationary to nonstationary. Summing the levels up to, and including level L^* will determine the nonstationary component of the original signal (as dictated by the ADF test) and the sum over the remaining levels will yield the stationary component. By applying the ADF test level-wise, one does not need to pre-determine the critical frequency f_c.

As before, for the assessments in this paper, the 99.73% confidence interval on the ADF test is used, which corresponds to a t-critical value of -3.045.

The decomposition proposed here can be made a little more informative, with a little more effort. First of all, the mean of the signal can be extracted before the wavelet decomposition; secondly, one can make a further separation within the stationary component in order to identify a 'noise' component.

2.2. Autocorrelation Functions (ACFs)

According to [9], for a stochastic process X_t for a finite sample, with mean μ and sample variance ${\sigma_x}^2$, which are both time-independent, the autocorrelation function (ACF) ϕ_{xx}, can be defined which determines the degree of dependence between any two samples in the process (signal). For stationary signals, the ACF depends only on the time interval τ between the pair of values considered and not their individual positions in time,

$$\phi_{xx}(\tau) = \frac{E\left[(X_t - \mu)\,(X_{t+\tau} - \mu)\right]}{\sigma_x^2} \tag{7}$$

where E again denotes the expectation operator.

Now, if one defines 'noise' as a component within the signal with *no* temporal correlations (i.e. one cannot predict any future values from past values), one can test for this property using the ACF. A signal will be termed noise, if it is *delta-correlated* i.e.

$$\phi_{xx}(k) = \delta_{k0} \tag{8}$$

with the index k now representing the sample index in terms of discrete time. With the normalisation given in equation (7), one can compute confidence intervals for a zero result; the intervals corresponding to 99.73% confidence are $\pm 3/\sqrt{N}$.

In order to establish the noise component in the signal, one computes the ACF for any levels in the stationary component (stationarity is required in the definition of the ACF) and moves into the noise component any levels which are delta-correlated.

This completes the description of how the new decomposition is defined. To summarise, the steps are as follows:

(i) Remove the mean from the signal $\{x_i : i = 1, \dots, N\}$. The mean μ is the first component.

(ii) Compute the wavelet transform, and determine the wavelet levels. (Note that this may require truncation or zero-padding in order to have 2^M points for the OWT, if N does not satisfy this condition already.)

(iii) Apply the ADF test level by level and determine L^*.

(iv) Resum the levels corresponding to the stationary and nonstationary parts.

(v) If desired, one can calculate the ACFs for the stationary levels and separate out the noise component.

The procedure will now be illustrated on a case study.

3. Case Study: The Z24 Bridge

As the motivation for this work came from the SHM context in the first place, the case study will be taken from data acquired in a major SHM campaign based on the monitoring of the Z24 Bridge. The second natural frequency from the bridge has already been encountered in Figure 1. The Z24 Bridge (overpass) was constructed in Switzerland in the beginning of the 1960s in order to link Koppigen and Utzenstorf by passing over the main national Highway A1 (Lausanne-Zurich). The bridge was superceded by a new construction in the late 1990s, and authorisation was given for an SHM campaign before its demolition in the autumn of 1998. The exercise, which included the introduction of realistic but controlled damage, is described in [10]. Various sensor signals were acquired, these included measurements of the ambient environmental variables like air and deck temperatures, humidity etc. The main features for SHM were modal quantities extracted automatically by stochastic subspace identification (SSI) at regular intervals [11]. The signal examined here will be the second natural frequency series.

From a first look at the signal (Figure 1) one can discern two major components. There is a high-frequency component, which is essentially the uncertainty associated with the SSI estimates; one would expect this component to be stationary. There is also a clear low-frequency trend in the time series that reflects the EOVs experienced by the bridge over the year. The most dramatic features in the signal are the peaks, which are associated with stiffening of the bridge asphalt layer during sub-zero temperature conditions in the winter months [11].

Moving now to the decomposition, Figure 2 shows the levels from the wavelet/MRA analysis. The Daubechies 4th order wavelet was used as a compromise between smoothness and support.

IOP Conf. Series: Journal of Physics: Conf. Series **1106** (2018) 012001 doi:10.1088/1742-6596/1106/1/012001

In total, 12 levels were extracted. Note that levels zero and one both seem to be copies of the mother wavelet; in fact the sum of the two levels carries the mean of the original signal, according to the implementation of the algorithm, which was taken from the book *Numerical Recipes* [12]. Furthermore, the algorithm is applied in such a way that the levels are numbered to increase with increasing frequencies. It is immediately obvious from the figure, that the trend behaviour – from the EOVs – is captured in the lower levels.

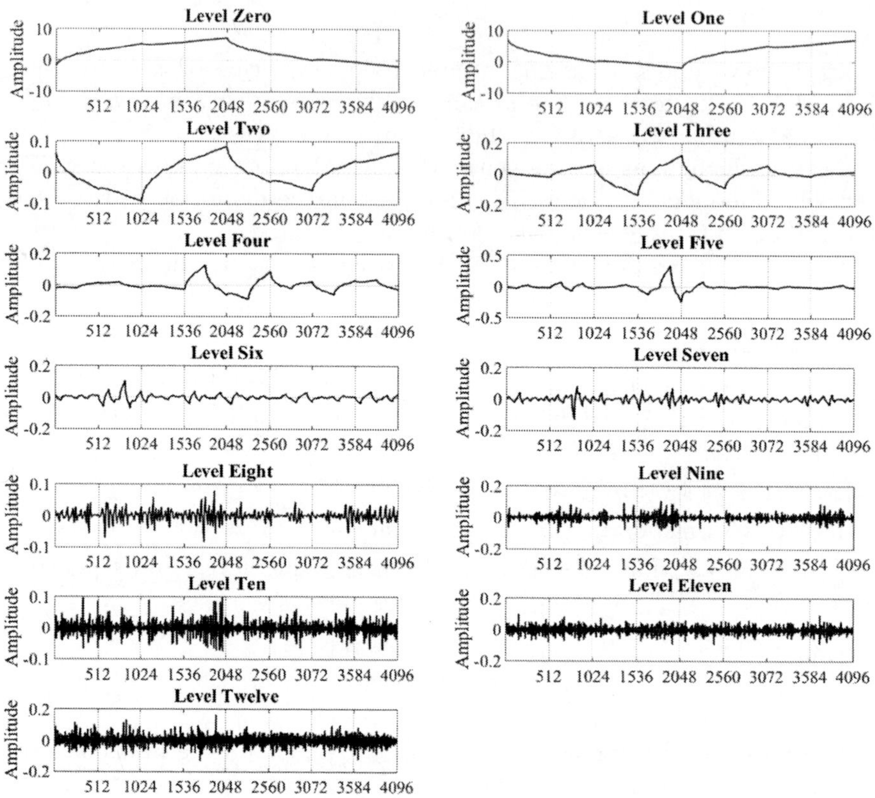

Figure 2. Wavelet decomposition/MRA levels for the Z24 second natural frequency.

The ADF test was then applied to each level in turn, and the statistic values obtained are summarised in Table 1. At the 99.73% confidence level applied here, the critical t-statistic is -3.045; this means that the critical level here is $L^* = 5$, and the sum of levels 0 to 5 is determined as the nonstationary component of the signal.

The final stage of the decomposition is to compute the ACFs for the stationary levels, so that one can separate out the delta-correlated levels as a noise component. The ACFs are given in Figure 3 (the ACFs for the nonstationary levels are also shown here for comparison); the red dashed lines in the figures are the 99.73% confidence levels. In this case, levels 10-12 are considered sufficiently delta-correlated and are separated out as a noise component.

The final decomposition is shown in Figure 4, where there is a clear separation into the nonstationary and stationary components. The further separation into a stationary and a noise component also appears meaningful, with the noise component resembling more the Gaussian white noise sequence one might associate with noise; however, one should recognise that the component is the result of essentially high-pass filtering the original signal, and therefore cannot

Table 1. T-statistics corresponding to different wavelet/MRA levels.

Level	t_{ADF}
0	-0.025
1	0.705
2	-0.1929
3	-1.103
4	-1.501
5	-2.724
6	-5.021
7	-9.0322
8	-15.177
9	-25.852
10	-42.085
11	-87.96
12	-93.58

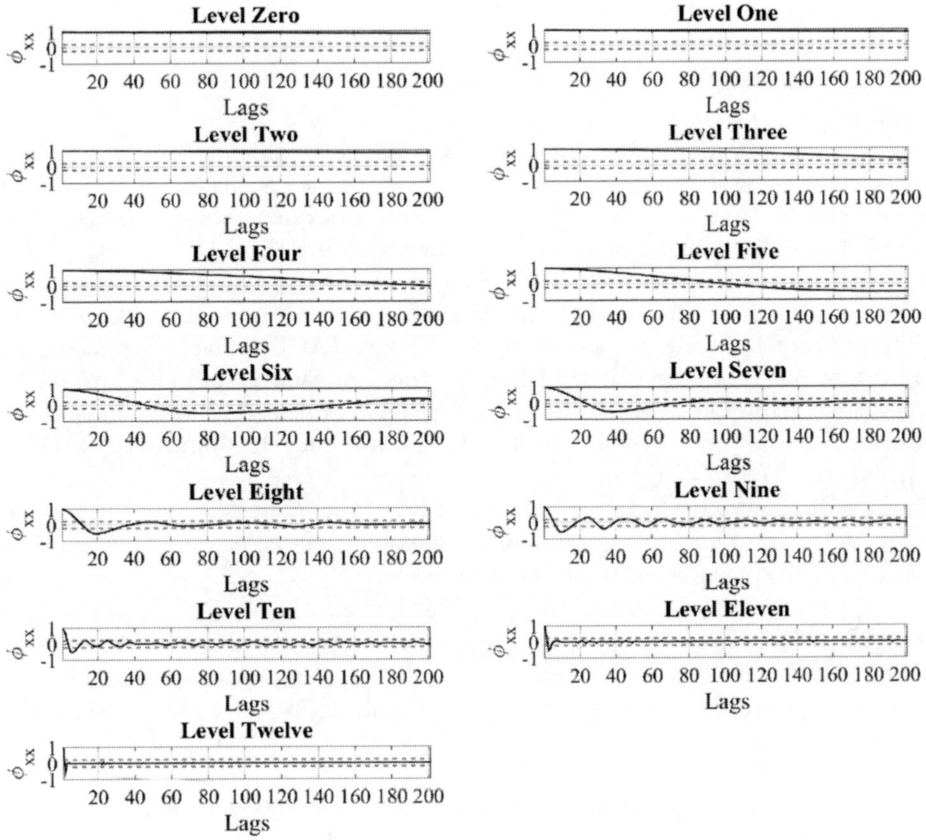

Figure 3. Autocorrelation functions corresponding to each wavelet/MRA level. The red dashed lines indicate the 99.73% confidence interval.

be truly white.

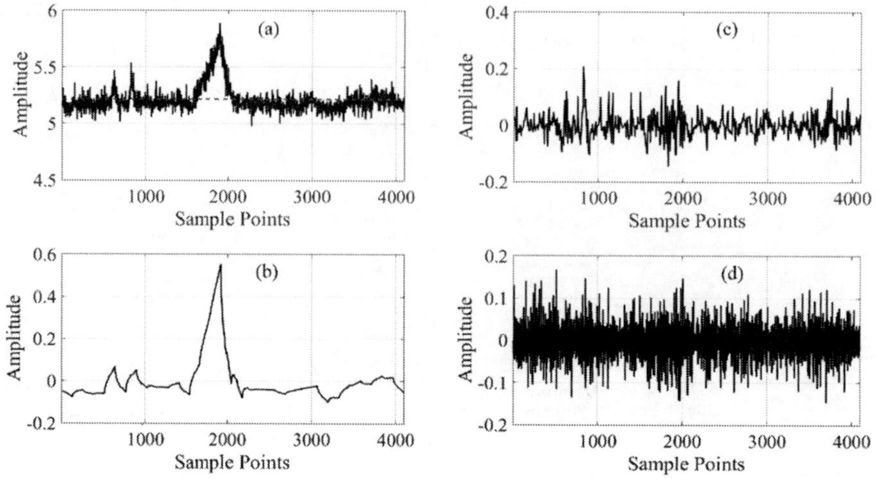

Figure 4. Decomposition of the Z24 second natural frequency: (a) original signal; (b) nonstationary component; (c) stationary component; (d) noise. The mean of the original signal is indicated using a blue dashed line in (a).

4. Conclusions

In this paper, a new decomposition method is presented, able to separate stationary and nonstationary components of signals. The basic idea is that a wavelet/MultiResolution Analysis is used to separate the time series of interest into a set of frequency band-limited wavelet levels. The levels are then individually tested for stationary by using the ADF statistic; the stationary levels are re-summed to give the stationary component of the original series and likewise the nonstationary levels. It is shown that one can also separate a 'noise' component from the stationary component by using autocorrelation functions (ACF). The 'noise' component is not a Gaussian white noise, but is distinguished by the fact that it is delta correlated, so that future values of the component are independent of past values. The decomposition method is demonstrated successfully on a signals from a Structural Health Monitoring (SHM) campaign on a real structure – the Z24 bridge.

Despite the SHM context selected here, the results of this paper are quite general and should be of interest, not only to engineers, but also to econometricians, or anyone seriously concerned with time series analysis where stationarity is an issue.

Acknowledgements

KW would like to acknowledge the support of the UK Engineering and Physical Sciences Research Council (EPSRC) through grant reference numbers EP/J016942/1 and EP/K003836/2.

References

[1] Farrar C and Worden K 2012 *Structural Health Monitoring: A Machine Learning Perspective* (John Wiley and Sons)
[2] Montgomery D 2009 *Introduction to Statistical Quality Control* (John Wiley and Sons)
[3] Mallat S 1999 *A Wavelet Tour of Signal Processing* (Academic Press)
[4] Worden K, Cross E, Antoniadou I and Kyprianou A 2014 *Mechanical Systems and Signal Processing* **47** 243–262

[5] Cross E, Worden K and Chen Q 2011 *Proceedings of the Royal Society of London A: Mathematical, Physical and Engineering Sciences* **467** 2712–2732

[6] Cross E, Manson G, Worden K and Pierce S 2012 *Proceedings of the Royal Society - Series A* **468** 4098–4122

[7] Worden K, Iakovidis I and Cross E 2018 *In Proceedings of the 36th International Modal Analysis Conference, Orlando, FL* (IEEE)

[8] Meyer Y 1993 *Wavelets – Algorithms and Applications* (Society for Industrial and Applied Mathematics Translation)

[9] Box G, Jenkins G, Reinsel G and Ljung G 2015 *Time Series Analysis: Forecasting and Control* (John Wiley & Sons)

[10] Kramer C, de Smet C and de Roeck G 1999 *Proceedings of IMAC 17, the International Modal Analysis Conference, Kissimmee, FL, USA* pp 1023–1029

[11] Peeters B and de Roeck G 2001 *Earthquake engineering & structural dynamics* **30** 149–171

[12] Press W, Teukolsky S, Vetterling W and Flannery B 1992 *Numerical recipes in C++: The Art of Scientific Computing* (Cambridge University Press)

IOP Conf. Series: Journal of Physics: Conf. Series **1106** (2018) 012002 doi:10.1088/1742-6596/1106/1/012002

Normalised curvature square ratio for detection of ballast voids and pockets under rail track sleepers

S Kaewunruen[1,2], R Janeliukstis[2,3], A Freimanis[2,3], K Goto[2,4]

[1]Department of Civil Engineering, School of Engineering, The University of Birmingham, U.K

[2]Birmingham Centre for Railway Research and Education, The University of Birmingham, U.K

[3]Department of Civil Engineering, Riga Technical University, Institute of Materials and Structures, Riga, Latvia

[4]Structural Mechanics Department, Railway Dynamics Division, Railway Technical Research Institute, Tokyo, Japan

E-mail: s.kaewunruen@bham.ac.uk

Abstract. After a railway track experiencing dynamic loading, the track settles and causes ballast to deform, spread and sometime damage. Without appropriate maintenance, void and pocket of ballast underneath railway sleepers can establish overtime and impair the ride quality of train services. In this study, the emphases will be placed on the application of non-destructive vibration-based technology, to investigate and evaluate dynamic characteristics of voided railway concrete sleepers, which are the fundamental element to provide track support to railway systems. The study has developed a curvature-based damage detention method to identify ballast voids under railway track sleepers. This method can be easily deployed in the field by using fibre bragg grating strain sensors to measure strains for curvature analysis. In this study, the assumption is that the time-dependent material degradation negligibly affects the curvature ratios. The dynamic finite element model has been established and validated for railway sleepers in the field. A variety of losses of ballast support have been simulated using the validated model. The dynamic mode shape has been analysed to evaluate curvature ratios under different types of ballast losses. Although the method provides positive outcomes, the advantages, disadvantages and limitation of the method are then identified and discussed.

1. Introduction

Modern railway tracks are constructed using several components grouped into two categories: substructure and superstructure. The substructure includes ballast, sub-ballast, and subgrade, while the superstructure includes sleepers (or sometimes called 'crosstie'), rail pads, fasteners, and rails. Figure 1 illustrates the typical ballasted railway tracks. Rolling stocks usually travel over the rails to transport passengers, goods, etc. The burden of vehicle and transported mass will transfer to axle, to wheel, and then to track structures. The loading conditions acting on railway tracks are normally time dependent since the wheels interacts with rails, causing dynamic effects. The dynamic loads often excite the railway track components with increased magnitudes at specific frequencies associated with such components. It is found that the railway concrete sleepers deteriorate greatly when they are subjected to dynamic loads at their resonant frequencies, especially in flexural modes of vibration [1-8].

Therefore, resonant characteristics of in-situ railway concrete sleepers are essential in their own analysis and design. Based on a number of research studies, it has been found that the resonant vibrations of sleepers affect not only the sleepers themselves, but also the wheel–rail interaction forces [9-16]. The vibrations of sleepers are also affected by the deterioration of ballast support conditions.

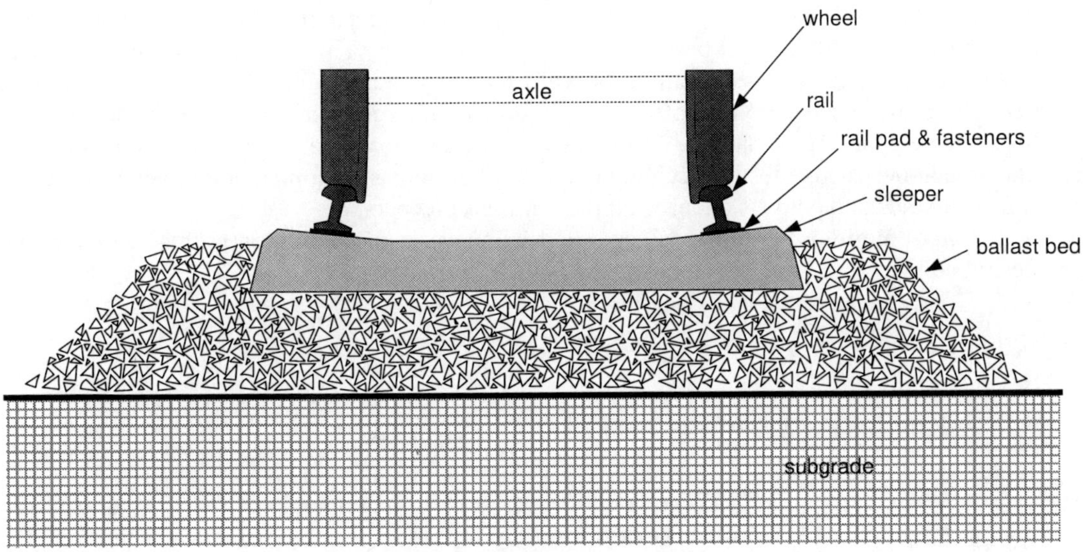

Figure 1. Typical railway tracks

Esveld [3] discovered that the ballast breakage increases substantially at a track resonance, so-called in-phase vibration. This phenomenon causes the voids and pockets, or even the poor packing of ballast support underneath railway concrete sleepers [17-23]. A sleeper having voids and pockets underneath is hereafter referred to as 'voided sleeper'. There are several investigations on free vibrations of in-situ railway sleeper in track system. Based on the literature survey, although the influence of voids and pockets on the dynamic behaviors of local railway concrete sleepers has been investigated, the damage detection and structural health monitoring of the ballast voids and pockets have not been fully established. In addition, although the sleepers are pressed down onto the ballast when a train passes, no one can guarantee whether under any circumstances the voids (at least most of them) are closed and the ballast fully supports the sleepers at the train passages. The fundamental reason of derailment and track overturning can be of numerous theories. One possible way is the high risk of overturning instability of railway tracks whereas the resonances due to void effects could enhance the overturning instability of global tracks [1-3].

This study is the world-first to investigate the potential of normalized curvatures and dynamic mode shapes of railway sleepers to detect ballast voids and pockets. Based on a study by Kliewer and Glisic [24], the normalized curvature ratio can be used to detect damages in concrete beams and girders over bridge. The potential of the method has inspired this study with the aim to provide an alternative method to detect ballast voids and pockets in the field. In this study, the nonlinear finite element model of ballast voids and pockets has been established and calibrated with experimental data. The simulation results are used to explore the possibility of the normalized curvature ratio method and to develop a novel curvature-based approach capable of detecting ballast voids and pockets using free vibration characteristics of the sleepers. The lowest mode of flexural vibration has been considered in this study since free vibration occurs primarily in the first or lowest mode in most cases in real structures [24-26]. The insight into this novel approach will pave itself as a supplement technique that helps rail track engineers evaluate the ballast support conditions, which cannot be visually inspected in practice [26-31].

2. Normalised curvature ratio

This study develops and explores a novel curvature-based method capable to detect ballast voids and pockets under railway sleepers using free vibration characteristics that can be predicted by simulations or measured in the field using either experimental modal testing or operational modal analysis. This novel technique has been inspired by a recent work [24] that depicts a potential to use this method for bridge girders. In general, the modeshapes and curvature modes under free vibration are not expected to change unless the structure experiences irregular behavior. This forms a basis for the key assumption of this curvature-based method. In real life, sleepers are embedded in the ballast and the support condition of the sleepers cannot be inspected or monitored. There is currently no sensor that could directly detect the ballast damage and voids underneath the sleepers. This study aims to develop an alternative non-destructive testing technique that can be used as a supplemental method in the field to detect ballast voids and pockets underneath the rail track sleepers.

The equation of motion of a railway sleeper under transverse free vibration with a small amplitude can be described by:

$$\frac{\partial}{\partial x}\left(kAG_r\left[\phi(x,t)-\frac{\partial w(x,t)}{\partial x}\right]\right)+m_r\frac{\partial^2 w(x,t)}{\partial t^2}=\bar{p}(x,t) \quad (1)$$

$$EI_r\frac{\partial^2\phi(x,t)}{\partial x^2}-kAG_r\left[\phi(x,t)-\frac{\partial w(x,t)}{\partial x}\right]-m_r r_r^2\frac{\partial^2\phi(x,t)}{\partial t^2}+P_a\phi(x,t)=0 \quad (2)$$

where: $w(x,t)$ is the vertical deflection of rail; $\phi(x,t)$ is the rotation angle of rail neutral axis; EI_r is the rail flexural rigidity; kAG_r is rail shear distortion rigidity; m_r is the rail mass per unit length; r_r is the radius of gyration of rail cross-section; P_a is the rail axial force; and $\bar{p}(x,t)$ is the generalized distribute force on the rail. Note that these equations are based on Timoshenko beam theory [32-33]. The curvature (κ) of a sleeper at a point along the sleeper can be written as:

$$\kappa = EI\frac{\partial^2 w(x,t)}{\partial x^2} \quad (3)$$

Based on equation (3), for any sleeper under free vibration in a single mode, the ratio of the curvature of the sleeper at one location and the curvature of the sleeper at another location should remain a constant value. This is true regardless of the boundary conditions of the system and is independent of the amplitude of motion [24]. This ratio between curvatures as normalized curvature ratio (NCR) can be defined by:

$$\mathrm{NCR}_{i,j} = \kappa_i/\kappa_j \quad (4)$$

where κ_i is the curvature of the sleeper at a location i and κ_j is the curvature of the sleeper at a location j. The curvature of the sleeper is based on the lowest bending mode of vibration of the sleepers since free vibration occurs primarily in the first or lowest mode in most cases in a real structure. In this study, a more sensitive curvature square is also considered. The ratio between normalized curvature square (NCSR) can be defined by:

$$\mathrm{NCSR}_{i,j} = \kappa_i^2/\kappa_j^2 \quad (5)$$

For this curvature-based method, either analytical or numerical model can be used for the NCR and NCSR evaluations. If there is a preexisting sensor network installed on the sleepers, the real-time data can be used to detect the ballast voids and pockets. Note that a benefit of the NCR and NCSR methods is that they do not require any correction of data related to temperature change [24].

IOP Conf. Series: Journal of Physics: Conf. Series **1106** (2018) 012002 doi:10.1088/1742-6596/1106/1/012002

3. Nonlinear Simulation of In-situ sleeper

A nonlinear finite element simulation of the in-situ railway concrete sleeper, which was developed earlier in two dimensions and successfully validated [1-2], has been adopted for this study. The railway concrete sleeper is modelled using fifty Timoshenko beam elements with a trapezoidal cross-section while rail pads and ballast systems were modelled as a spring-dashpot element and the elastic beam support features in STRAND7 [34]. The elastic support feature allows the changes in any element along the beam, making it possible to investigate any type of potential void configurations. The standard input data and engineering properties of those components have been adopted from previous investigations [1-2]. The in-situ boundary condition is simplified using spring elements as rail pads connected to rails. Verifications of the model have been done earlier in both free-free and in-situ conditions. The results of either eigenvalues or eigenmodes are in very good agreement with the previous findings in [1, 2]. The results have clearly proven that the finite element model and the simplified approach are capable of predicting the dynamic characteristics of in-situ railway concrete sleepers in a track system [35-36]. Figure 2 shows an example of simulation results.

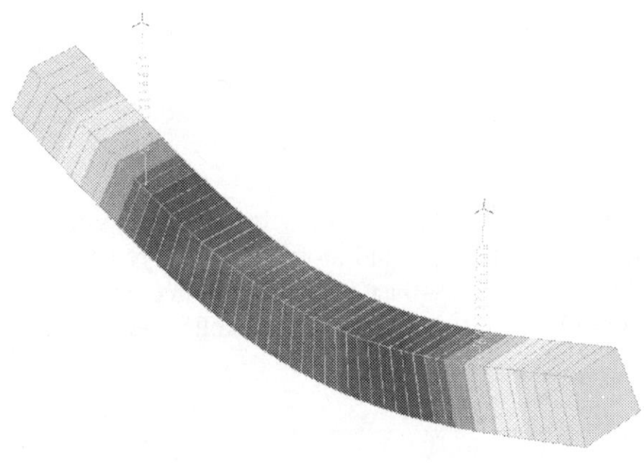

Figure 2. The lowest mode of flexural vibration of a sleeper (135 Hz)

Figure 3. Experimental setup [1]

4. Experimental Study

Commonly, experimental modal analysis (EMA) or modal testing is a non-destructive testing strategy based on vibration responses of the structures. Over the past decade, the modal testing has become an effective means for identifying, understanding, and simulating dynamic behaviour and responses of structures. One of the techniques widely used in modal analysis is based on an instrumented hammer impact excitation. By using signal analysis, the vibration response of the structures to the impact excitation is measured and transformed into frequency response functions (FRFs) using Fast Fourier Transformation (FFT) technique. Subsequently, the series of FRFs are used to extract such modal parameters as natural frequency, damping, and corresponding mode shape. In a wide range of practical applications the modal parameters are required to avoid a resonance in structures affected by external periodic dynamic loads. Practical applications of modal analysis span over various fields of science, engineering and technology. In this study, the modal testing is sound and has been employed in the experiments as shown in Figure 3 [1].

IOP Conf. Series: Journal of Physics: Conf. Series **1106** (2018) 012002 doi:10.1088/1742-6596/1106/1/012002

5. Results and discussion

The comparison between the nonlinear finite element simulations and the experimental results can be demonstrated in Figure 4. In the small range of ballast voids (a: void length; L: the total length of sleeper), only 1-2% difference between experimental and numerical results can be observed. Over the larger range of ballast voids, up to 3-4% of discrepancy can be observed. On this ground, the numerical model is considered in excellent agreement with the experimental studies.

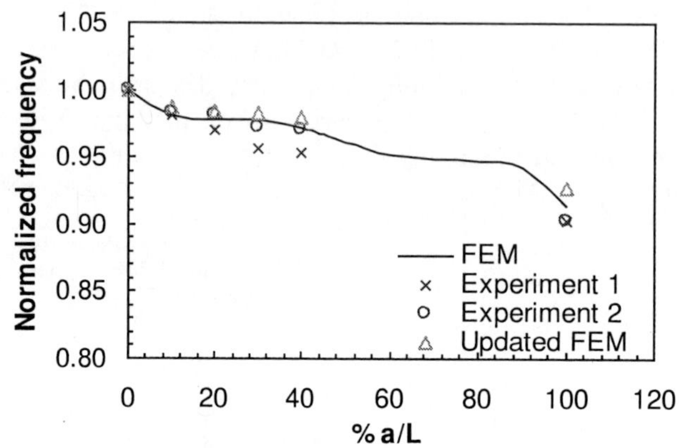

Figure 4. Comparison between numerical and experimental results of voided sleepers

The numerical studies have been further extended to evaluate the free vibration behavior of the sleepers under 5 cases: full ballast support condition; minor ballast void at railseat; severe ballast void at railseat; severe ballast void over midspan; and total loss of ballast (or hanging sleeper). Figure 5 shows the dynamic modeshapes of the sleeper under various cases. It can be observed that small variation of the shapes can be observed.

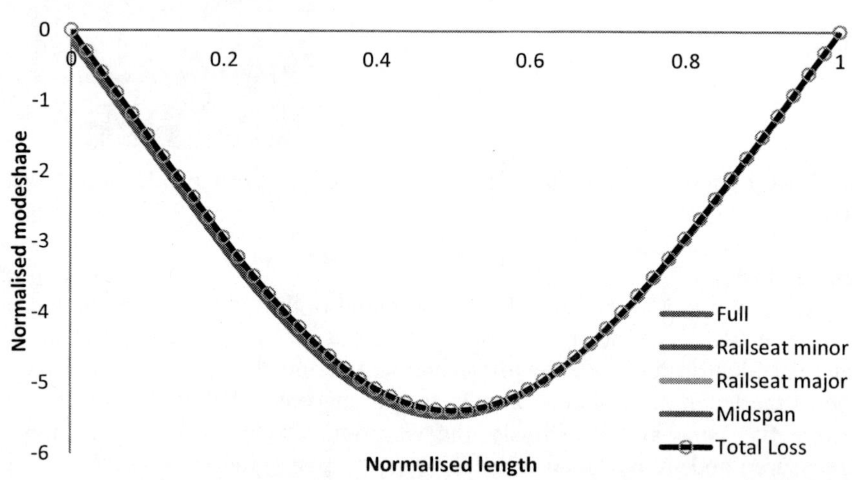

Figure 5. Lowest modeshapes of voided sleepers

The curvatures and curvature squares of the sleepers can be seen in Figure 6. It is clear that the voids and pockets can slightly affect the normalised curvature. However, the curvature squares show relatively more sensitive to the effect of ballast voids and pockets.

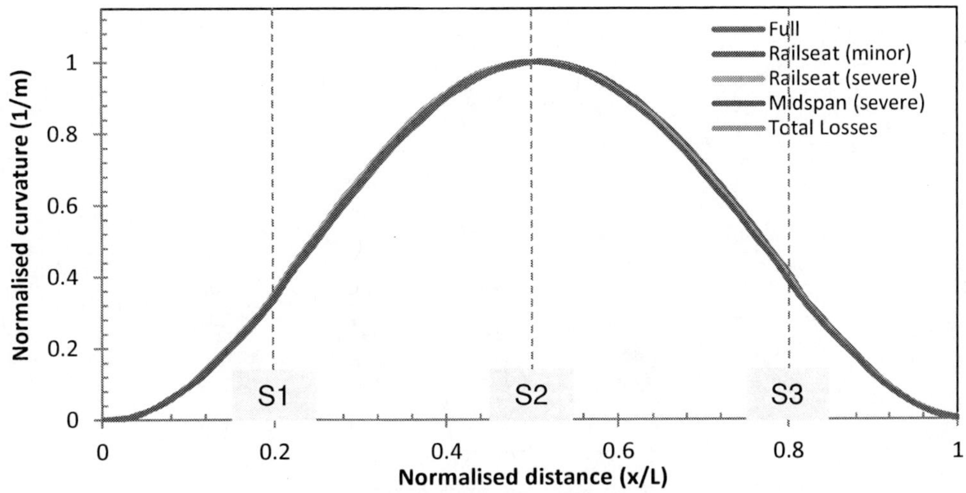

a) normalised curvature

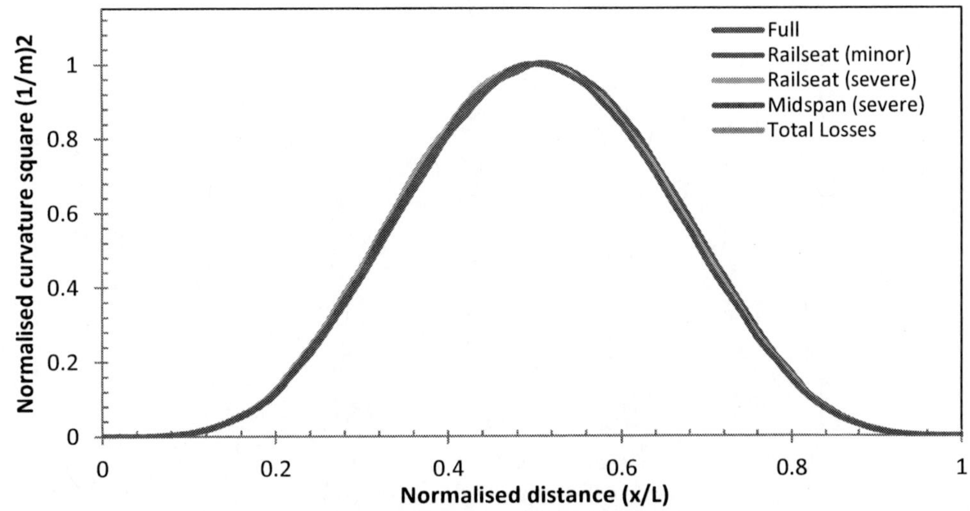

b) normalised curvature square

Figure 6. Lowest bending curvatures of voided sleepers (Location indicators: S1 at railseat; S2 at midspan; and S3 at the other railseat)

It can be observed that the ballast conditions do not significantly affect the curvature or the curvature square of the *in-situ* sleeper. NCR and NCSR can be evaluated as shown in Figure 7. It can be seen that NCR is not sensitive to the ballast voids, when compared with NCSR. Based on this study, it is apparent that NCR and NCSR do not present major precursor for the detection of ballast voids and pockets. It is because the sleepers tend to behave symmetrically under free vibrations. It is clear from the case of total loss of ballast contact. In the total loss case, the sleepers vibrate symmetrically so the NCR and NCSR become less sensitive. This indicates that a given configuration of sensors is not equally sensitive to the damages occurring symmetrically.

However, for the cases of asymmetrical ballast voids (e.g. midspan and railseat loss), NCSR can be seen a more suitable ratio as the precursor used to detect the ballast voids and pockets. The dominate precursor for detecting ballast voids and pockets is S2/S3 (midspan over railseat). This NCSR ratio can be used for direct detection of the ballast voids at midspan and at railseat.

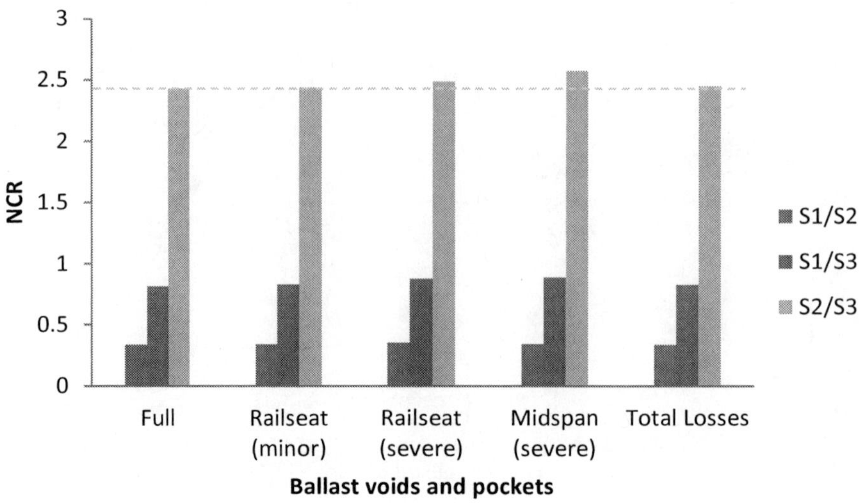

a) normalised curvature ratio (NCR)

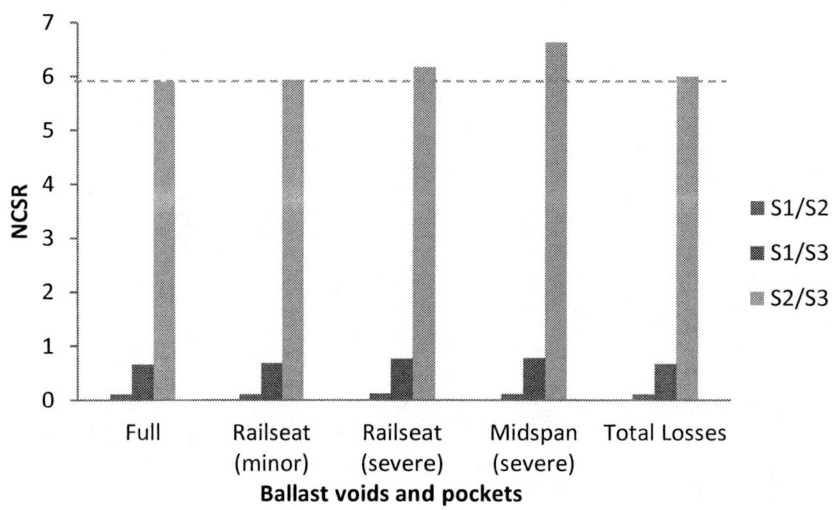

b) normalised curvature square ratio (NCSR)

Figure 7. Lowest bending curvature ratios of voided sleepers

6. Conclusion

This paper presents a novel and simplistic condition monitoring method based on dynamic curvature approach. The method uses the normalized curvature ratio (NCR) and the normalized curvature square ratio (NCSR) as the precursor for detection of ballast voids and pockets underneath the rail track sleepers. The method is initially presented through the analytical study of the railway sleeper using Timoshenko beam theory. The experimental and numerical studies have been performed to validate the nonlinear finite element model. An excellent agreement between the experimental and numerical results can be found. The model has been further extended to evaluate the effect of ballast support conditions for this curvature-based method. The aim of this study is to develop a novel alternative method that could be used as a supplement technique to detect ballast voids and pockets, which cannot be normally inspected visually in practice.

The study found that the curvature-based method is sensitive to asymmetrical damage of ballast conditions. It does not well respond to symmetrical vibrating behavior of sleepers (such as in the case of full support or total loss of ballast condition). In fact, it can be found that the NCSR is a more suitable precursor for damage detection of ballast voids and pockets underneath the rail track sleepers. The insight into this free vibration characteristics can be simply used to guide railway and track engineers to identify voided sleepers in the field using either modal testing or operational responses of the sleepers.

Acknowledgement

The first author wishes to gratefully acknowledge the Japan Society for Promotion of Science (JSPS) for his JSPS Invitation Research Fellowship (Long-term), Grant No L15701, at Track Dynamics Laboratory, Railway Technical Research Institute and at Concrete Laboratory, the University of Tokyo, Tokyo, Japan. The JSPS financially supports this work as part of the research project, entitled "Smart and reliable railway infrastructure". Special thanks to European Commission for H2020-MSCA-RISE Project No. 691135 "RISEN: Rail Infrastructure Systems Engineering Net-work" (www.risen2rail.eu) [37]. In addition, the sponsorships and assistance from CEMEX, Network Rail, RSSB (Rail Safety and Standard Board, UK) are highly appreciated. Financial support from BRIDGE Grant (Collaboration between University of Birmingham and University of Illinois at Urbana Champaign) is also grate-fully acknowledged.

References

[1] Kaewunruen, S. and Remennikov, A.M., "Application of vibration measurements andfinite element model updating for structural health monitoring of ballasted railtrack sleepers with voids and pockets", in Sapri, R.C.(ed), Mechanical Vibration: Measurement, Effect, and Control, New York USA, Nova Science Publishers, 2009, pp.621-644. [Available Online URL http://ro.uow.edu.au/engpapers/502/]

[2] Kaewunruen S., and Remennikov A.M., "Investigation of free vibrations of voided concrete sleepers in railway track system," Proceedings of the Institution of Mechanical Engineers Part F Journal of Rail and Rapid Transit, 221(4), 495-508, 2007. doi: 10.1243/09544097JRRT141

[3] Esveld C.,. Modern Railway Track, second edition, The Netherlands, MRT-Productions, 2001.

[4] Kaewunruen S., and Remennikov A.M., "Applications of experimental modal testing for estimating dynamic properties of structural components," Proceedings of Australian Structural Engineering Conference 2005, Newcastle, Australia, 2005 [CD Rom]

[5] Kaewunruen S., and Remennikov A.M., "Sensitivity analysis of free vibration characteristics of an in-situ railway concrete sleeper to variations of rail pad parameters," Journal of Sound and Vibration 298, 453-461, 2006.

[6] Kaewunruen S., and Remennikov A.M., "Post-failure mechanism and residual load-carrying capacity of railway prestressed concrete sleeper under hogging moment," Proceedings of International Conference of Structural Integrity and Failure 2006, Sydney, Australia, pp. 331-336, 2006.

[7] Kaewunruen S., and Remennikov A.M., "Rotational capacity of railway prestressed concrete sleeper under static hogging moment," Proceedings of 10th East Asia Pacific Conference of Structural Engineering and Construction, Bangkok, Thailand, Vol. 5, pp. 399-404, 2006.

[8] Indraratna, B., Rujikiatkamjorn C., and Salim W., Advanced Rail Geotechnology – Ballasted Track, CRC Press, London, UK, 2011.

[9] Kaewunruen S., and Remennikov A.M., "Effect of improper ballast packing/tamping on dynamic behaviours of on-track railway concrete sleeper," International Journal of Structural Stability and Dynamics 7, 167-177, 2007.

[10] Kaewunruen S., and Remennikov A.M., "Nonlinear transient analysis of a railway concrete sleeper in a track system," International Journal of Structural Stability and Dynamics 8(3), 505-520, 2008.

[11] Kaewunruen S, Remennikov AM. "Influence of ballast conditions on flexural responses of railway concrete sleepers in track systems", Concrete In Australia, *Journal of Concrete Institute of Australia*, 35(4): 57-62, 2008.
[12] Plenge M., and Lammering R., "The dynamics of railway track and subgrade with respect to deteriorated sleeper support," In: K. Popp und W. Schiehlen (Ed.), System Dynamics and Long-term Behaviour of Railway Vehicles, Track and Subgrade, (Springer Verlag, pp. 295-314), 2003.
[13] Remennikov A.M., and Kaewunruen S., "Experimental investigation on dynamic sleeper/ballast interaction," Experimental Mechanics, 46, 57-66, 2006.
[14] Dahlberg T., and Nielsen J., "Dynamic behaviour of free-free and in-situ concrete railway sleepers," Proceedings of International Symposium on Precast Concrete Railway Sleepers, Madrid, Spain, 1991.
[15] Cai Z., "Modelling of rail track dynamics and wheel/rail interaction," Ph.D. Thesis, Department of Civil Engineering, Queen's University, Ontario, Canada, 1992.
[16] Barke D.W., and Chiu W.K., "A review on the effect of out-of-round wheels on track and vehicle components," Proceedings of the Institution of Mechanical Engineers Part F, 219, 151-175, 2005.
[17] Kaewunruen S., Remennikov A.M., Experimental simulation of the railway ballast by resilient materials and its verification by modal testing, Experimental Techniques, 32(4): 29-35, 2008.
[18] Kaewunruen S., Remennikov A.M., Nonlinear finite element modeling of railway prestressed concrete sleeper, Proceedings of the 10th East Asia-Pacific Conference on Structural Engineering and Construction, EASEC 2010, 4, 323-328, 2006.
[19] Gamage, E.K.; Kaewunruen, S.; Remennikov, A.M.; Ishida, T. Toughness of Railroad Concrete Crossties with Holes and Web Openings. Infrastructures 2017, 2, 3.
[20] Ngamkhanong, C.; Kaewunruen, S.; Costa, B.J.A. State-of-the-Art Review of Railway Track Resilience Monitoring. *Infrastructures* 2018, 3, 3. doi:10.3390/infrastructures3010003
[21] Kaewunruen, S.; Wu, L.; Goto, K.; Najih, Y.M. Vulnerability of Structural Concrete to Extreme Climate Variances. Climate 2018, 6, 40.
[22] S. Kaewunruen, "Impact Damage Mechanism and Mitigation by Ballast Bonding at Railway Bridge Ends", International Journal of Railway Technology, 3(4), 1-22, 2014. doi:10.4203/ijrt.3.4.1
[23] Friswell M.I., Mottershead J.E., Finite Element Model Updating in Structural Dynamics, The Netherlands. Kluwer Academic Publishers, 1995.
[24] Kliewer K and Glisic B, Normalized Curvature Ratio for Damage Detection in Beam-Like Structures. Front. Built Environ. 3:50. 2017. doi: 10.3389/fbuil.2017.00050
[25] Remennikov, A., Kaewunruen, S., Determination of dynamic properties of rail pads using instrumented hammer impact technique. Acoustics Australia 33(2): 63-67, 2005.
[26] Kaewunruen, S., Remennikov, A.M. Field trials for dynamic characteristics of railway track and its components using impact excitation technique. NDT & E International 40 (7), 510-519, 2007.
[27] Kaewunruen, S., Remennikov, A.M., Experimental determination of the effect of wet/dry ballast on dynamic railway sleeper/ballast interaction, ASTM Journal of Testing and Evaluation 36 (4), 412-415, 2008.
[28] Kaewunruen, S., Tang, T., Ngamkhanong, C., Aikawa, A., Dynamic properties of flooded railway ballast, International Conference on Modern Practice in Stress and Vibration Analysis (MPSVA 2018), University of Cambridge, 2018.
[29] Kaewunruen, S. and Remennikov, A.M. "Current state of practice in railway track vibration isolation: an Australian overview", Australian Journal of Civil Engineering 14 (1), 63-71, 2016. doi: 10.1080/14488353.2015.1116364
[30] Kaewunruen, S. "Monitoring in-service performance of fibre-reinforced foamedurethane sleepers/bearers in railway urban turnout systems", Structural Monitoring and Maintenance,

1(1): 131-157, 2014. doi: 10.12989/smm.2014.1.1.131

[31] Sae Siew, J., Mirza, O., Kaewunruen, S., Torsional effect on track support structures of rail turnouts crossing impact, ASCE Journal of Transportation Engineering, 2017. doi: 10.1061/JTEPBS.0000004

[32] Kaewunruen, S., Remennikov, A.M. Effect of a large asymmetrical wheel burden on flexural response and failure of railway concrete sleepers in track systems, Engineering Failure Analysis, 15(8): 1065-1075, 2008.

[33] Remennikov, A.M., Kaewunruen, S., Experimental load rating of aged railway concrete sleepers, Engineering Structures, 76: 147-162, 2014.

[34] G+D Computing, STRAND7 Theoretical Manual, G+D Computing Pty Ltd, 2005.

[35] Ngamkhanong, C., Li, D., Remennikov, A.M., Kaewunruen, S. Dynamic capacity reduction of railway prestressed concrete sleepers due to surface abrasions considering the effects of strain rate and prestressing losses, International Journal of Structural Stability and Dynamics, in press, 2018. https://doi.org/10.1142/S0219455419400017

[36] Chiengson C., Kaewunruen S., Aikawa A., Coupling dynamic vehicle-track interaction over coupled short and long wavelength rail defects, The 4th Thailand Rail Academic Symposium (TRAS 2017), Khao Yai, Pakchong, Nakhon Ratchasima, Thailand, August 31 – September 1, 2017. https://www.tras2017.com/

[37] Kaewunruen, S,, Sussman, J.M. and Matsumoto, A. “Grand Challenges in Transportation and Transit Systems”, Frontiers in Built Environment, 2:4, 2016. doi: 10.3389/fbuil.2016.00004

IOP Conf. Series: Journal of Physics: Conf. Series **1106** (2018) 012003 doi:10.1088/1742-6596/1106/1/012003

Long Range Guided Wave Propagation Experimental Analysis in Overhead Power Line Cables under Different Axial Load Levels.

Sergio Malo[1], Makis Livadas[1], Shu-Mei Tan[1], Jamil Kanfoud[1], Tat-Hean Gan[1], Cristinel Mares[2]
[1]Brunel Innovation Centre, Brunel University London, Uxbridge, Middlesex, UB83PH, United Kingdom

[2]Dept. Mechanical, Aerospace and Civil Engineering, Brunel University London Uxbridge, Middlesex, UB8 3PH, United Kingdom

Tat-Hean.Gan@brunel.ac.uk

Abstract. Over the past hundred years, overhead power line cables have been widely installed around the world. These cables are commonly exposed to adverse environmental conditions that can affect their structural integrity and over time, could lead to the complete failure of the structure. This research presents the use of guided waves for the inspection of the structural integrity of the overhead power line cables. The proposed system relies on permanently installing on power line cables a multiple transducers collar as well as a pulse receiver. The system is installed on the cable and performs automated regular inspections. One of the key features of this technique is its ability to inspect a long section of the cable from a single inspection point. To achieve this objective, the wave propagation features have been studied in a wide range of frequencies in 50m long cables, where different collar configurations were used. In addition, power line cables, when installed, are subjected to different axial loads depending on the type of cable and the distance between pylons. The effects of the axial load on the wave propagation have been studied. To assess the defect detection capabilities of the system over a wide range of distances without damaging the cable samples, a metallic clamp is used to introduce a cross-section change on the cables, simulating the effect of a fault on the cables (i.e. corrosion). The experimental results show the highly attenuative effect of the axial load on the wave propagation for most of the frequency spectrum. However, it was found that at low frequencies the system performance allowed the inspection of long distances. This was further proven with the experimental results for the clamp detection study.

1. Introduction

Multi-wire cables are used in many industrial applications in different fields. One of them is their use as overhead power line cables. For decades, these cables have been installed globally and currently the grid covers thousands of kilometers. These cables are exposed to adverse weather conditions and many times they are located close to salty or polluted atmospheres. Over time, these conditions can affect the structural integrity of the cables leading to corrosion or broken strands, and eventually this could lead to the complete failure of the structure.

To avoid unpredicted failures and in order to obtain information about the health of the cables, different inspection techniques have been used. The most common approach has been visual inspection, where an operator assesses the status of the cables looking for broken strands or corrosion. This task is normally performed on foot or with the use of a helicopter. This is a time-consuming and costly task that could represent a hazard for the operator due to the difficulty in accessing the location of some of the cables. Due to this limitation, in recent years, different solutions have been investigated for the inspection and monitoring of the cables. The use of UAV has been introduced to substitute the visual inspection by the HD camera or the use of infrared cameras. The defects or corrosion in the cables increase the temperature of the conductor and the infrared cameras are used to find these hot spots. The battery life of the UAV and the possibility of only inspecting the external wires are two of the main limitations of this technique.

As an alternative, the use of ultrasonic guided waves has been investigated as a possible solution for the inspection or monitoring of these structures. Guided waves testing technique is commonly used for the inspection of the pipeline and other plate like structures. The waves have the ability to travel long distances within the boundaries of the structure. This is an advantage over other techniques because a large length of the structure can be investigated from a single point of inspection.

Based on this, the INTEL-LINE project aims to produce a power line cable monitoring solution using guided waves inspection. The INTEL-LINE system (Figure 1) includes a compact electronic design to generate the high voltage pulses with the necessary wireless communication hardware and software to process and send the collected data. The waves are generated by PZT transducers mounted on a collar specially designed to enhance the generation and propagation of the waves. In addition, an energy harvesting system has been specifically designed to use the magnetic field generated by the high current in the cables as a power source for the system. The system has been integrated in a spherical metallic encapsulation to resist the adverse conditions derived from the power line cables operation, as well as the weather conditions. The experimental results presented in this research were performed as part of the INTEL-LINE project.

Figure 1 INTEL-LINE system

2. Guided waves applied to multi-wire cables

Different investigations have been carried out to study the propagation of the guided waves into the multi-wire cables. These studies focused on different types of cables depending on their applications varying from steel strands [1], bridge stand cables [2], or power line cables [3]. The complexity of the cables makes the study of the wave propagation a challenging topic. Two main factors that highly affect the wave propagation have been identified; their multi-element characteristic, with multiple points of contact between the strands, and the complex helical shape of each of these elements [4]. The effect of these two factors in the wave propagation is being studied for different types of cables [5-6]. As a result of these investigations, it was first concluded that the L(0,1) is the optimal wave mode thanks to its lower attenuation and dispersion properties compared to F(1,1).

In addition to the multi-element condition and helical shape of the cables, another parameter that affects the propagation of the waves for the power line case is their operation under external axial load. Once these cables are installed, the axial load is decided according to a safety factor for their maximum operational axial load. Therefore, this value varies depending on the type of cables. Different investigations have been carried out taking into account the axial load effect on the wave propagation in the multi-wire cables [1]. These results show the attenuation suffered by the waves at different frequencies, however these studies were not carried out on long length power line cables [6-7].

3. Experimental setup

The present experimental case shows the effect of the axial load on the wave propagation in a BEAR cable sample of 50 meters length. The cable is formed of 30 aluminum strands in the two external layers and a core of 7 steel strands, all of them with a diameter of 3.35 mm. The experimental test rig was designed in order to allow the axial load applied to the power line cable sample to be controlled, for up to 600 kg force.

Figure 2 50 meters cable test-rig

In order to study the wave propagation in these structures, two different transducers' configuration were used: pitch-catch to study and pulse-echo. For the pitch-catch configuration, two prototypes of the INTEL-LINE system were used, with a distance between the transmitter (Tx) and the receiver (Rx) of 5 and 20 meters.

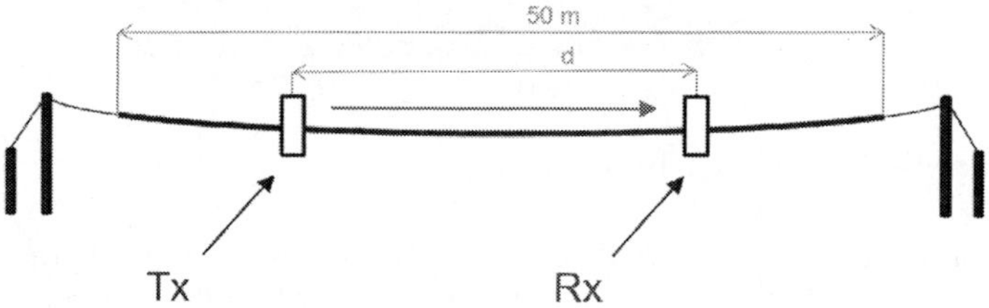

Figure 3 Pitch-catch configuration

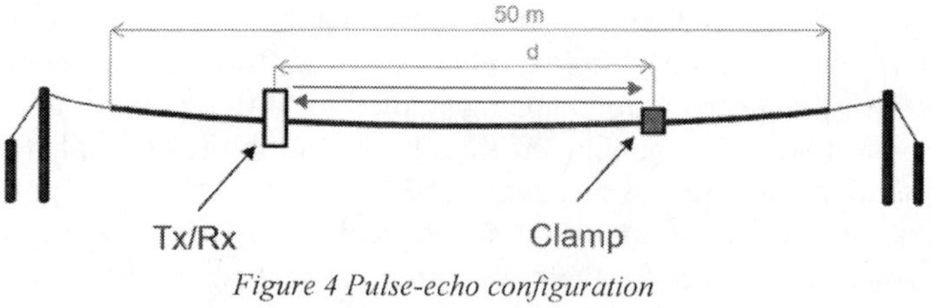

Figure 4 Pulse-echo configuration

In contrast, for the pulse-echo configuration a single unit of the INTEL-LINE system was used (Tx/Rx). To simulate the presence of defects on the cables, without damaging the cables and therefore affecting the structural integrity of the samples, a metallic clamp was used as shown in Figure 5. This clamp also provided flexibility in the experiments due to the fact that it allowed the defects to be located at different positions.

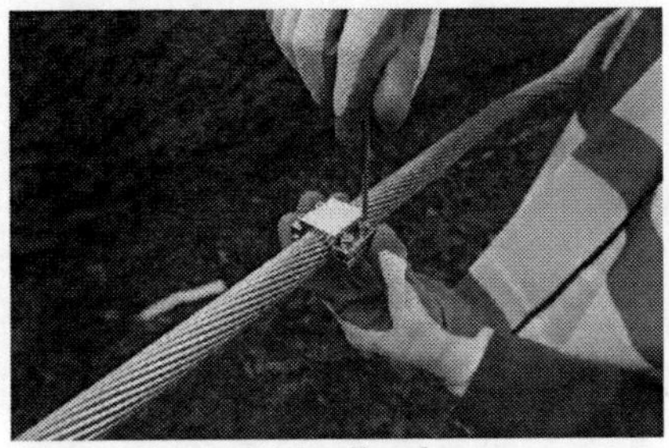

Figure 5 Metallic clamp

4. Results

4.1. Case 1 Wave propagation study under different axial loads with a pitch-catch configuration.

The main objective of this case is to study the amplitude and SNR of the waves travelling throughout the cables. For that a pitch-catch configuration was used, as shown in Figure 3. The initial distance between Tx and Rx was 5 meters. Figure 6 gives the results of the first passing signal for 5 and 10 kHz under different axial loads, [0 – 600 kg force] every 100 kg force. These signals show the high influence of the axial load on the propagation properties of the cables.

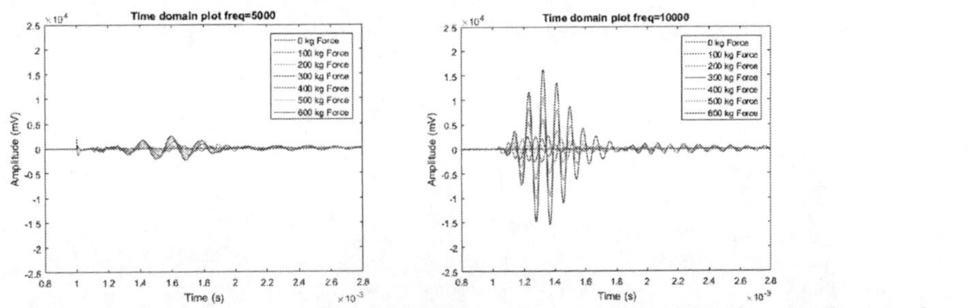

Figure 6 First passing signal at 5 kHz (left) and at 10 kHz (right) for axial loads of [0 – 600 kg force] at 5 meters distance between Tx and Rx.

In order to illustrate this effect, the maximun values of the first passing signals, at 10 and 15 kHz are shown in the following Figure 7. In addition the SNR of the signals is included. Both cases demonstrate how the axial load reduces the amplitude and SNR of the waves up to 200 kg force. For higher values of axial load, the amplitude and SNR increases.

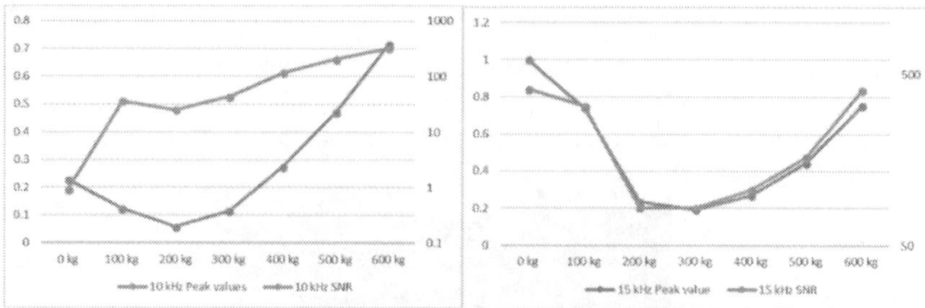

Figure 7 Peak values and SNR of the first passing signals at 10 kHz (left) and 15 kHz (right) at axial load of [0 – 600 kg force]

The following 3D images (Figures 8 and 9) display the effect of the of the different axial load levels on the amplitude and SNR values respectively at axial loads of 0 to 600 kg force with 100 kg force step and a frequency range of 5 to 150 kHz with 5 kHz step.

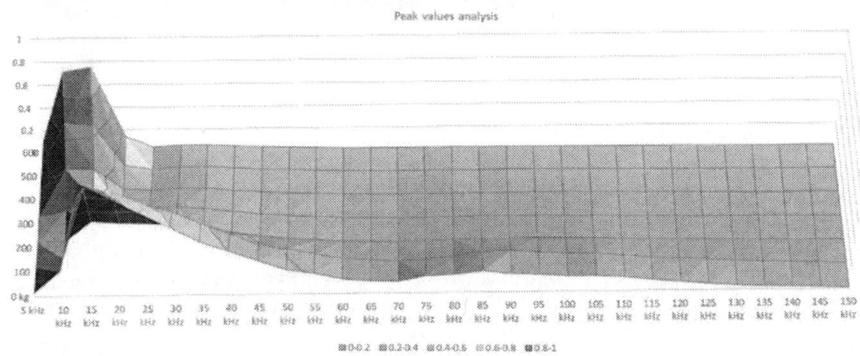

Figure 8 Amplitude values of the first passing signals for axial loads of [0 – 600 kg force] at 5 meters distance between Tx and Rx.

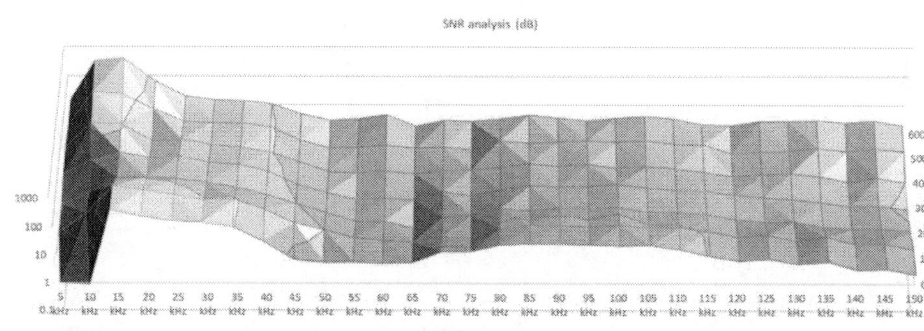

Figure 9 SNR of the first passing signals for axial loads of [0 – 600 kg force] at 5 meters distance between Tx and Rx.

According to Figures 8 and 9 the previous effect shown for 10 and 15 kHz is common for frequencies between 5 to 40 kHz. In addtion, these images also show a shift into the frequency with the highest amplitudes and SNRs to lower frequencies. While under no axial load, the maximum values are obtained between 10 kHz and 35 kHz. Under 600 kg force, this range is concentrated between 5 to 20 kHz. These results compared the first passing signals for a distance of 5 meters, which allowed a wide range of frequencies to be recorded. For a distance of 20 meters it was found that the waves were only visible under an axial load higher than 400 kg force. In order to compare both distances, the peak values and SNR of the first passing signals, under an axial load of 600 kg force, have been included in Figure 10. In this case the analysis focuses on frequencies below 13 kHz, where the maximum amplitudes and SNR levels were found.

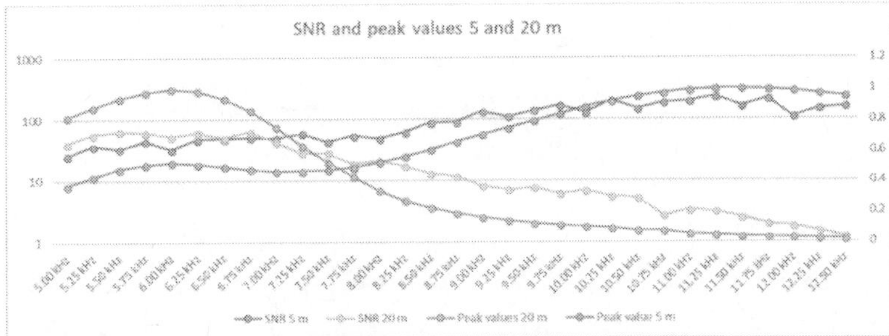

Figure 10 Normalize peak values (left axis) and SNR (right axis) of the first passing signals for axial loads of 600 kg force at 5 and 10 meters distance between Tx and Rx.

Figure 10 shows how, while with a distance of 5 meters the maximum amplitude and SNR levels correspond to the range between 11 to 12 kHz, for 20 meters distance the maximum level is found at 6 kHz.

4.2. Case 2 Wave propagation study for defected cables simulation with a pulse-echo configuration.

After the previous analysis where the most convenient frequency range was found regarding the amplitude and SNR, this second case of study investigates the defect detection capabilities of the technique. To do so, a pulse-echo configuration is used where the waves are reflected by a metallic clamp located at different distances from the INTEL-LINE system. The frequency range used was [6 – 13 kHz] and the axial load was fixed to 600 kg force. One of the limitations of the pulse-echo configuration is that part of the waves generated by the transducers are reflected and collected as part of the received signal. This dead-zone represents a limitation in terms of the study of defects located a short distance from the source of excitation. In order to mitigate the effect of this noise, a baseline subtraction approach was used where the signal from a sample without the clamp attached was used as the baseline. In order to illustrate this method, Figure 11 demonstrates the baseline used and the signal received from the system with the clamp at 8 meters before and after baseline subtraction.

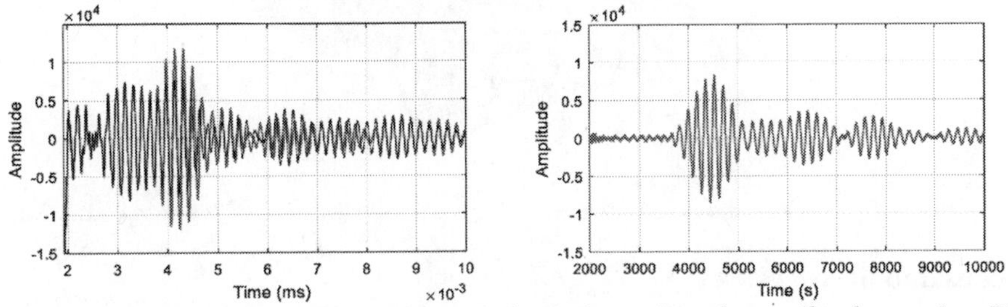

Figure 11 Left: Received signal at 5.75 kHz before (blue) and after (red) attaching the metallic clamp at 8 m. Right: Received signal after baseline subtraction.

This method was used with the clamp attached at different distances for up to 18 m. The results are shown in Figure 12 for a frequency of 10 kHz. This figure demonstrates the attenuation of the reflected waves with the increase in the distance between the excitation source and the clamp.

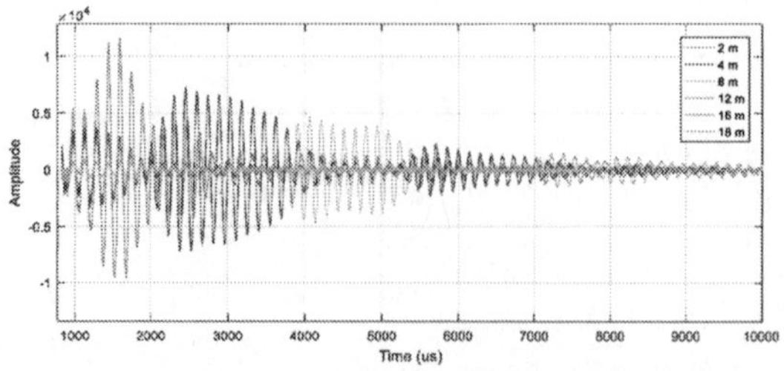

Figure 12 Received signals after baseline subtraction with the clamp at different distances from the Tx at 6.5 kHz

This method was reproduced for the following range of frequencies [6.0 – 13.5 kHz] every 0.5 kHz. The following Figures 13 and 14 show the results of these experiments, the peak value of the reflection, and their SNR respectively.

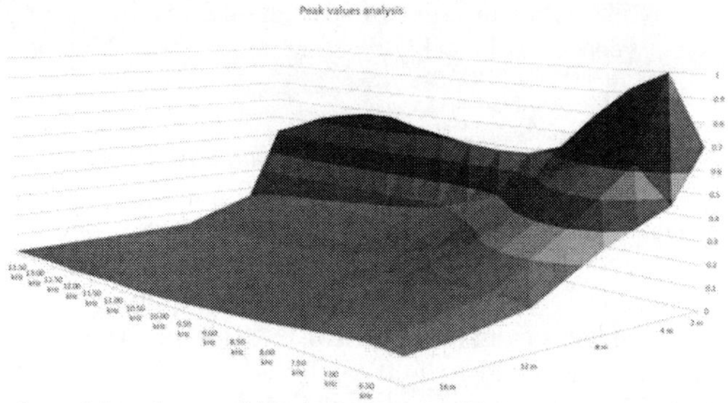

Figure 13 Peak values of the echoes at different distances and frequencies at 600 kg force axial load

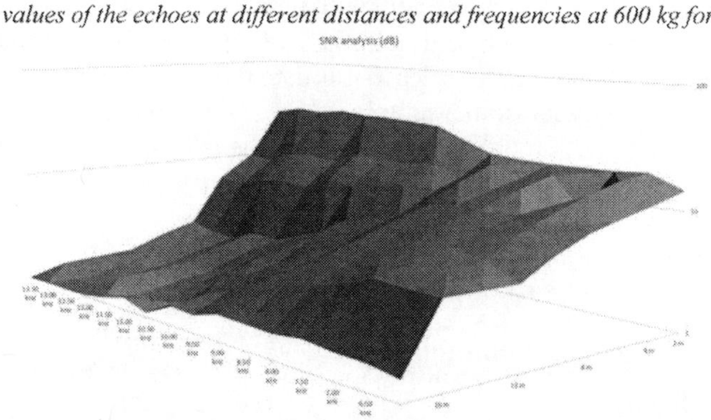

Figure 14 SNR of the echoes at different distances and frequencies at 600 kg force axial load

5. Conclusions and future work

A series of experiments were performed in order to study the wave propagation in power line cables. Two different cases of study were presented. First, we studied the effect of the axial load on the wave propagation. A pitch-catch configuration was used with two different distances between the excitation and reception systems, 5 and 20 meters. For 5 meters distance, the results showed how the axial load have a big influence on the propagation of the waves, heavily attenuating for tensions between 100 – 200 kg forces. For higher tensions of 500 to 600 kg force the opposite effect was demonstrated, where the amplitude values were similar or higher compared to the results without axial load. In addition, with no axial tension, it was found that the maximum amplitudes and SNR corresponded to frequencies between 10 kHz and 35 kHz. However, at 600 kg force these maximum values shifted to lower frequencies [5 - 15 kHz]. Within this range of frequencies, at 5 meters the higher values corresponded to 11 to 12 kHz. In contrast, for a distance of 20 meters, the best frequency was 6 to 7 kHz according to the SNR and amplitude values. These results prove the long range propagation capabilities of the technique.

The second case of investigation's main objective was to study the defect detection capabilities of the technique. In order to study the reflection of the waves at different distances and under axial load without damaging the cables, a metal clamp was used to simulate the defects. Due the use of a pulse-echo configuration and the characteristics of the power line cables with multiple strands, a large dead-zone appeared in the signals. In order to remove this effect, a baseline subtraction approach was used. The results show how the technique was able to increase the SNR of the reflected signals for distances of up to 18 meters. It was demonstrated that the attenuation of the waves varied depending on the frequency. This attenuation is especially severe at higher frequencies [10 – 13 kHz]. In contrast, this frequency range presented higher SNR for distances below 4 meters. This information will be useful when selecting the most convenient frequencies depending on the location of the defects.

The system shows that the waves generated have the ability to travel long distances in cables under axial load and the ability of the technique to detect changes in cross-section characteristic of corrosion. This technique will need to be validated for higher axial loads to reproduce the operation conditions of the cables with the introduction of saw cuts or corrosion into the cables.

Acknowledgments

The experimental results presented in this research are part of the investigation performed as part of the INTEL-LINE project. This work has been carried out in collaboration with different members of the INTEL-LINE consortium, including Kena Jolley from Pi Ltd., Pierre Jost from MCI and Hilal Tusal from Nesne. INTEL-LINE is a fast tracked innovation project funded by the EU under the H2020 framework with a duration of two years. For further information about the project please visit http://www.intel-line.eu.

References

[1] S. Chaki and G. Bourse, "Guided ultrasonic waves for non-destructive monitoring of the stress levels in prestressed steel strands.," *Ultrasonics*, vol. 49, no. 2, pp. 162–71, 2009.

[2] J. Xu, X. Wu, C. Cheng, and A. Ben, "A magnetic flux leakage and magnetostrictive guided wave hybrid transducer for detecting bridge cables.," *Sensors (Basel)*, vol. 12, no. 1, pp. 518–33, 2012.

[3] M. Legg, M. K. Yücel, V. Kappatos, C. Selcuk, and T.-H. Gan, "Increased range of ultrasonic guided wave testing of overhead transmission line cables using dispersion compensation," *Ultrasonics*, vol. 62, pp. 35–45, 2015.

[4] S. Malo Peces, M. Livadas, C. Selcuk, T.-H. Gan, and C. Mares, "Wave Scattering Analysis in an ACSR Cable Using Piezoelectric Transducers," in *EWSHM-8th European Workshop on Structural Health Monitoring*, 2016.

[5] P. Rizzo and F. L. di Scalea, "Wave propagation in multi-wire strands by wavelet-based laser ultrasound," *Experimental Mechanics*, vol. 44, no. 4, pp. 407–415, 2004.

[6] A. Baltazar, C. D. Hernandez-Salazar, and B. Manzanares-Martinez, "Study of wave propagation in a multiwire cable to determine structural damage," *NDT \& E International*, vol. 43, no. 8, pp. 726–732, 2010.

[7] A. Frikha, F. Treyssede, and P. Cartraud, "Effect of axial load on the propagation of elastic waves in helical beams," *Wave Motion*, vol. 48, no. 1, pp. 83–92, 2011.

Parameter selection for model updating based on the global sensitivity method

Zhaoxu Yuan[1], Kaiping Yu[1] and John E Mottershead[2*]

[1] Harbin Institute of Technology (HIT), Department of Astronautical Science and Mechanics, P.O. Box 304, No.92 West Dazhi Street, Harbin 150001, Peoples' Republic of China.
[2] University of Liverpool, Department of Mechanical, Materials and Aerospace Engineering, The Quadrangle, Liverpool L69 3GH, United Kingdom.

E-mail: *J.E.Mottershead@liverpool.ac.uk

Abstract. A new model-updating parameter selection method based on global sensitivity analysis is presented in this work. A specifically designed evaluation function is used for the probability that the sample fits the distribution of test data. In contrast to other parameter selection methods the test-data information is introduced to the parameter selection procedure. Global sensitivity analysis is performed and a set of composite indices for parameter selection is calculated. The parameters are selected based on the values of these composite indices. The method is validated using simulation data from a pin-jointed truss structure model. The cases of independent and correlated parameters are studied and the presented method is shown to be effective for both.

1. Introduction

Finite element model updating [1], [2] is now a widely used technique in many industries, but the problem of selecting a suitable set of candidate updating parameters remains to be fully solved. There have been many studies carried out on this topic [3-5] often making use of local sensitivity data. These approaches are dependent upon initial parameter estimates and are not able to provide an assessment of the global influence of a candidate parameter.

Statistical methods were introduced to overcome this problem, including Bayesian evidence [6] and the F-test [7]. Global-sensitivity based methods are among the most popular [8], [9], but none of these involve the test data, which carries information on the modelling error. In [10] a local sensitivity-based technique was introduced which selects parameters that contribute most to the output test-data covariance terms, but it still needs an initial guess of the parameters and is limited by an assumption of uncorrelated parameters.

The purpose of the present research is to propose a global sensitivity-based parameter selection method, which uses both information from the model and from the test data. This is done by designing an evaluation function, with built-in test data, and based on the multivariate normal distribution probability density function (PDF). Then by using the global sensitivity method, the evaluation function is analysed and the parameters that have a higher contribution to the uncertainty of the model error are chosen as updating parameters. Effectiveness and robustness are validated in two numerical examples of a truss structure with independent parameters and correlated parameters.

2. Parameter selection by using global sensitivity

The parameter selection method is described by the flowchart shown in Figure 1 and details are provided in the following sections.

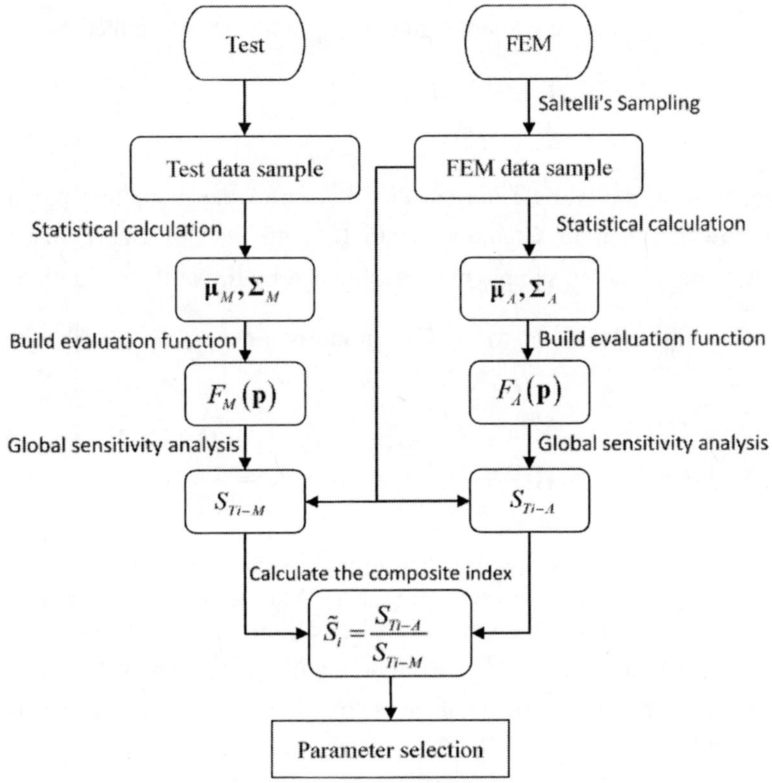

Figure 1. Flowchart of the method.

2.1. Building the evaluation function

Consider an analytical finite element model with input parameter vector $\mathbf{p} = (p_1, p_2, \ldots, p_n)^T$ and output modal frequencies $\boldsymbol{\omega}_A = (\omega^1, \omega^2, \ldots, \omega^k)_A^T$, where n is the number of parameters to be selected and k is the number of output modal frequencies. The model can be described as a function,

$$\boldsymbol{\omega}_A = f_A(\mathbf{p}) . \tag{1}$$

The model output is chosen here to be the set of modal natural frequencies, but could include mode shapes, frequency response functions or any other model output.

The corresponding test data measurement can be described as a random vector for each of k natural frequencies $\left(\boldsymbol{\omega}_M^i\right)^T = (\omega_1^i, \omega_2^i, \ldots, \omega_m^i)_M$; $i = 1, 2, \ldots, k$ obtained typically from tests carried out on m nominally identical structures. To show no preference for any particular output, the samples from finite element analysis and test are scaled with the mean of the test data $\mu_M^i = E[\boldsymbol{\omega}_M^i]$ such that,

$$\mathbf{y}_A^i = \frac{\boldsymbol{\omega}_A^i}{\mu_M^i} = \frac{f_A^i(\mathbf{p})}{\mu_M^i} . \tag{2}$$

$$\mathbf{y}_M^i = \frac{\boldsymbol{\omega}_M^i}{\mu_M^i}. \tag{3}$$

Then the normalized mean $\bar{\boldsymbol{\mu}}_M$ and the covariance matrix $\boldsymbol{\Sigma}_M$ may be determined as,

$$\begin{aligned} \bar{\boldsymbol{\mu}}_M &= \mathbf{I} = (1,1,\ldots,1)_n^T \\ \boldsymbol{\Sigma}_M &= E[(\mathbf{y}_M - \mathbf{I})(\mathbf{y}_M - \mathbf{I})^T] \end{aligned}. \tag{4}$$

Similarly, by sampling from a multivariate normal distribution of the modelling parameters, and in preparation for a later requirement in §2.3, the mean $\bar{\boldsymbol{\mu}}_A$ and the covariance matrix $\boldsymbol{\Sigma}_A$ of the normalized finite element model output sample may be obtained by forward propagation.

Then, from (4) an evaluation function in the form of a log multivariate normal distribution PDF may be constructed as,

$$F_M(\mathbf{p}) = \log \mathrm{PDF}_M(\mathbf{y}_A(\mathbf{p})) = \log\left(\frac{\exp(-\frac{1}{2}(\mathbf{y}_A(\mathbf{p}) - \mathbf{I})^T \boldsymbol{\Sigma}_M^{-1}(\mathbf{y}_A(\mathbf{p}) - \mathbf{I}))}{\sqrt{(2\pi)^k |\boldsymbol{\Sigma}_M|}}\right). \tag{5}$$

The log form of multivariate normal distribution PDF gives a higher gradient when the sample is far from the mean value. This can be helpful in preventing the sensitivity from becoming too small when the samples are far from the mean of the test data. The physical meaning of this evaluation function is the probability that the simulation result $\mathbf{y}_A$ calculated with given parameter vector $\mathbf{p}$ fits the distribution of the test data. Therefore, the parameters that should be updated may be selected by analysing the sensitivity of this evaluation function.

2.2. Global sensitivity analysis

There are many ways of calculating the global sensitivity indices including Fourier Amplitude Sensitivity Test (FAST), the method of Morris, the Derivative-based Global Sensitivity Measure (DGSM) and the Delta Moment-Independent Measure. Here Sobol's sensitivity analysis method (originally developed by Sobol [11] and improved by Saltelli [12]) is adopted for reasons of robustness and widespread usage. All these methods share the same concept of global sensitivity analysis, so any of them may be used in the research presented here.

In the present research, the parameters show an obviously high order effect brought by the interaction of parameters in the finite element analysis. Therefore, the total sensitivity indices will be mainly used. Fortunately, there is a powerful open source Python library (SALib [13]) which provides a convenient way to access the most popular global sensitivity methods [12] and is validated to be effective and efficient. The sampling procedure and the sensitivity evaluation were carried out using the functions available in this toolbox.

2.3. Eliminating the effect of the model

The evaluation function Eq. (5) gives the probability that a sample finite-element output falls within the distribution of the test data. But just as it contains the information from the test data, it also contains information from the finite element model itself, which will affect the selection and must be removed. It has been found intuitively that this can be achieved by the use of a baseline (or reference) function in the same form as the evaluation function, but with finite element data as $F_A(\mathbf{p})$.

Then, to eliminate unwanted model effects a new composite index may be created by dividing the baseline total sensitivity index by the same index from the evaluation function,

$$\tilde{S}_i = \frac{S_{Ti-A}}{S_{Ti-M}}. \tag{5}$$

where S_{Ti-A} and S_{Ti-M} denote the total global sensitivity of the baseline function and of the evaluation function respectively. Then $\tilde{S}_i$ is the composite index that is used for parameter selection. From the geometric characteristics of multivariate normal distribution, the curve tends to be flat when the variance is high. This means that those parameters with high uncertainty will tend to have low sensitivity. This is why the baseline-function sensitivity appears in the numerator and the test function in the denominator, with the result that the parameters to be selected appear with higher values of the composite index than the other parameters.

3. Numerical case study

3.1. Pin-joined truss

As shown in Fig.2, a pin-jointed truss structure of overall dimensions 5m×1m and composed of 21 bar elements, will be used as a simulated numerical case study.

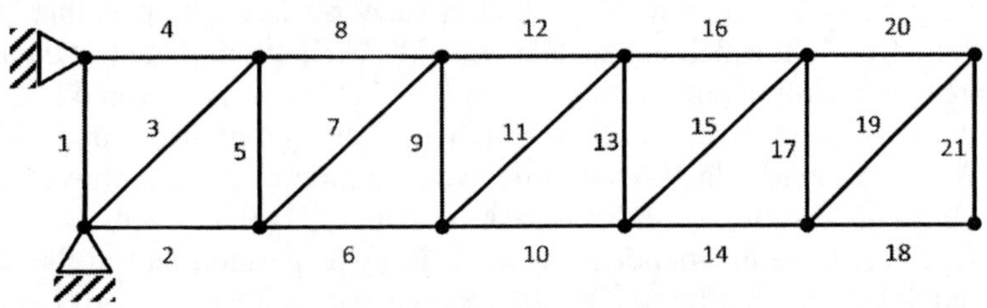

Figure 2. Pin-joined truss.

Each bar element is randomised and considered as a candidate for updating. Therefore, the whole structure is an uncertain system with 21 randomized parameters. Then, in what follows, the mean and standard deviation values of chosen parameters are changed to simulate highly erroneous parameters in the structure. Modal frequency data are used as the nominal measured output and using the parameter selection method presented in §2, the parameters that contribute most to the uncertainty of the structure will become apparent.

3.2. Selection result for independent and correlated parameter cases

The first case is when the parameters to be selected are independent of each other. Then, using numbering of elements as in Figure 3, the diagonal bars listed as [3,7,11,15,19] are chosen as target parameters. The mean and standard deviation values of parameters in this list are changed when generating the nominal test data.

Then in the correlated parameters case, we used the same list of target parameters. The only difference is when generating the nominal test data, the parameters in the list are correlated with each other. The result of the parameters selection is presented in Fig.3 and Fig.4. respectively. We can see in both cases, the parameters in the target list are correctly selected. But in the correlated case, the result is still affected by the correlation of the parameters. As shown in Fig.4, parameters p_3 and p_7 have a very strong correlation, which makes them have lower index values than other parameters. Also, as for the effect of correlation, parameter p_9 may also be mistakenly selected. This is not exactly what we want, but for most model updating problems, selecting a limited extra number of parameters than needed will not have much effect on the updating result. Thus, correlated parameters may also be selected by the method presented in this paper.

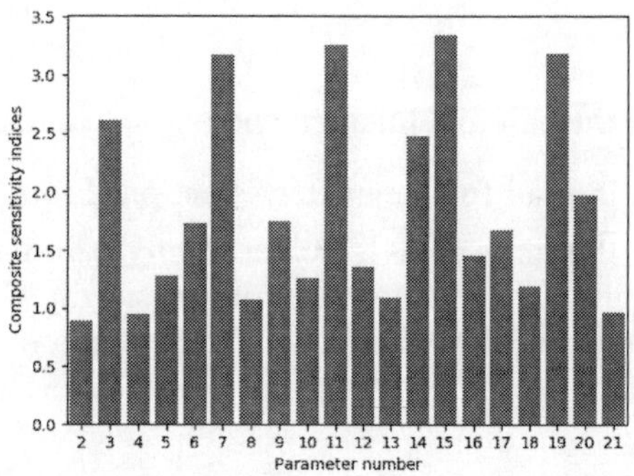

Figure 3. Composite sensitivity indices calculation result.

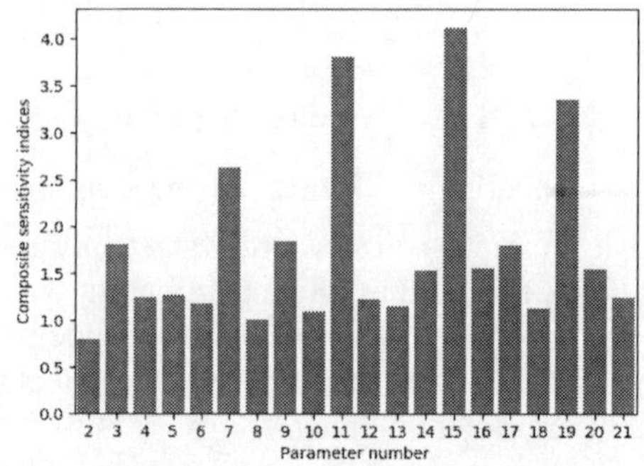

Figure 4. Composite sensitivity indices calculation result for correlated parameters.

4. Conclusions

In this paper, a parameter selection method for stochastic model updating is presented. The method is based global sensitivity analysis and a new evaluation function is developed that makes use of the probability that a sample from finite element analysis fits the distribution of the test data. Two simulation cases studies were presented with parameter selection carried out on a pin-jointed truss structure. In the first study, the uncertain parameters were taken to be independent and in the second the parameters were correlated. All stiffness parameters in the structure were randomized with chosen mean values and standard deviations, but only a subset of these were erroneous. Excellent results were obtained in both cases. When the parameters were independent the selection was perfect, and in the case of correlated parameters, it was found that, in addition to all the erroneous parameters, the correlation resulted in one additional parameter being falsely selected.

Acknowledgments

The research was supported by the National Science Foundation of China (NSFC) under Grant No. 11372084 and the China Scholarship Council (CSC). It was carried out during a visit to the University of Liverpool by Mr Zhaoxu Yuan and was supervised by Professor John E Mottershead.

References

[1] Mottershead J E and Friswell M I 1993 Model updating in structural dynamics: A survey *J. Sound Vib.* **167** 347–75

[2] Friswell M I and Mottershead J E 1995 *Finite element model updating in structural dynamics* vol 38

[3] Friswell M I, Mottershead J E and Ahmadian H 1998 Combining Subset Selection and Parameter Constraints in Model Updating *J. Vib. Acoust.* **120** 854

[4] Mottershead J E, Mares C, Friswell M I and James S 2000 Selection and updating of parameters for an aluminum space-frame model *Mech. Syst. Signal Process.* **14** 923–44

[5] Abu Husain N, Haddad Khodaparast H and Ouyang H 2012 Parameter selection and stochastic model updating using perturbation methods with parameter weighting matrix assignment *Mech. Syst. Signal Process.* **32** 135–52

[6] Mthembu L, Marwala T, Friswell M I and Adhikari S 2011 Model selection in finite element model updating using the Bayesian evidence statistic *Mech. Syst. Signal Process.* **25** 2399–412

[7] Fang S-E and Perera R 2011 Damage identification by response surface based model updating using D-optimal design *Mech. Syst. Signal Process.* **25** 717–33

[8] Saltelli A, Ratto M, Andres T, Campolongo F, Cariboni J, Gatelli D, Saisana M and Tarantola S 2008 *Global Sensitivity Analysis. The Primer*

[9] Sudret B 2008 Global sensitivity analysis using polynomial chaos expansions *Reliab. Eng. Syst. Saf.* **93** 964–79

[10] Silva T A N, Maia N M M, Link M and Mottershead J E 2016 Parameter selection and covariance updating *Mech. Syst. Signal Process.* **70–71** 269–83
[11] Sobol I M 2001 Global sensitivity indices for nonlinear mathematical models and their Monte Carlo estimates *Math. Comput. Simul.* **55** 271–80
[12] Saltelli A, Annoni P, Azzini I, Campolongo F, Ratto M and Tarantola S 2010 Variance based sensitivity analysis of model output. Design and estimator for the total sensitivity index *Comput. Phys. Commun.* **181** 259–70
[13] Herman J and Usher W 2017 SALib: An open-source Python library for Sensitivity Analysis *J. Open Source Softw.* **2**

Machine Learning for Energy Load Forecasting

David Scott[1], **Tom Simpson**[1], **Nikolaos Dervilis**[2], **Timothy Rogers**[2], **Keith Worden**[2],

[1] Department of Mechanical Engineering, University of Sheffield, Mappin Street, Sheffield, S1 3JD, UK

[2] Dynamics Research Group, Department of Mechanical Engineering, University of Sheffield, Mappin Street, Sheffield, S1 3JD, UK

E-mail: `n.dervilis@sheffield.ac.uk`

Abstract. With an increasing penetration of renewables into energy markets, it is desirable to have a flexible grid in order to match large fluctuations in supply to a volatile power output typical of renewable supply. Hence, it is imperative to accurately forecast power load demand. The recent emergence of big data analytics and machine learning techniques have shown great success in a wide range of regression problems in varied industries and various data can be harnessed by the energy industry to better understand likely energy loads placed upon the system. This paper presents a comparison of several regression models which can be used for accurate predictions of energy load given environmental feature data. Here we show that dynamic Gaussian Processes can be used as a powerful tool taking into account the non-stationarity of the data under analysis. This regression model was compared Neural Networks, used most extensively in the industry, and linear regression models to give an idea of their comparable accuracy. However, it was noted that the dynamic Gaussian Process were inferior to a Neural Network when training for huge datasets due to their high relative computational cost, increased uncertainty with projection time, and large memory usage. Though primarily used for dynamics problems, there are a range of non-stationary problems which could benefit from the use of a dynamic Gaussian Process of which this paper just presents one. It also considers online learning models be used for real time forecasting.

1. Introduction

1.1. Energy Load Forecasting

Accurate prediction of grid electricity demand is key to the operation of a power utility company. Load forecasting is essential to ensure that sufficient capacity is available to maintain system frequency whilst minimising the amount of redundant generation and the use of backup peaker plants [1]. The increasing penetration of variable renewable energy sources in utility generation is resulting in an increased requirement for grid flexibility and accurate predictives, especially if the eventual goal is the complete phasing out of fossil fuel generation [2].

Load forecasting falls into two distinct categories: long term forecasting and short/medium term forecasting [1]. Long term forecasting is principally interested in predicting multi-year trends in electricity demand which are to influence decisions regarding infrastructure investment. This requires a lot of contextual information regarding governmental policy, economic growth and industrial activity etc. within the country of interest. Conversely, short-medium term forecasting focuses on the prediction of power demand on a day to day basis in order to inform the management of operational assets. This short/medium term forecasting is of greater importance when it comes to maximising the performance of renewable energy assets and is the principal interest of this paper.

1.2. Machine Learning

Machine learning is the name for a set of mathematical techniques which can be applied to data to find patterns within it. Advances in computers have allowed these computationally expensive methods to be used on vast datasets where patterns may be obscured by the noise or scale of the data. Patterns can be used to form models to make predictions on new data.

In supervised learning, one is given $(x_1, y_1), (x_2, y_2)(x_n, y_n)$ and must find the relationship $f(x) = y$ which takes a new datapoint $\mathbf{x}_i$ and produces a prediction of y_i [3]. In regression, y_i is a point whilst in classification problems, it is a class. In this paper, regression techniques are used to predict the energy load of an area in the UK given environmental data and other information relevant to energy use such as school holidays.

2. Models

2.1. Linear Regressions

A linear regression model fits a function to data using the squared euclidean distance metric to optimise the functions fit to training data. In the most simple model, $y = \beta_1 x + \beta_0$, a straight line is fitted to the data with gradient β_1 and y-intercept β_0. With larger numbers of features, this becomes a hyperplane.

$$y = \beta_0 + \beta_1 x_1 + \ldots + \beta_k x_k \tag{1}$$

Interaction terms $x_i x_j$ can be included to map the relationship between terms x_i and x_j. This yields better results than a simple linear regression if the effect of x_i on y depends on the other predictors $x_{j,j\neq i}$.

$$y = \beta_0 + \beta_1 x_1 + \ldots + \beta_k x_k + \beta_{12} x_1 x_2 + \ldots + \beta_{1k} x_1 x_k + \ldots + \beta_{(k-1)k} x_{k-1} x_k \tag{2}$$

In a quadratic model, a quadratic function is fit to the data and optimised in a least squares manner. Consequently, the model contains an intercept, linear terms, interactions, and squared terms as visualised in Equation 3.

$$y = \beta_0 + \beta_1 x_1 + \ldots + \beta_k x_k + \beta_{12} x_1 x_2 + \ldots + \beta_{1k} x_1 x_k + \ldots + \beta_{(k-1)k} x_{k-1} x_k + \beta_{11} x_1^2 + \ldots + \beta_{kk} x_k^2 \tag{3}$$

2.2. Gaussian Processes

Gaussian process regression is a non-parametric regression technique widely used for approximating arbitrary non-linear functions. It has several desirable characteristics that make it rather suitable for the problem studied herein. Firstly, being non-parametric it is not required that any structure be imposed on the data as is necessary in polynomial regression with regards to the number of terms considered or in neural-networks in which network architecture must be selected. Secondly, Gaussian process regression is derived purely from well known rules for combining normal distributions using Bayes theorem. As a result it naturally provides error bounds for the predictions made by the algorithm, for any practical application the error bounds are likely almost equally important as the accuracy of the prediction itself.

A Gaussian process regression predicts a probability distribution for a given input vector based on previous training input vectors and associated responses. The predicted distribution is given by Equation 4 as derived in detail by Rasmussen [4]. That is the probability distribution of the predictor function f_s at points of interest X_s given the training values of X and y, is a Gaussian distribution with mean vector $\sigma_n^{-2} X_s^T A^{-1} X y$ and covariance matrix given by

$X_s^T A^{-1} X_s$. In these equations σ_n is the assumed standard deviation of the noise in the measurement values and Σ_p is the covariance function.

$$P(f_s|X_s, X, y) = \mathcal{N}(\sigma_n^{-2} X_s^T A^{-1} X y, X_s^T A^{-1} X_s) \tag{4}$$

$$A = \sigma_n^{-2} X X^T + \Sigma_p^{-1} \tag{5}$$

Gaussian process regression models can also be naturally extended to take previous function values as input known as auto-regressive exogenous Gaussian process models. Further details on auto-regressive exogenous models are given in Section 2.4.

2.3. Neural Networks

Loosely based on the brain, the neural network describes an iterative procedure by which a collection of units known as neurons transmit and process signals in order to create an associative mapping of $X \mapsto Y$. The method has become popular with the advent of huge quantities of computational power which have enabled large combinations of equations to be solved quickly.

Each node is a non-linear activation function, Equation 6, each of which are arranged in layers through which a signal moves through an input layer, passes through various hidden layers, and finally is transmitted to an output layer. The activation function describes the movement from layer j to layer i with a weight w_{ij} which are summed at node i to be passed through the activation function.

$$x_i = f\left(\sum_j w_{ij} x_j\right) \tag{6}$$

A learning procedure is carried out to optimise the mapping by changing the weights at each neuron to best describe the general mapping of $X \mapsto Y$ without overfitting. There are various methods of doing this; the most simple of which is Gradient Descent. After a training procedure, the neural network should accurately predict new y values given new x values. More details about the process of neural networks can be found in [3].

2.4. Auto-regressive with Exogenous Variables Models

An auto-regressive model is a regression model which takes as input the previous n values of the system response, where n is the order of the model, and predicts the system value at the next time step as a function of these previous responses. These models can be very powerful in the case of dynamic systems in which current and previous system states can have considerable impact on future values.

$$y(t_i) = f(y(t_{i-1}), y(t_{i-2}), \ldots, y(t_{i-n})) \tag{7}$$

Auto-regressive, exogenous models combine a more traditional regression model with auto-regressive models. These models use both normal input vectors, for example in this case: the temperature data, and previous predictions made by the model.

$$y(t_i) = f(x_1, x_2, \ldots, x_m, y(t_{i-1}), y(t_{i-2}), \ldots, y(t_{i-n})) \tag{8}$$

A key parameter of auto-regressive models which must be stipulated is the number of previous model predictions taken as inputs, the order of the model. This must be done with care as similarly to most regression algorithms there is a trade-off between model flexibility and over-fitting of the training data.

3. Methods

3.1. Preprocessing

The data were scaled to have a zero mean and unit variance to ensure that scaling effects of data with different magnitudes did not affect regression performance. Categorical data were converted to numerical data and the date information was separated into three features indicating day of the week, day of the year, and year reflecting the seasonality of the data. Analysis showed that this separation doubled performance accuracy when using a dynamic Gaussian Process $n = 7$, reducing the prediction performance error from 0.45% to 0.26%.

3.2. Optimisation of NARX Gaussian Process

An investigation into the optimal number of previous time steps used was performed on the data varying n between 1 and 10, visualised in Figure 1. It was found that 7 time points was the optimal number for the dataset under investigation.

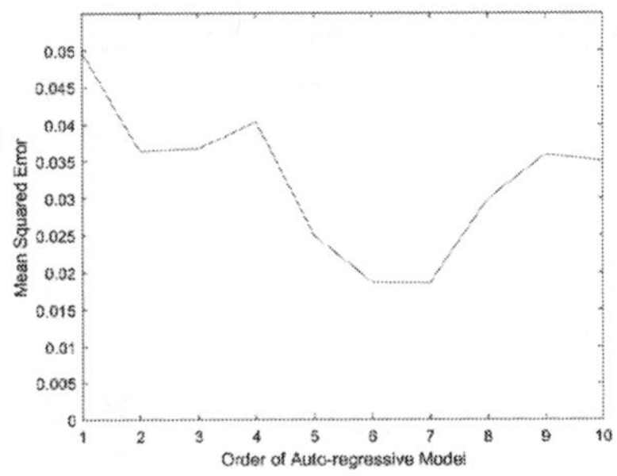

Figure 1. Optimisation of number of previous time steps used in model.

3.3. Optimisation of Neural Network

The neural network's shape was difficult to optimise given the high variability of the function to fall into local minima given random initial weights. To mitigate these circumstances, a Bayesian neural network was used with a single layer of 10 neurons.

4. Results

Table 1. Results of different regression methods on energy load data.

Method	Accuracy (%)
Linear Regression	13.78
Linear Regression with Interactions	10.80
Linear Regression (Quadratic)	6.36
Neural Network	28.15
NARX Gaussian Process	0.26

It is clear from this table that the NARX Gaussian process significantly outperformed all other algorithms trained on the training data due to its ability to use the previous states as predictors for future states. A detailed examination of the performance of all the algorithms is given in the following sections.

4.1. Linear Regression

The three linear regression models increased in accuracy as their complexity increased, a predictable result given the clear non-linearities in the dataset.

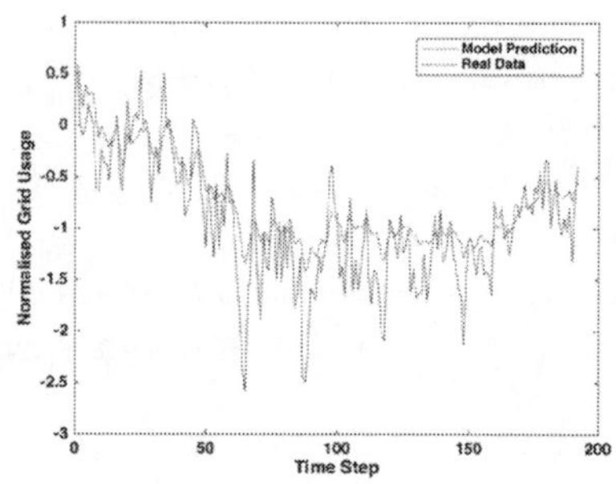

Figure 2. Linear regression.

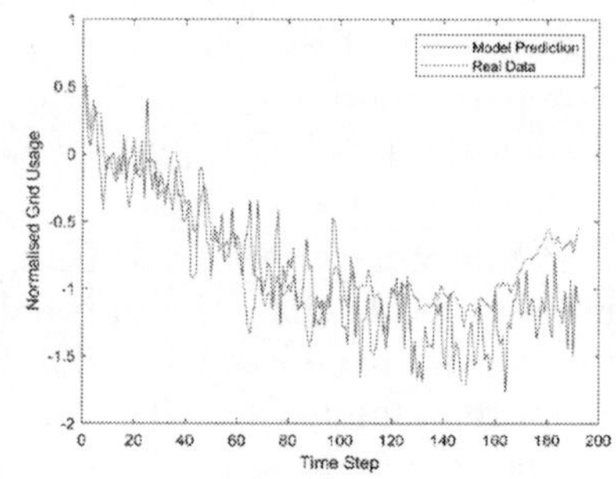

Figure 3. Linear regression with interaction terms.

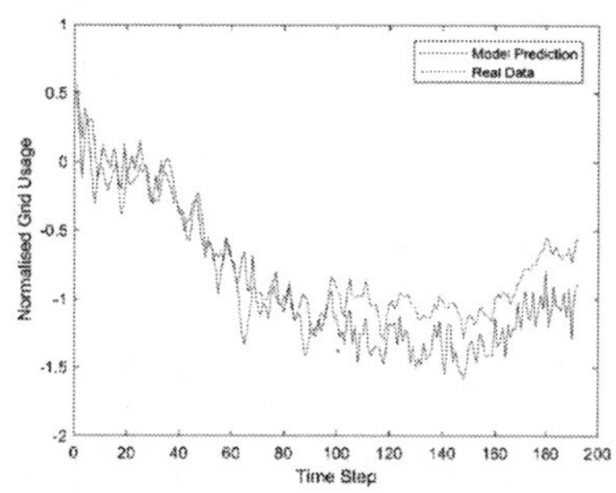

Figure 4. Linear regression (quadratic).

4.2. Neural Network

The neural network had a large degree of variability due to random initial weights as shown by Figure 6. With a larger training data set, the network could likely generate more accurate predictions. Using an auto-regressive network would also be likely to boost performance.

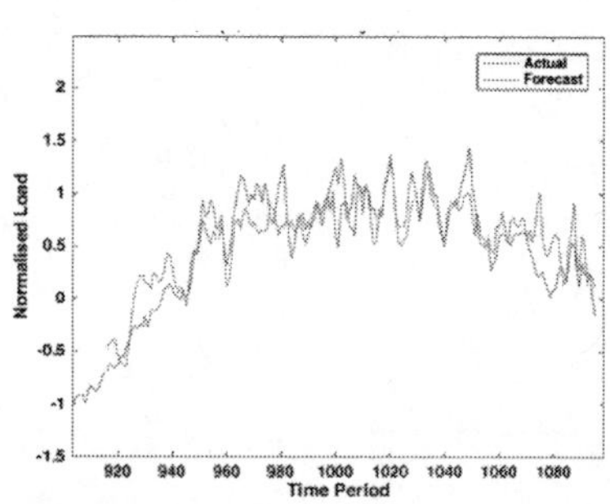

Figure 5. Neural Network regression.

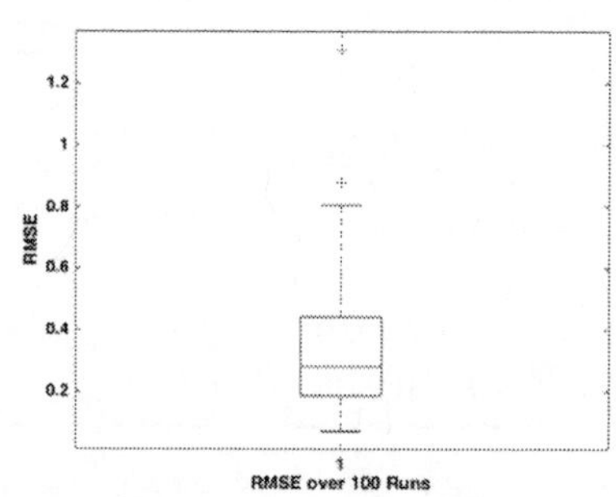

Figure 6. RMSE of Neural Network over 100 runs.

4.3. NARX Gaussian Process

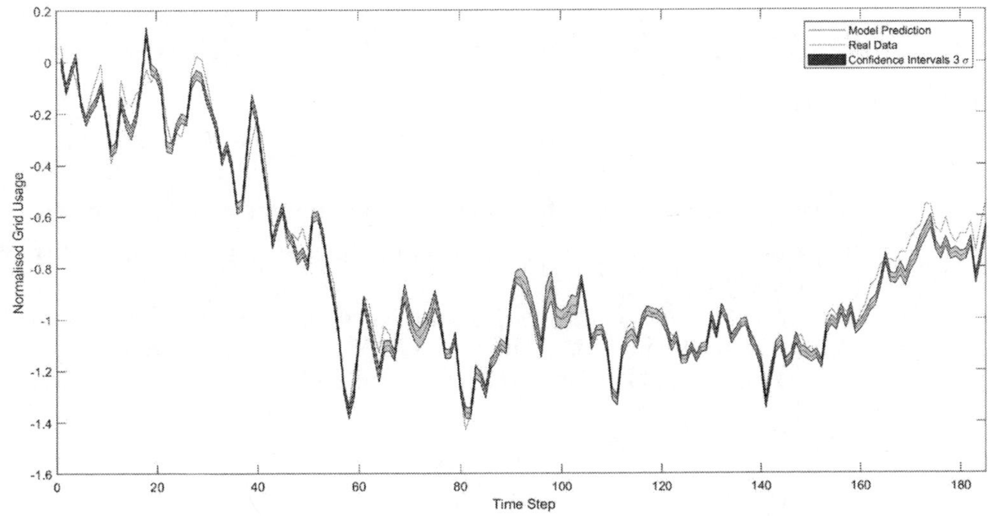

Figure 7. NARX Gaussian Process regression with 3σ confidence intervals.

The best results came from the NARX Gaussian Process which had a prediction error of 0.26%, far outperforming the other algorithms under investigation.

4.4. Evaluation

The linear regression was less effective than the Gaussian process in its forecasting power however it was very quick and simple to run, making it a useful tool for lower accuracy forecasting with a large quantity of data.

The Neural Network for the low number of data points had low prediction accuracy relative to the other methods tested, was slow to run and was difficult to optimise. With more data, it may be more effective especially when compared to a Gaussian process which could run prohibitively slowly on large datasets. NARX Neural Networks could be used and optimised to determine whether their enhancement to be auto-regressive with exogenous variables significantly increases their prediction accuracy.

It was seen in Figure 7 that the Gaussian accurately found the physics of the system, demonstrating the power of auto-regressive gaussian processes for non-stationary regression problems. In an industrial application, the method may be less well suited as it has significant problems of computational time for large numbers of datapoints, being an $\mathcal{O}\left(n^3\right)$ algorithm where n is the number of data points.

References

[1] Soliman S A and Alkandari A M 2010 *Electrical load forecasting: modeling and model construction* (Butterworth-Heinemann) ISBN 0123815444

[2] Stram B N 2016 *Energy Policy* **96** 728–734 ISSN 0301-4215 URL `https://www.sciencedirect.com/science/article/pii/S0301421516302646`

[3] Bishop C M 2006 *Pattern recognition and machine learning* (Springer) ISBN 9780387310732

[4] Rasmussen C E and Williams C K I 2006 *Gaussian processes for machine learning* (MIT Press) ISBN 026218253X URL `https://dl.acm.org/citation.cfm?id=1162254`

IOP Conf. Series: Journal of Physics: Conf. Series **1106** (2018) 012006 doi:10.1088/1742-6596/1106/1/012006

Dynamic responses of railway ballasted track considering rail pad deterioration

Chayut Ngamkhanong[1,2], Keiichi Goto[2,3] and Sakdirat Kaewunruen[1,2]

[1]Department of Civil Engineering, School of Engineering, University of Birmingham, Birmingham B152TT, United Kingdom

[2]Birmingham Centre for Railway Research and Education, School of Engineering, University of Birmingham, Birmingham B152TT, United Kingdom

[3]Railway Dynamics Division, Railway Technical Research Institute, Japan

cxn649@student.bham.ac.uk

Abstract. Rail pads, which can be made of polymeric compound, rubber, composite materials etc., are resilient components installed on rail seats between the rail and sleeper in order to attenuate the impact loads and moderate track stiffness at the special locations. Based on literature, rail pad can be deteriorated over time. This can change the dynamic characteristics of rail pads: stiffness and damping. This paper presents the dynamic responses of railway track with new rail pads and aged rail pads. It should be noted that the parameters used of rail pads were analysed using impact excitation technique and then adapted for this study. To study the dynamic responses, D-track software is used based on Winkler foundation principle. Wheel flat is considered as imperfect wheel for impact case. The results obtained show that rail pads deteriorations seem to have significant effects on the increasing of rail/sleeper contact force rather than wheel/rail contact force. It is know that softer pads normally have a better performance on reducing dynamic responses. Although, the stiffness of rail pads reduce over time, damping coefficient also decrease. These lead to a poor performance of rail pad resulting in the reducing of rail/sleeper contact force.

1. Introduction

Rail pads are resilient elastic pads installed on rail seats between the rail and sleeper. It is noticeable that rail pads, which can be commonly made of polymeric compound, rubber, composite materials etc., are the important component of railway track The main duties of rail pads are to adjust the track stiffness at the differential track stiffness locations such as bridge transitions[1-2], switches and crossings [3] etc. It should be noted that these areas are prone to dynamic loading. This is used to reduce the differential track stiffness. In the stiffer portion of transition zone or bridge end, rail pads were used to reduce track stiffness [4].

In fact, railway track often experience high intensity impact load due to the irregularities of wheel or rail [5-6]. This is a high shock load applied in short duration with the frequency range between 0Hz and 2000Hz depending on vehicle velocities, types of wheel or rail defects etc [5]. While, the impact capacity of track components can be reduced due to material degradations and long term behaviour [7-13]. Another main duty of rail pads is attenuate the dynamic responses induced by wheel impacts [14]

The use of softer pad provided a better damping effects resulting in the reductions of forces acting and dynamic responses. As shown in previous study, the use of stiffer rail pad can induce cracking in concrete sleepers as the stress on sleeper was increased. It was clearly seen that the replacement of stiffer pads by softer pads indicated a reduction of sleeper loads resulting in better ride comfort [1]. Thus, it should be noted that the performance of rail pads depends on the properties of rail pads used in those particular areas. Hence, it is important to consider the pad stiffness when selecting and applying pads [15].

According to previous studies, it is evident that deterioration rate has significant influence on the dynamic rail pad properties that affect the dynamic responses of railway tracks [16-17]. The experimental techniques using instrumented hammer coupled with modal analysis have been developed in order to extract the dynamic characteristics of rail pad [18-19]. In this study, it should be noted that the material properties of aged rail pad have been collected and extracted from previous studies at different years in service [20-22].

This paper presents the dynamic responses at the contact between rail and sleeper with rail pads. The study case of worn wheel flat was considered as an impact case to compare with the perfect wheel/rail contact. The material degradations of rail pads were taken into account based on previous experiments. The dynamic amplification factors are highlighted. The outcome of this study will help railway engineer in improving the predictive maintenance and inspection regimes.

2. Methodology

2.1. Simulation scenarios

D-Track software was used to simulate the static and dynamic responses of railway track based on Winkler foundation principal which cross section of track dynamic responses was considered symmetrically. It should be noted that Timoshenko beams were used to model the rails and sleepers as this take into account shear characteristics which have a significant effects on results [23]. The Timoshenko coefficients used for rails and sleepers were 0.833 and 0.34, respectively [24]. The numerical simulations were carried out using 106t freight wagon with wheel radius of 0.46m and Hertzian spring constant of 0.87×10^{11} N/m$^{3/2}$.

Simulation scenarios and vehicle/track types are presented in table 1. It should be noted that worn wheel flat is considered as wheel/rail contact irregularity case. This causes high frequency impact force on railway track. It is noted that wheel flat is the most common local surface defect of wheels [5, 25-27].

Table 1. Simulation scenarios

Vehicle type	Track type	Train speed	Wheel/rail contact condition
106t Coal Wagon	Standard Gauge 1435mm 60kg/m rail, Prestressed concrete sleepers	20-100	No irregularity Worn wheel flat (Length = 75 mm, Depth = 1 mm)

2.2. Rail pad properties

An alternative rail pad tester based on the SDOF vibration response measurement for determining the dynamic properties of rail pads has been devised, as well as the state-dependent model of rail pads [19]. Total pad stiffness including both frequency-dependent stiffness and frequency-independent stiffness has been calculated based on modal testing with the consideration of state dependent model. HDPE (High Density Polyethylene) pads were used. The new rail pad and worn rail pads in the years in services of 18 and 20 were taken into account [19]. The damping coefficient and stiffness of rail pad are shown in figure 1. It was assumed that the deterioration rate was linear. The degradation equation of both damping and stiffness are noted in the Figure.

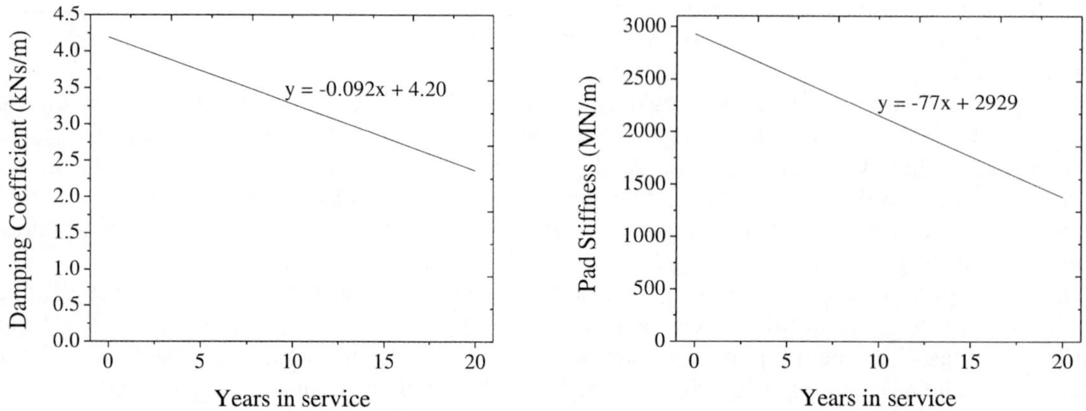

a) Damping coefficient b) Rail pad stiffness

Figure 1. Dynamic properties of rail pad.

3. Results and Discussions

Figures 2-3 show the time histories of wheel/rail and rail/sleeper contact forces subjected to the train speed of 100 km/h. It should be noted that the analysis positions were at rail seat. The pulse durations were about 0.7 s. It can be seen from both figures that wheel flat defect play a role on the increasing of dynamic force. It is observed that railway track with new and aged rail pads have the same magnitude of wheel/rail contact forces. However, aged rail pads have significant effects on the rail/sleeper contact force in comparison with new rail pads, as seen in figure 3.

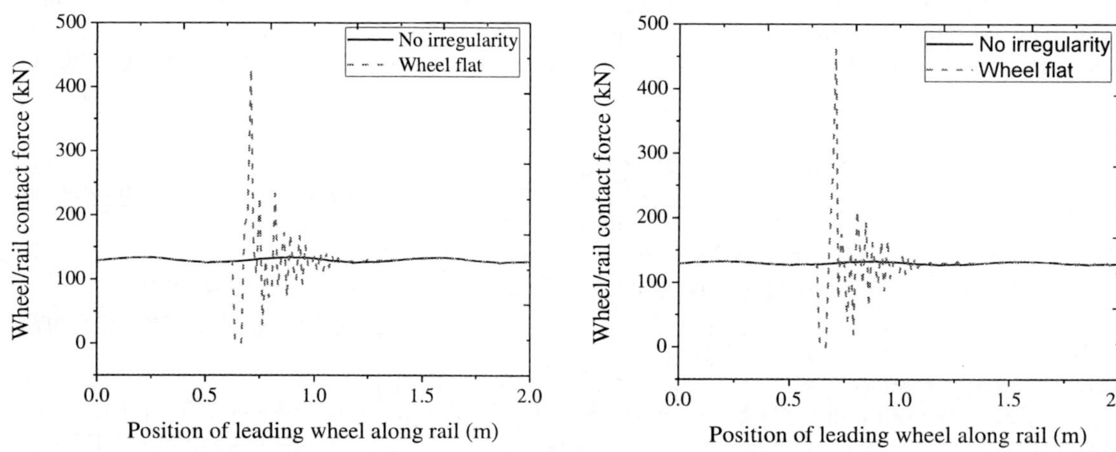

a) Year 0 b) Year 20

Figure 2. Wheel/rail contact force.

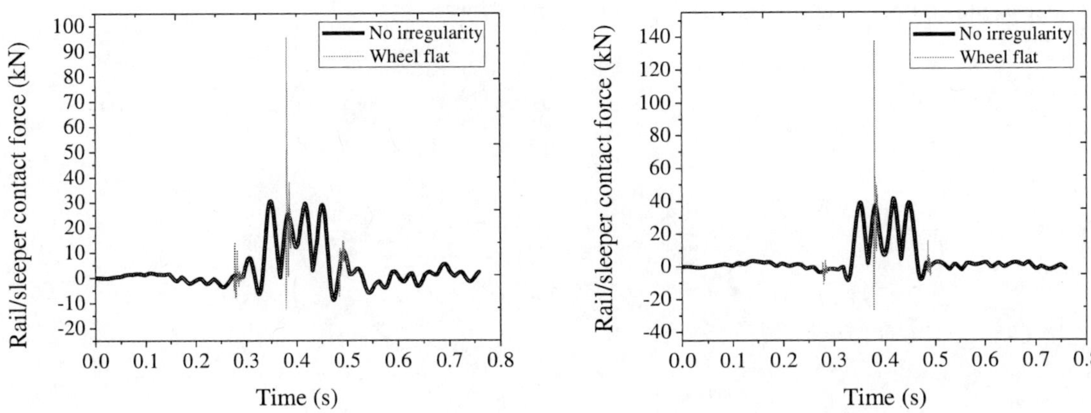

a) Year 0 b) Year 20

Figure 3. Rail/sleeper contact force at rail seat.

Wheel/rail contact force at different train velocities and years in service of rail pads are presented in figure 4a. It should be noted that for the perfect wheel/rail interaction, wheel/rail contact forces are about 140 kN. As for worn wheel flat case, contact forces increase nonlinearly with the increase of vehicle velocity. It is noted that rail pad deterioration does not play a role on wheel/rail contact force. It should be noted that wheel flats would create the forces varying from 0 to 2000 Hz, depending on the train speeds [5].

Figure 4b presents the rail/sleeper contact forces considering worn wheel flat. As for static case (no irregularity), the contact forces slightly linearly increase with the increasing of velocity and years in service. However, the increases are less than 5%. In case of worn wheel flat, the contact forces significantly increase with the increasing of train velocity. It is also noted that aged rail pad play a role in rail/sleeper contact force reduction especially at the higher speed.

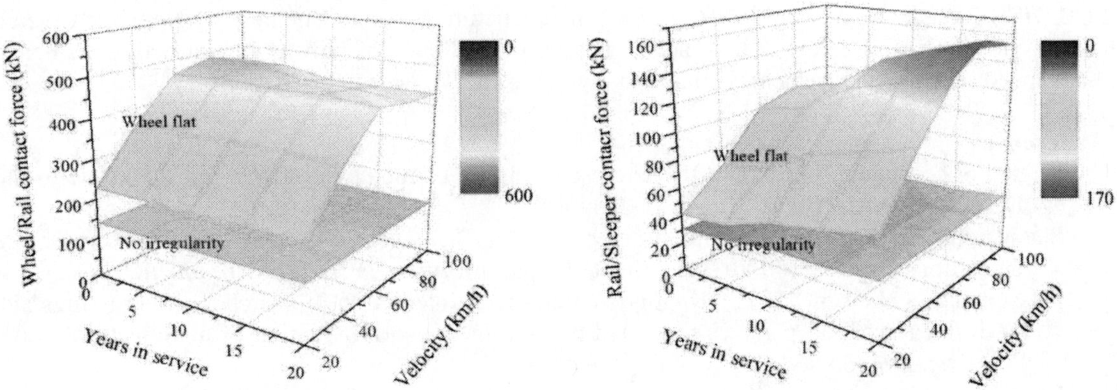

a) Wheel/rail b) Rail/sleeper

Figure 4. Contact force contour at different velocities over time.

The dynamic factor can be evaluated from rail/sleeper dynamic over static forces ratio, as illustrated in figure 5. Dynamic factors are in the range between 1.3 and 3.6 at year 0 and between 1.7 and 3.9 at year 20. Interestingly, rail pad degradations does not seem to play a role on wheel/rail contact force as they show quite the same dynamic factor at different ages of rail pads. Nevertheless, they instead play a significant role on rail/sleeper contact force.

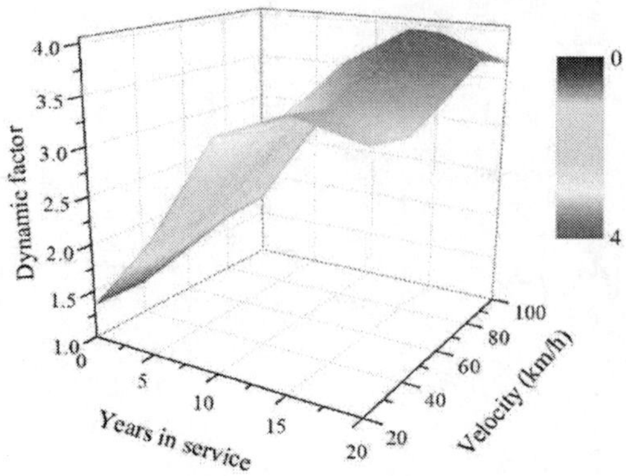

Figure 5. Dynamic factor.

4. Conclusions

This paper presents the dynamic response between vehicle and track considering degradations of rail pads. It is well known that rail pads are generally used in railway track in order to attenuate the dynamic load and adapt the overall track stiffness. The new and aged HDPE pads have been previously studied in terms of properties and then used in this study. It was noted that the dynamic stiffness and damping coefficient tended to reduce over the time. The results show that rail pads deteriorations seem to have significant effects on the increasing of rail/sleeper contact force rather than wheel/rail contact force. The increases of dynamic factor of rail/sleeper force of about 25% are noted. Although, the softer rail pads can significantly make benefit to railway track as seen in previous studies, damping properties, which are observed over time, reduce the damping performance which induce the higher rail/sleeper contact force.

5. Acknowledgements

The authors are sincerely grateful to the European Commission for the financial sponsorship of the H2020-RISE Project No. 691135 “RISEN: Rail Infrastructure Systems Engineering Network”, which enables a global research network that tackles the grand challenge of railway infrastructure resilience and advanced sensing in extreme environments (www.risen2rail.eu) [28].

6. References

[1] Miguel Sol-Sánchez F M N G 2015 The use of elastic elements in railway tracks: a state of the art review *Constr. Build. Mater.* **75** 293–305

[2] Setsobhonkul S, Kaewunruen and Sussman J M 2017 Lifecycle Assessments of Railway Bridge Transitions Exposed to Extreme Climate Events *Frontiers in Built Environment* **3** 2017

[3] Palsson B and Nielsen J 2015 Dynamic vehicle-track interaction in switches and crossings and the influence of rail pad stiffness field measurements and validation of a simulation model *Veh. Syst. Dyn.* **53** 734–755

[4] Lund H and Åswärdh A 2014 *Transition Zones between Ballasted and Ballast less Tracks.* Lund: LTH School of Engineering at Campus Helsingborg, Department of Technology and Society, Lund University.

[5] Remennikov A M and Kaewunruen S 2008 A review on loading conditions for railway track structures due to wheel and rail vertical interactions *Structural Control and Health Monitoring*, **15** 207-34

[6] Ngamkhanong C, Kaewunruen S and Costa B J A 2018 State-of-the-Art Review of Railway Track Resilience Monitoring *Infrastructures* **3**

[7] Ngamkhanong C, Li D and Kaewunruen S 2017 Impact capacity reduction in railway prestressed concrete sleepers with vertical holes *IOP Conference Series: Materials Science and Engineering* **236**

[8] Ngamkhanong C, Kaewunruen S and Remennikov A M 2017 Static and dynamic behaviours of railway prestressed concrete sleepers with longitudinal through hole *IOP Conference Series: Materials Science and Engineering* **251**

[9] Ngamkhanong C, Li D and Kaewunruen S 2017 Impact capacity reduction in railway prestressed concrete sleepers with surface abrasions *IOP Conference Series: Materials Science and Engineering* **245**

[10] Ngamkhanong C, Li D, Kaewunruen S 2017 and Remennikov A M Capacity Reduction in Railway Prestressed Concrete Sleepers due to Dynamic Abrasions *International Journal of Structural Stability and Dynamics* (Accepted)

[11] Li D, Ngamkhanong C and Kaewunruen S 2017 Influence of vertical holes on creep and shrinkage of railway prestressed concrete sleepers *IOP Conference Series: Materials Science and Engineering* **236**

[12] Li D, Ngamkhanong C and Kaewunruen S 2017 Time-dependent topology of railway prestressed concrete sleepers *IOP Conference Series: Materials Science and Engineering* **245**

[13] Li D, Ngamkhanong C and Kaewunruen S 2017 Influence of Surface Abrasion on Creep and Shrinkage of Railway Prestressed Concrete Sleepers. *IOP Conference Series: Materials Science and Engineering* **245**

[14] Nyström P and Prokopov A 2011 Spårkonstruktioner för 320 km/h – Övergångskonstruktioner. Borlänge, Stockholm: Trafikverket.

[15] Jenks C W 2006 Design of Track Transitions. (Washington DC, USA: Transportation Research Board).

[16] Grassie S L and Cox S J 1984 The dynamic response of railway track with flexible sleepers to high frequency vertical excitation *Proceedings of Institute of Mechanical Engineering Part D* **24** 77-90

[17] Wu T X and Thompson D J 1999 The effects of local preload on the foundation stiffness and vertical vibration of railway track *Journal of Sound and Vibration* **215** 881-904

[18] Kaewunruen S and Remennikov A M 2005 Monitoring structural degradation of rail bearing pads in laboratory using impact excitation technique *1st International Conference on Structural Condition Assessment, Monitoring, and Improvement* Perth Australia December 14-12 .394-389.

[19] Kaewunruen S and Remennikov A M 2009 State dependent properties of rail pads *Transport Engineering in Australia* **12** 17-24

[20] Kaewunruen S and Remennikov A M 2006 Sensitivity analysis of free vibration characteristics of an in situ railway concrete sleeper to variations of rail pad parameters *Journal of Sound and Vibration* **298** 453-461

[21] Remennikov A M and Kaewunruen S 2006 Laboratory measurements of dynamic properties of rail pads subjected to incremental preloads *19th Australasian Conference on the Mechanics of Structures and Materials* 29 Nov 29 – 1 Dec Christchurch New Zealand 319-324

[22] Oregui M, Nunez A, Dollevoet R and Li Z 2017 Sensitivity Analysis of Railpad Parameers on Vertical Railway Track Dynamics *J. Eng. Mech.* **143** 04017011

[23] Steffens D M 2005 Identification and development of a model of railway track dynamic behaviour Master thesis Queensland University of Technology Australia

[24] Cai Z 1992 Modelling of rail track dynamics and wheel/rail interaction PhD Thesis Department of Civil Engineering Queen's University Ontario Canada

[25] Kanoje N K, Sharma S C, and Harsha S P 2014 Wheel-rail and wheel-flat as a coupled system: Contact dynamics modeling with finite element analysis *J. Coupled Syst. Multiscale Dyn.* **2** 20-27

[26] Knothe K L and Grassie S L 1993 Modelling of railway track and vehicle/track interaction at high frequencies *Vehicle System Dynamics* **22** 209–262

[27] Sadeghi J 1997 Investigation of characteristics and modelling of railway track system PhD Thesis School of Civil, Mining, and Environmental Engineering University of Wollongong Australia

[28] Kaewunruen S, Sussman J M and Matsumoto A 2016 Grand challenges in transportation and transit systems *Frontiers in Built Environment* **2**

Vertical vibrations of rail track generated by random irregularities of rail head rolling surface

P Koziol and D Kudla

Faculty of Civil Engineering, Cracow University of Technology,
ul. Warszawska 24, 31-155 Krakow, Poland

pkoziol@pk.edu.pl

Abstract. It is known that geometrical irregularities of the rail head rolling surface produce additional force when the train runs on track. This force can be quite significant and should not be neglected in the analysis, especially when one deals with high speed railways. In this paper, an analytical method of modelling of such irregularities is presented, along with procedure of their stochastic simulation. The detailed description of this approach is associated with its practical application to the analysis of the rail track dynamic response to moving train. The type of imperfections is chosen randomly, leading to stochastic function which describes their distribution along the track. The system response to moving train is described by two layer model which allows to consider all important elements of rail track such as rails, sleepers, rail pads or under sleeper pads. This previously developed model allowed to analyse the dynamic response of rail track in the case of four forces, mainly due to complex computational procedure needed to apply. In this paper, the improved procedure is used, which allows to calculate the track response to the whole passing train.

1. Introduction

The modelling of load generated by train is of importance in the rail track dynamic response analysis, especially when it comes to changes of foundation stiffness or quite significant degradation of track [1]. Various imperfections or damages of rails can be found in operational lines. Some of them need expensive reparations while some can be marginalized by careful maintenance. The influence of such damages, if not critical, can be sometimes omitted. However, in many cases these imperfections lead to the increase of noise and dynamically varying vibrations that can propagate to surroundings, including buildings near rail track.

Geometrical irregularities of the rail head rolling surface belong to such imperfections. They are dangerous for the environment protection mainly in the case of high speed rails. Modelling of the load generated by moving train is quite difficult. Except of the quasi-static load produced by the weight of train, other additional loads are generated. These regular ones, as e.g. the load coming from the rail deflection between sleepers, are relatively easy to model by harmonic functions with the use of experimentally determined additional static force. Others, like the rail head rolling surface damages, can be easily described only if their regular shape is assumed on the whole length of rails [2]. In reality, these imperfections vary along track structure and, unluckily, their nature changes in time during everyday use of rail track. Even more difficult task is to build the dynamic model of rail track which takes into account nonlinear properties of structure [2], which are confirmed experimentally, and, additionally, stochastic characteristics of railway [3].

In this paper, the problem of modelling of stochastically distributed rail imperfections is discussed. In the final stage of this preliminary work, the load including the effects of such irregularities will be represented by random function, leading to the stochastic differential equation of the fourth order. For this purpose, one must develop methodology of analytical modelling of such imperfections. Having this procedure, one might solve classical dynamic system of differential equations and then make extended statistical analysis of functions representing the track response to moving train.

The two layer model described in this paper allows to consider mechanical properties of construction elements of rail track and also to analyse interactions between rails and sleepers [1, 4]. This model assumes that the rail can be represented by the Euler-Bernoulli beam with parameters constant along the track, whereas sleepers are modelled as a rigid body. The fasteners are described by the viscoelastic continuous constrains, i.e. stiffness and damping, and the sleeper foundation is modelled as a viscoelastic subgrade. The improved computational procedure to represent the dynamic response of the track to moving train is used in this paper. This procedure allowed to analyse the track response to the whole passing train, however in simpler case than this associated with stochastic load. One should underline that this ability of the developed technique is very important for the analysis of systems with nonlinear and stochastic properties [5]. In the case of such models, it is not enough to consider separate forces generated by train axles [6].

Each force associated with axles of train has three components: the force constant in time, produced by weight of train, the part varying in time, produced by the rail deflection between sleepers, and the part of load generated by rail head rolling surface irregularities, which are modelled randomly by using the developed stochastic procedure. Vertical vibrations of rails and sleepers can be obtained by applying the Fourier transform and its approximation based on wavelet filters [2].

2. Modelling of the rail track

The rail track can be modelled in various ways. It is usually done by numerical approaches based on FEM methods. Analytical solution to mathematical models are relatively rare, although they are valued for their ability to perform parametrical analysis with relatively low cost.

Two main approaches to analytical modelling of rail track can be found in the literature. The first one is based on so called one layer model. It assumes that the rail track can be represented by either the Euler-Bernoulli beam or the Timoshenko beam supported by viscoelastic foundation which represents track bed. The second approach uses the two layer model which consists of two layers associated with rails and sleepers. The upper layer, associated with rails, is represented by a beam, usually the Euler-Bernoulli one, mainly because it is computationally easier. The lower layer, associated with sleepers, is modelled as a rigid body. The fasteners between rails and sleepers are modelled by the viscoelastic continuous constrains, i.e. stiffness and damping coefficients. The sleepers foundation is considered as a viscoelastic layer. The moving train load is represented by a set of forces in various configurations related to train axles [1, 2].

The one layer dynamic model can be described by the following equation [1, 2]:

$$EI\frac{\partial^4 y_r}{\partial x^4} + N\frac{\partial^2 y_r}{\partial x^2} + m_r\frac{\partial^2 y_r}{\partial t^2} + c_r\frac{\partial y_r}{\partial t} + k_r y_r - k_N {y_r}^3 = P(x,t) \tag{1}$$

The parameters appearing in this equation are: EI [Nm2] – bending stiffness of rail steel, N [N] – axial force, m_r [kg/m] – rail unit mass, k_r [N/m^2] – stiffness of rail foundation and c_r [Ns/m^2] – viscous damping of rail foundation. The parameter k_N [N/m^4] represents nonlinear part of rail track foundation stiffness

The two layer dynamic model of rail track is represented by the system of two coupled differential equations [1, 4]:

$$EI\frac{\partial^4 y_r}{\partial x^4} + N\frac{\partial^2 y_r}{\partial x^2} + m_r\frac{\partial^2 y_r}{\partial t^2} + c_r(\frac{\partial y_r}{\partial t} - \frac{\partial y_s}{\partial t}) + k_r(y_r - y_s) - k_N {y_s}^3 = P(x,t) \tag{2}$$

$$m_s \frac{\partial^2 y_s}{\partial t^2} + c_s \frac{\partial y_s}{\partial t} + k_s y_s + k_N {y_s}^3 = c_r (\frac{\partial y_r}{\partial t} - \frac{\partial y_s}{\partial t}) + k_r (y_r - y_s) \tag{3}$$

The additional parameters appearing in this equation are: m_s [kg/m] – unit mass of sleepers uniformly distributed along the track, k_s [N/m^2] – stiffness of sleepers foundation and c_s [Ns/m^2] – viscous damping of sleepers foundation.

3. Modelling of the train load

The load generated by train can be represented by a set of forces concentrated or distributed with density $d(x, t)$:

$$P(x,t) = P_C(x,t) + P_V(x,t) + P_R(x,t) \tag{4}$$

$$P_C(x,t) = \sum_{l=0}^{L} \frac{P_0}{a} d(x,t) H(a^2 - (x - Vt - s_l)^2) \tag{5}$$

$$P_D(x,t) = \sum_{l=0}^{L} \frac{\Delta P}{a} e^{i\omega} d(x,t) H(a^2 - (x - Vt - s_l)^2) \tag{6}$$

and the term $P_R(x, t)$ denotes factor associated with random irregularities of rail head surface. *H(.)*, $2a$, s_l, L, ω and V are the Heaviside function, the span of single load, the distance between consecutive forces produced by wheels, the number of axles, the frequency of load and the speed of train, respectively. The axles configuration can be introduced by inclusion of the phase shift associated with the wheel position on the surface of regular imperfection (e.g. cosine shape linked with rail deflection between sleepers). In the case of regular spacing between sleepers, the amplitude of this deflection is constant and its shape is regular, having a cosine form.

The density $d(x, t)$ gives possibility of analysis of different types of loads. The load can be e.g. rectangular, Gaussian [1], cosine type or this one used in previous publications [2, 4]:

$$d(x,t) = cos^2 (\frac{x - Vt - s_l}{2a}) \tag{7}$$

which gives high regularity of the solution.

The random part $P_R(x, t)$ is generated by irregularities appearing on the rolling surface of the rail head and modelling of this effect is described in the next section. Other factors influencing the load are omitted, e.g. sudden changes of foundation stiffness or characteristics of the train structure.

4. Random imperfections of the rail track

Random approach to modelling the dynamic response of the train track can be recognized in two different ways:

1. First approach (classical in Engineering):
   To perform an appropriate number of output realisations and then carry out „*statistical*" analysis of the obtained results. This means: to solve the system for deterministic input which is chosen randomly from a set of possible scenarios. The system itself still remains deterministic.
2. Second approach (based on stochastic analysis):
   To perform an appropriate number of input realisations, carry out „*statisical*" analysis and build the system of stochastic equations (in particular differential ones) which might be solved by using tools from *stochastic analysis*. The system becomes stochastic from the beginning.

Both approaches need computational solution to mathematically describe random changes of rolling surface along rail. For this purpose, a special algorithm is constructed and then implemented in Excel software in such a way that gives possibility to include the result in Mathematica code. Mathematica software was used previously to solve and to analyse the analytical systems described by equations (1-3). Therefore, it is crucial to keep the original path of investigations which will allow to follow analytical approach and, finally, to reach solution to the second approach mentioned above, i.e.

formulate random system described by the stochastic differential equations. This could be solved by using tools coming from mean-square analysis and hybrid approximation methods.

5. The draw procedure

The assumptions made for irregularity of rail are as follows:

1. The length of the investigated area of beam with „surface imperfections" is finite, although the length of beam remains infinite.
2. The step of draw in x direction is fixed, although it can be changed in separated sections.
3. The amplitude of imperfection is limited (finite).
4. The amplitude increase between two consecutively selected points can be limited (and fixed) either unlimited (although remaining within the limits specified by the condition 3).
5. The amplitude of imperfection and the direction of its change at consecutive points is randomly chosen as many times as all conditions are fulfilled and, additionally, realistic physical character of imperfection is kept (all irregularities are below horizontal line – there are no elevations above the zero level).

Figures 1 and 2 show examples of imperfections chosen by using the above procedure without limitation of the single draw amplitude (Fig. 1) and with limitation (Fig. 2). The condition 5 is omitted in these examples.

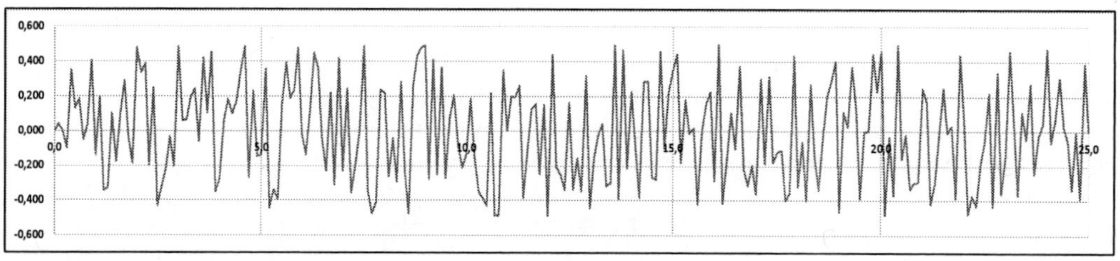

Figure 1. Case 1: random imperfection without limits of randomly chosen amplitude increase.

Conditions for imperfections shown in Fig. 1 are as follows:

1. The length of the investigated area of beam with „surface imperfections": 25 m;
2. The step of draw in x direction: 0.1 m;
3. The amplitude of imperfection: 0.5 mm;
4. The amplitude increase: without limits.

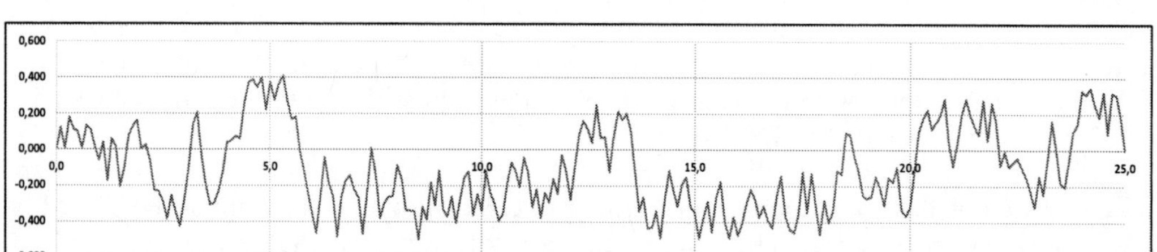

Figure 2. Case 2: random imperfection with fixed limit of randomly chosen amplitude increase.

Conditions for imperfections shown in Fig. 2 are as follows:

1. The length of the investigated area of beam with „surface imperfections": 25 m;

2. The step of draw in x direction: 0.1 m;
3. The amplitude of imperfection: 0.5 mm;
4. The amplitude increase: 0.25 mm.

Another question is: how many times the amplitudes should be randomized? Is it enough to make only one draw or more realisations should be made in order to obtain real representation of physical situation?

Figures 3-4 shows the spectra amplitudes of imperfections related to cases 1 and 2. Figure 4 shows that indeed the increase in the number of draws leads to more realistic representation which is justified by several measurements done on operational rail lines. On the other hand, this result strongly depends on a number of parameters of the developed procedure, so high caution must be kept already at this stage in order to stay within limits of the model applicability.

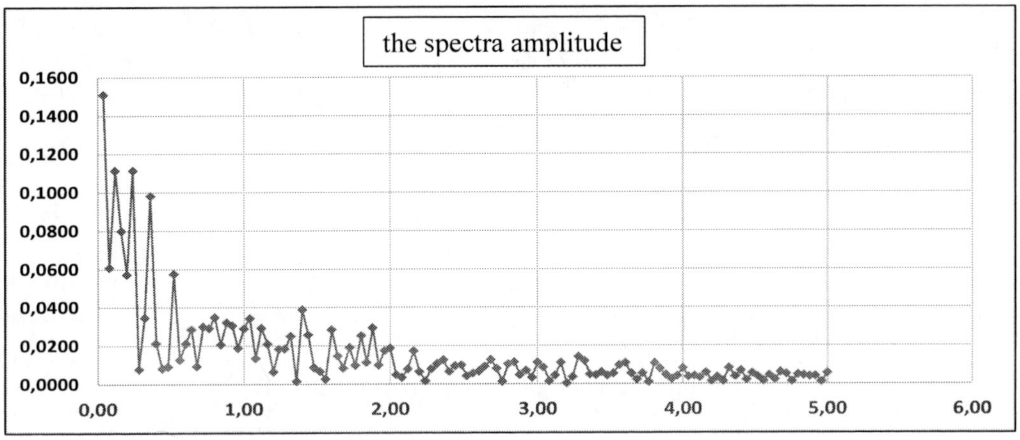

Figure 3. Case 2, the length of rail: 5 m, the number of draws: 1.

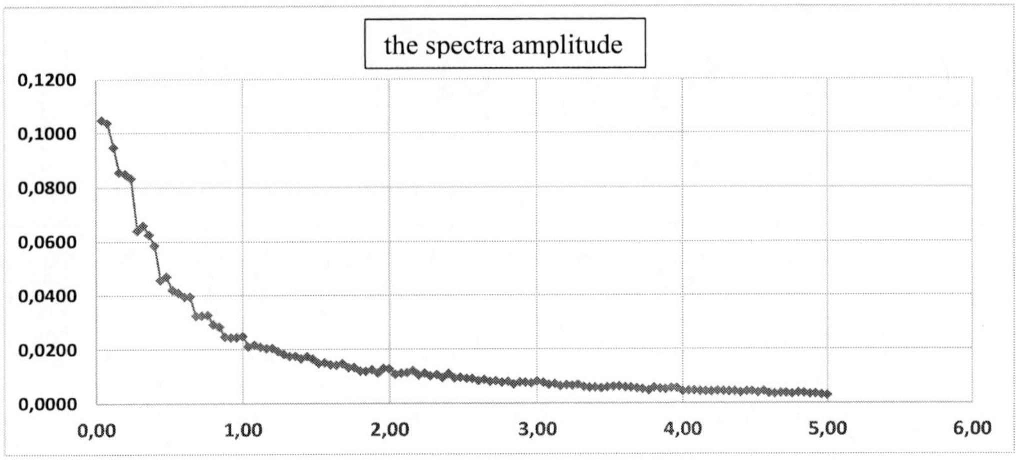

Figure 4. Case 2, the length of rail: 5 m, the number of draws: 100.

6. Approximation of imperfections

The main idea presented in this paper deals with the problem of approximation of randomly chosen imperfections in a way that allows to use it in analytical computations leading to solution of the systems defined by equations (1) and (2-3). A very natural approach can be sought in application of the Fourier series. However, this kind of approximation leads to a very complex computations which are highly inefficient. Because of this, parametrical analysis of the system can be computationally impossible.

Figures 5-7 shows approximation of randomly modelled irregularities by using the Fourier series. One can see relatively good agreement of shape and amplitudes between approximated and chosen imperfection. Nevertheless, as expected, this kind of approximation gives formula which is ineffective when used in analytical computations, especially when one deals with the whole train loading. Therefore other ways of approximation must be sought.

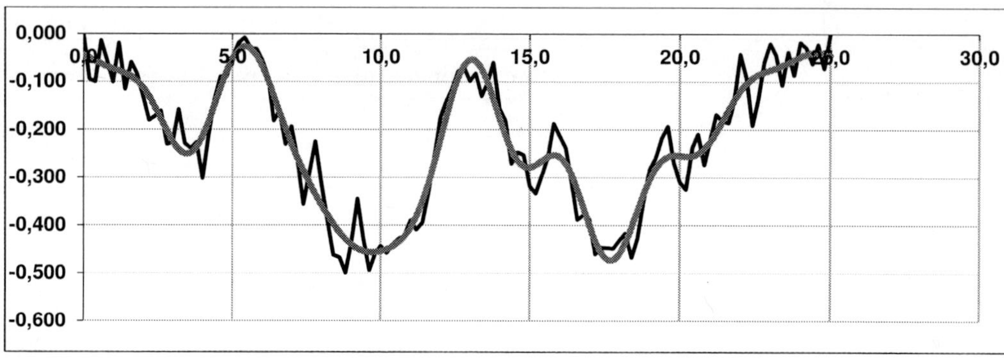

Figure 5. Case 2; the length of rail: 25 m, random data - black line, Fourier approximation - red line, number of Fourier coefficients: 10.

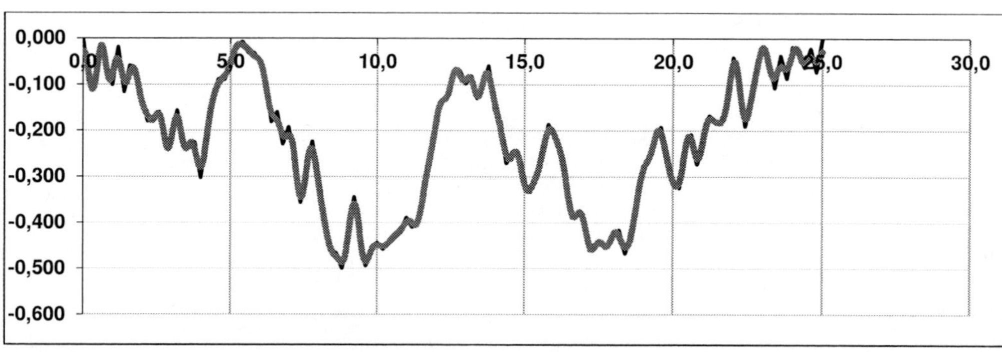

Figure 6. Case 2; the length of rail: 25 m, random data - black line, Fourier approximation - red line, number of Fourier coefficients: 100.

One can see that the higher number of Fourier coefficients gives better approximation. In the same time, the approximate solution becomes extremely difficult to apply in analytical calculations related to the solution of the system described by equation (1) and even more in the case of two layer model: equations (2-3).

Figure 7 shows in details approximation applied to a short length of the beam. One can see another disadvantage of the proposed method: the approximate solution might not fit with zero level at the ends of area with random irregularities.

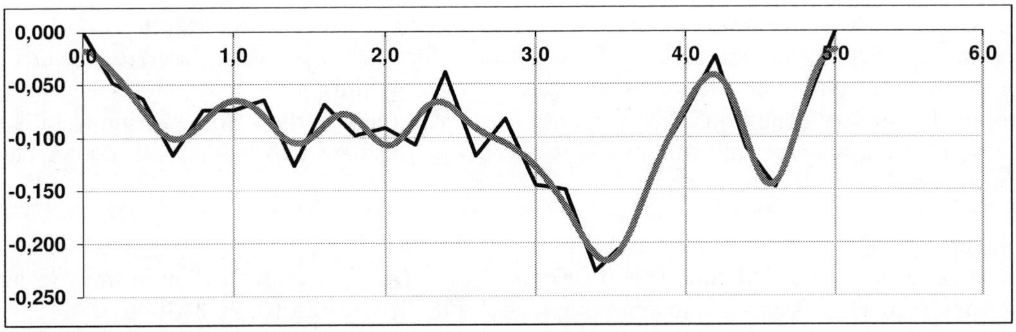

Figure 7. Case 2; the length of rail: 5 m, random data - black line, Fourier approximation - red line, number of Fourier coefficients: 10.

7. The track response

The found approximation procedure is still not good. Due to a high number of Fourier coefficients needed for exact enough approximation of randomly generated irregularities, the output function cannot be effectively used in analysis of the rail track response to passing train. Such a response can be relatively easy obtained in the case of linear and deterministic rail track models defined by equations (1-3) (with $k_N = 0$). Figure 8 presents example of rail deflection as a result of passing train EMU250 Pendolino, obtained for realistic parameters taken from operational lines of HSR in Poland. Inclusion of stochastic and nonlinear properties of rail track in this complex dynamic model still remains an open problem in the case of analytical solution.

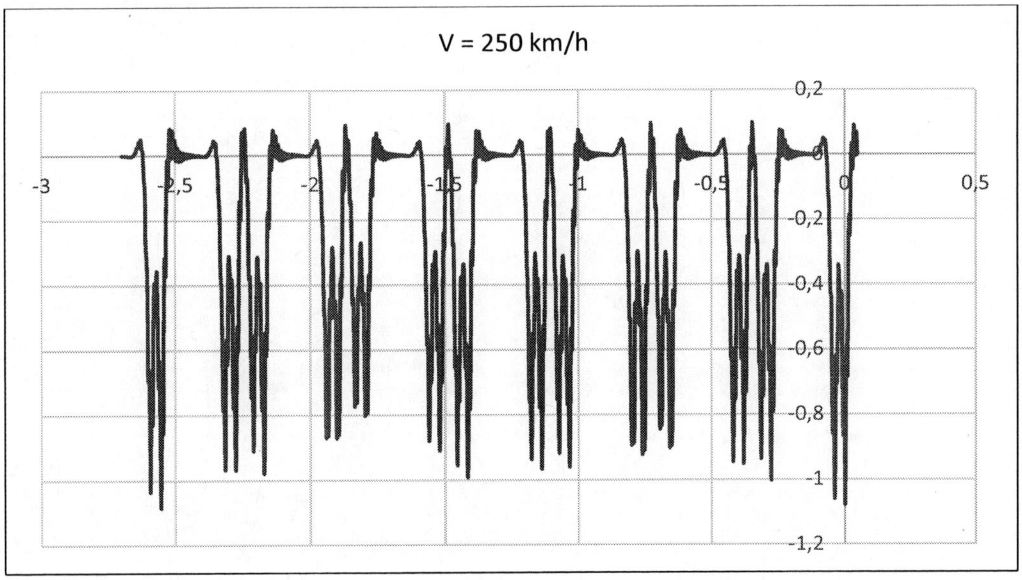

Figure 8. Rail track response to moving train: rail deflection under load generated by EMU250 Pendolino train.

8. Conclusions

Preliminary results with regard to modelling of random irregularities of the rail head rolling surface are presented. Although the obtained results are satisfactory in some cases, they are still difficult to implement in previously developed analytical systems, mainly due to complex formulas and inappropriate boundary conditions. The response of the rail track to the whole passing train is shown in the case of linear and deterministic model. Combination of these two models still remains an open problem and it is left for future work.

References

[1] Czyczula W, Koziol P, Kudla D and Lisowski S 2017Analytical evaluation of track response in the vertical direction due to a moving load *J. Vib. Control* **23** 18, 2989-3006

[2] Koziol P 2016 Experimental validation of wavelet based solution for dynamic response of railway track subjected to a moving train *Mech. Syst. Signal Process.* **79** 174-181

[3] Koziol P and Hryniewicz Z 2009 Vibrations of beams on a random foundation due to a moving load: wavelet approach *Proceedings of the Twelfth International Conference on Civil, Structural and Environmental Engineering Computing CC2009*, Civil-Comp Press, Stirlingshire, United Kingdom, paper 42, doi:10.4203/ccp.91.42

[4] Koziol P 2015 Vibrations of Railway Tracks Modelled as a Two Layer Structure *Proceedings of the Fifteenth International Conference on Civil, Structural and Environmental Engineering Computing*, Civil-Comp Press, Stirlingshire, UK, Paper 199, doi:10.4203/ccp.108.199

[5] Koziol P 2014 Wavelet approximation of the Adomian's decomposition applied to a nonlinear problem of a double-beam response subject to a series of moving loads *J. Theor. Appl. Mech.* **52** 3, 687-697

[6] Kargarnovin M H, Younesian D, Thompson D J and Jones C J C 2005 Response of beams on nonlinear viscoelastic foundations to harmonic moving loads *Comput. Struct.* **83** 1865-77

IOP Conf. Series: Journal of Physics: Conf. Series **1106** (2018) 012008 doi:10.1088/1742-6596/1106/1/012008

Impact Load Response of PC Rail Joint Sleeper under a Passing Train

K Goto[1,2], S Minoura[2], T Watanabe[2], C Ngamkhanong[1] and S Kaewunruen[1]

[1] The University of Birmingham, Edgbaston, UK
[2] Railway Technical Research Institute, Tokyo, Japan

E-mail: goto.keiichi.90@rtri.or.jp

Abstract. The purpose of this study was to clarify the influence of an impact load caused by a train passing over a rail joint on a prestressed concrete (PC) sleeper. In this study, we conducted a three-dimensional finite element (FE) model of the PC sleeper, which was able to simulate its nonlinear behaviour. Using this numerical analytical model, the influence of the impact load caused by the rail joint on the maximum bending moment and stress generated in the sleeper for various irregularities and train speeds were quantitatively evaluated. This study clarified the following items: 1) No cracks occurred in the PC sleeper under the rail surface irregularity actually measured in the field; 2) The crack generated owing to a train running at a speed of 87 km/h when the rail surface irregularity was three times the measured value (the maximum dimension of the irregularity was 7.05mm); 3) The bending moment of the PC sleeper had the characteristic of speed dependency, which was regarded as an influence of the rail surface irregularity.

1. Introduction

The design of prestressed concrete (PC) sleepers is carried out in consideration of the influence of impact loads. Impact loads are caused by rail joints, welded rail joints, track irregularities on the rail surface and wheel flats (damage generated on the wheel treads when the wheels slide on the rail running surface). Among them, impact loads caused by rail joints and wheel flats tend to be relatively large. In Japan, since the beginning of the development of the PC sleeper, wheel flats have been recognized as one of the leading causes of impact loads. Studies on the influence of the impact load caused by the wheel flats on PC sleepers have also been actively conducted [1]. On the other hand, there have been few studies on the influence of the impact load caused by a rail joint on a PC sleeper.

In a previous study, a numerical analysis with a beam model was conducted, and the effect of the impact load caused by a rail joint on a PC sleeper was evaluated [2]. This previous study indicated that the supporting condition of the PC sleeper has a significant influence on the generated bending moment in the sleeper.

However, studies on the cracks generated in a PC sleeper due to the impact load and the influence of wear on the bottom surface of the sleeper, the pre-stressing force, corrosion of the PC steel wire and deterioration of the concrete are inadequate. To evaluate these influences using numerical analysis, it is necessary to construct a three-dimensional finite element (FE) model, taking account of the nonlinearity of the sleeper material.

Based on the above background, we focused on the rail joint and constructed a nonlinear three-dimensional FE model to evaluate the dynamic and nonlinear behaviours of a PC sleeper for a rail joint. Using this model as a fundamental study, the influences of an irregularity on the rail surface and train speed on the bending moment and stress in a sleeper for a rail joint were quantitatively evaluated.

2. Methodology

2.1. Target PC sleeper

Figure 1 shows the outline of the target track structure. In this section, standard rails of 25m in length and 60 kg/m in weight were used and there was a rail joint at every 25m. PC sleepers in the plane section were Type-6 PC sleepers and those at each rail joint were specifically designed for rail joints. Both of these types are specified in the Japanese Industrial Standards (JIS). The number of PC sleepers per 25-m rail was 41.

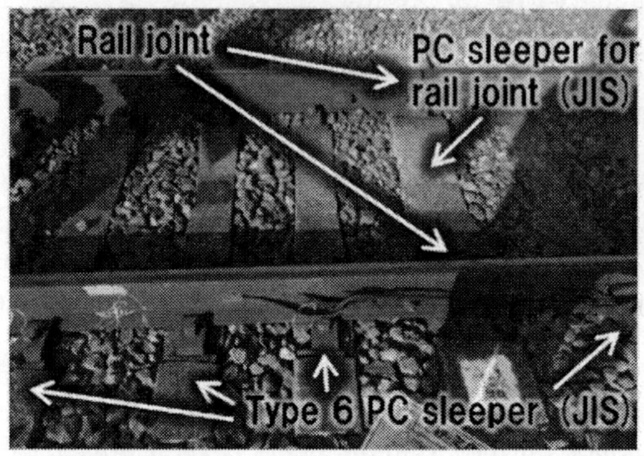

Figure 1. Target track structure.

Figure 2 shows the attachment positions of strain gages for field measurement to obtain the bending moment of the PC sleeper under a passing train, and Fig. 3 shows the attachment positions of the acceleration meters to evaluate the vibration mode of the sleeper. In order to compare the measurement results of the actual sleeper with the results of the numerical analysis, the bending moment and vibration mode of the sleepers during a passing train were measured. For the measurement of the bending moment, the strain gauges were attached to the cross-section at the rail position and the cross- section at the middle of the sleeper. The bending moment at each cross-section was calculated from the measured strains. For the measurement of the vibration mode, five piezoelectric accelerometers (PV-85, Rion) were placed at the positions shown in Fig. 3, and the free vibration response was measured using an impulse hammer. The identification method of the vibration mode was the eigensystem realization algorithm (ERA) method [3]. Trains running on this line were commuter trains for the 1067-mm gauge line, and the wheel load is approximately 50 kN.

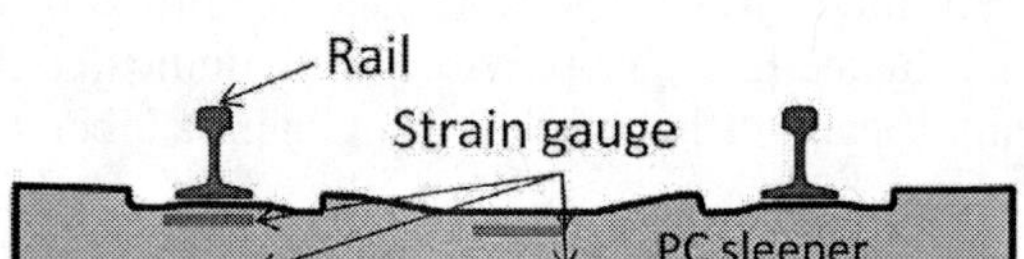

Figure 2. Attachment positions of the strain gauges.

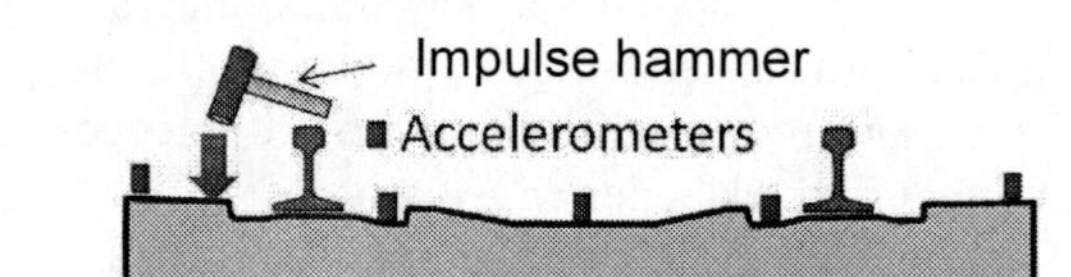

Figure 3. Attachment positions of the acceleration meters.

2.2. Analysis model

Figure 4 shows the outline of the numerical analysis model, and Fig. 5 shows the analysis model of the PC sleepers around the rail joint. In this numerical analysis, a three-dimensional nonlinear FEM analysis was performed using the LS-DYNA (ver. R 8.0.0). For each vehicle, the car, two bogies and four wheel axles were modelled as rigid body. These rigid bodies were connected with springs and dampers. The rail was modelled with beam elements. Around the rail joint, the rails were divided into 30 elements between rail fastening devices. For other sections, the rails were divided into 30 elements between rail fastening systems. The concrete of the PC rail joint sleeper was modelled with hexahedral solid elements. The PC steel wires and stirrups were modelled with beam elements. The connection between the concrete elements, PC steel wire elements and stirrup elements was assumed to be a perfect bond condition. The Type-6 PC sleeper was modelled as beam elements. The pre-stressing force of the PC rail joint sleeper was reproduced by giving an initial tensile stress to each PC steel wire. The number of elements was 10792, and the number of nodes was 13271.

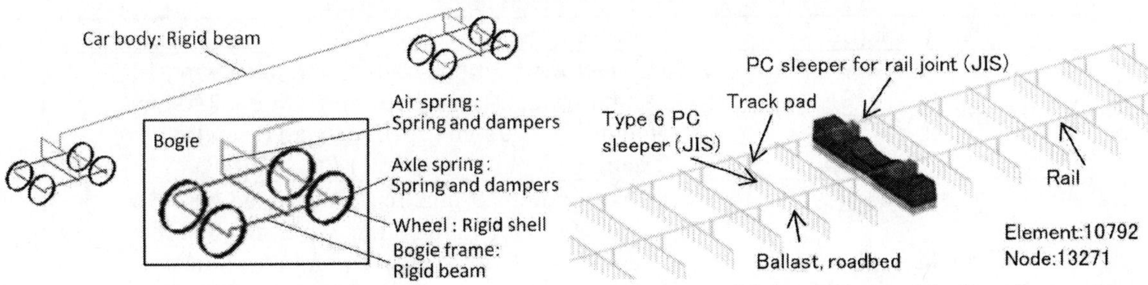

(a) Analysis model of train (b) Analysis model of track structures

Figure 4. Outline of numerical analysis model.

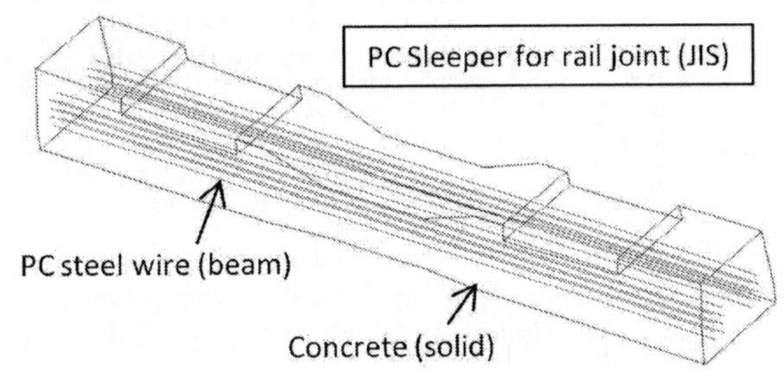

Figure 5. Analysis model of PC sleeper for rail joint.

Figure 6 shows the surface irregularities at the rail joint section. In order to study the influence of the impact force generated by them, the irregularities on the rail surface around the joint were measured using a device with a measurement length of 1.0 m. The measured irregularities were input to the analysis model.

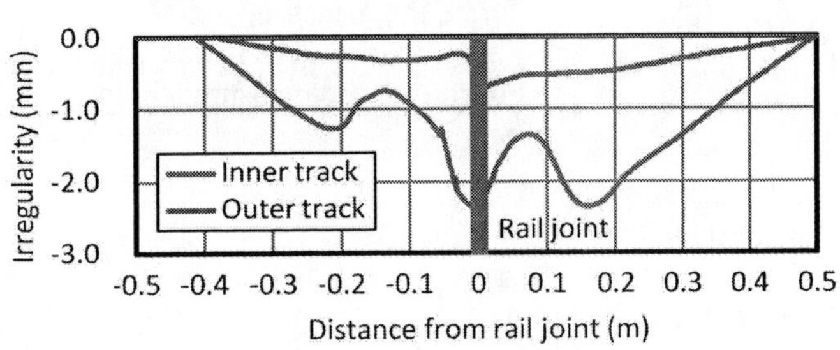

Figure 6. Rail surface irregularities at the rail joint section.

IOP Conf. Series: Journal of Physics: Conf. Series **1106** (2018) 012008 doi:10.1088/1742-6596/1106/1/012008

Table 1 shows the material properties used in the numerical analysis. The properties of the respective materials were based on the design standards for Japanese railway structures and track structures [4] [5]. Based on the vibration modes of the PC sleeper and the measurement results of the generated bending moment, the spring constant of the track pad was set to twice the nominal value and the spring constant of the support of the sleeper was set to a value corresponding to the centre-supported state.

Table 1. Material properties used in the analysis.

Rail	Type: 60-kg rail (JIS) Young's modulus E_S: 200 MPa
Track pad	Spring constant D_P: 220 MN/m
Type 6 PC sleeper (JIS)	PC steel wire: 2.9-mm dai. *3-stranded wire, 12 Length L_P: 2000 mm, bottom width B_P: 240 mm Height H_P: 170 mm (rail), 150 mm (middle) Young's modulus (concrete) E_C: 49.5 GPa
PC rail joint sleeper (JIS)	PC steel wire: 2.9-mm dia. *3-stranded wire, 16 Length L_P: 2000 mm, bottom width B_P: 300 mm Height H_P: 170 mm (rail), 145 mm (middle) Young's modulus (concrete) E_C: 49.5 GPa Tensile strength of concrete: 3.08 MPa
Ballast	Ballast thickness h: 250 mm Spring constant D_B: 180 MN/m (for 1 rail)
Roadbed	Ground reaction force coefficient K_{30}: 110 MN/m^3 Spring constant D_S: 111 MN/m (for 1 rail)

2.3. Analysis case

Table 2 shows the details of the analysis case. In this study, the influences of the rail surface irregularity and train speed were evaluated. The rail surface irregularity was based on the measured values (the maximum dimension of the irregularity was 0.72 mm for the outer rail and 2.35 mm for the inner rail). In addition, the analysis was executed for cases which the rail surface irregularity doubled and was three times as large as the measured value. The train running speed in the base case was set to 87 km/h, which was the speed at the time of measurement.

Table 2. Analysis case.

Train speed V	30–130 km/h (87 km/h for basic case)
Rail surface irregularity δ	• Measured value (maximum 0.72 mm for outer rail, maximum 2.35 mm for inner rail), • Twice the measured value • Three times the measured value • Four times the measured value

3. Analysis result and discussion

3.1. Vibration modes of the PC sleeper

Table 3 shows a comparison of the natural frequencies of the PC sleeper for a rail joint. Figure 7 shows the vibration mode forms of such a sleeper. Comparing the natural frequencies of the measurement results with the numerical analysis results, excluding the rotation mode, the difference between them was within about five percent.

Table 3. Comparison between the measurement and analysis of vibration modes of the PC rail joint sleeper.

Vibration mode	Vertical translation	Rotation	1st mode	2nd mode	3rd mode
Measured (Hz)	101	74	202	534	907
Analysis (Hz)	99	89	199	513	925

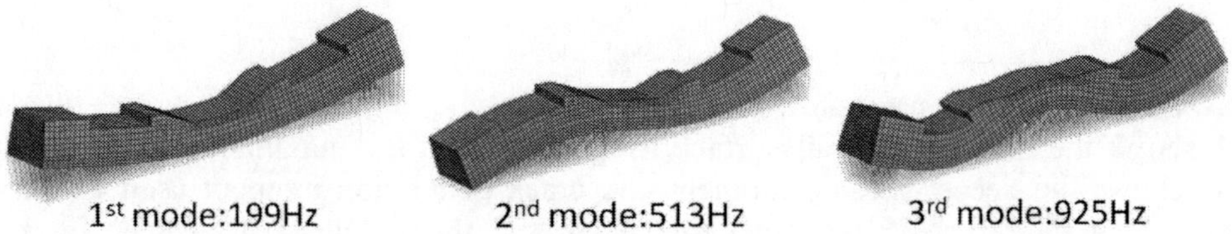

Figure 7. Vibration mode form of the PC sleeper for the rail joint.

3.2. Bending moment on the PC sleeper

Figure 8 shows a comparison between the measurement results and numerical analysis results by means of the time history waveform of the bending moment on the PC rail joint sleeper when a train was passing. As shown in the figure, the maximum value of the negative bending moment occurred at the middle of the PC sleeper is larger than that of the positive bending moment at the rail position of the outer rail. It was also confirmed that although a negative moment instantaneously occurred at the rail position in the numerical analysis, the numerical analysis results and measurement results generally agreed with each other.

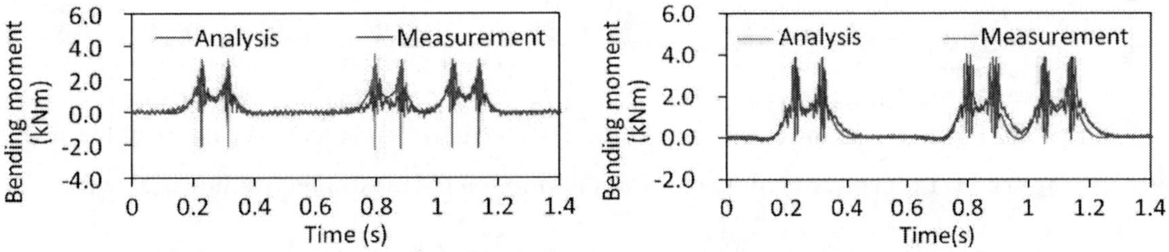

(a) Positive bending at rail position (outer rail) (b) Negative bending at the middle of the sleeper

Figure 8. Comparison between the measurement and the numerical analysis.

3.3. Stress in the concrete and the PC wire

Figure 9 shows the time history of the concrete element stresses in the longitudinal direction of the PC rail joint sleeper. From the figure, it was confirmed that each concrete element has an initial compressive stress owing to the pre-stressing force and that no tensile stress generates on any concrete element even when a train passes.

Figure 10 shows the time history of the axle stress of the PC steel wire. The axle stress is the stress of the element in which the maximum value was observed. It was confirmed that the initial stress generated owing to added initial tensile stress and that the maximum stress was less than the yield stress (1619 MPa), the stress amplitude was approximately 20 MPa.

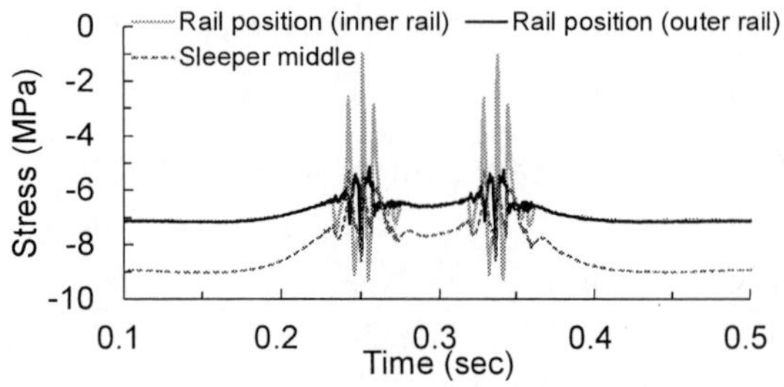

Figure 9. Time history of the concrete element stresses in the longitudinal direction of the PC sleeper.

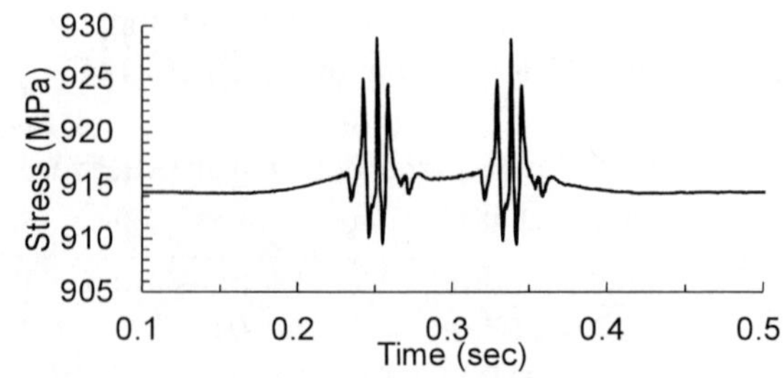

Figure 10. Time history of the axle stress in the PC steel wire.

3.4. Influence of rail surface irregularities and train speed

Figure 11 shows the influence of rail surface irregularities on the maximum bending moment. The figure also shows the decompression moment and crack generation moment used in the design for positive bending at the rail position and negative bending at the middle of the sleeper. From the figure, it was confirmed that the maximum bending moment increased as the irregularity became larger in all the cases. Moreover, the generated moment at the rail position of the inner rail side exceeded the crack generation moment to be considered in the design when the rail surface irregularity is three or more the measured value. For the bending moment at the rail position cross section on the inner rail side, the generated moment in the case when the rail surface irregularity is four times the measured value is smaller than that in the case when the irregularity is three times the measured value. This is thought to be the influence of the train speed, as described below.

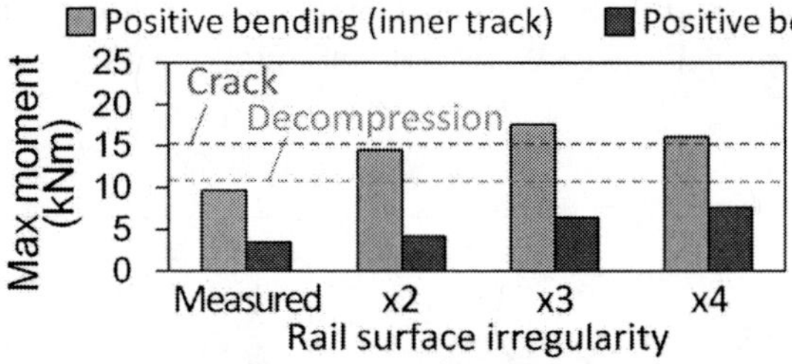

(a) Positive bending at rail position

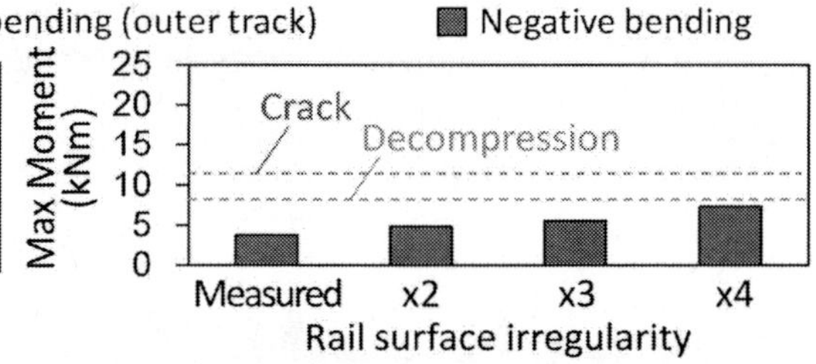

(b) Negative bending at the middle of the sleeper

Figure 11. Influence of rail surface irregularities on the maximum bending moment

Figure 12 shows the influence of the train running speed on the maximum bending moment for various amounts of the rail surface irregularity. From the figure, the maximum bending moment tends to increase with the increase of the rail surface irregularity and train speed. However, for the maximum bending moment at rail position, there is a peak observed at a running speed of around 70 km/h in the base case and there are peaks observed at running speeds of around 30 km/h, 70 km/h and 120 km/h in the case where the rail surface irregularity is four times the measured value. This confirms that there was speed dependence with respect to the bending moment. Although the profile of the measured irregularity on the surface of the inner rail was not a simple shape, it could be seen that rail surface dropped 2.35 mm of the end of the length of the rail of about 0.4 m. Therefore, during the time in which the wheel took to fall freely by 2.35 mm and the time in which the train took to move 0.4 m along the rail coincided with each other, the amount of wheel drop became maximum and the generated maximum moment was expected to increase. The train speed corresponding to this case was about 65km/h for the measured rail surface irregularity, about 46 km/h for the case in which the rail surface irregularity was twice the measured value and about 33 km/h for the case in which the rail surface irregularity was four times the measured value. In each analysis case, since a peak appeared at these speeds, the maximum generated moment was considered to be influenced by the shape of the rail surface irregularity.

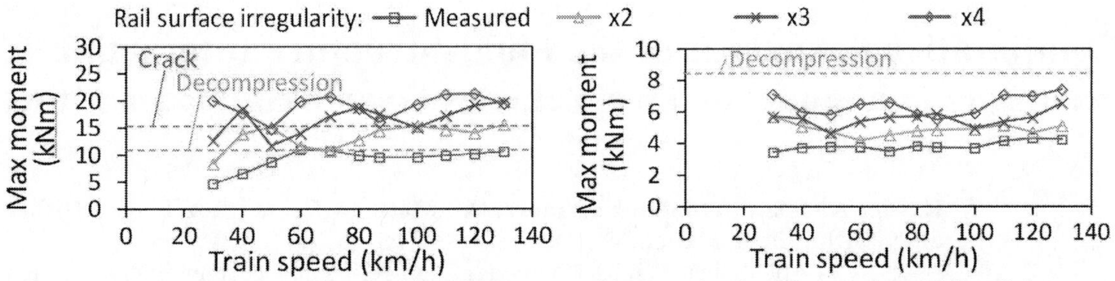

(a) Positive bending at rail position (b) Negative bending at the middle of sleeper

Figure 12. Influence of train running speed on the maximum bending moment for various amounts of rail surface irregularity.

4. Conclusions

In this study, we conducted a nonlinear three-dimensional FE model reproducing the PC rail joint sleeper. Using this model, the influences of the rail surface irregularities and train speed on the bending moment and the stress of the PC rail joint sleeper were quantitatively evaluated.

- According to the numerical analysis using the measured irregularity of the rail surface, no crack generated owing to the train running at a speed of 87 km/h when the rail surface irregularity was as large as the measured value.
- According to the numerical analysis using the measured rail surface irregularity, the crack generated owing to the train running at a speed of 87 km/h when the rail surface irregularity was three times the measured value (the maximum amount of the irregularity on the inner rail side was 7.05 mm).
- The bending moment of the PC sleeper had the characteristics of speed dependency, which is regarded as an influence of the rail surface irregularity.

We will take these findings into the consideration in the design of the PC sleeper and build an optimal design method for PC rail joint sleeper to cope with the impact load caused by rail joints in the future research.

References

[1] Wakui H and Okuda A 1999 A study on limit state design method for prestressed concrete sleepers *Concrete library of JSCE* **33** 1-25

[2] Watanabe T, Matsuoka K and Minoura S 2016 Numerical analysis for the dynamic response characteristics of the prestressed concrete sleeper *VII European Congress on Computational Methods in Applied Sciences and Engineering (ECCOMAS Congress)*

[3] Matsuoka K, Kaito K and Ishii H 2012 Statical consideration regarding the vibration characteristics and vibration factors of 24-span steel railway bridge in 86 years in service *Journal of Japan Society of Civil Engineers F4* 68 3 157-74

[4] Railway Technical Research Institute 2004 *Design Standards for Railway Structures and Commentary (Concrete Structures)* (Tokyo: Maruzen)

[5] Railway Technical Research Institute 2012 *Design Standards for Railway Structures and Commentary (Track Structures)* (Tokyo: Maruzen)

IOP Conf. Series: Journal of Physics: Conf. Series **1106** (2018) 012009 doi:10.1088/1742-6596/1106/1/012009

Computational Analysis of the Fluid-Structure Interaction occurring in a model of Two Vehicles Overtaking Each Other

Hayder Al-Jelawy[1], Stefan Kaczmarczyk[1], Matthew Cross[2], Rob Lewis[2], Nishant Singh[1], Dhirgham Alkhafaji[3]
[1] Faculty of Arts, Science and Technology, Engineering and Technology Department, The University of Northampton, UK.

[2] TotalSim Ltd. UK.

[3] College of Engineering, The University of Babylon, Iraq.

The corresponding author's e-mail: hayder_sabah56@yahoo.com

Abstract. A computational study has been conducted to investigate the transient aerodynamic forces experienced by two vehicles overtaking each other. Ahmed body model has been used to represent the vehicles. The transient effects during the overtaking scenario have been determined by using a Computational Fluid Dynamics (CFD) model. This has been achieved by developing an open source computational code. The aerodynamic effects have been investigated by emulating this event in a virtual wind tunnel. One of the generic vehicles was set to be stationary, while the other was allowed to be moving at constant speed. The Delayed Detached Eddy simulation (DDES) turbulence approach has been applied in this study based on the finite volume analysis (FVA). The computational results such as fluid forces and moments acting on the vehicle structures have been compared against published experimental results. Encouraging correlations between those results are observed. In the present work, the fluid-structure interaction (FSI) phenomenon has also been studied by developing a flexible body dynamic model. The modal and dynamic responses are determined from FEM analysis by the application of fluid transient loads obtained from the CFD model. This investigation is important to address the issue of stability and performance of the vehicle system. This work provides a significant understanding into the complex aerodynamics of an overtaking process and gives the foundation for further analysis in the area of fluid-structure interactions for more complex geometries and scenarios.

1. Introduction

On one hand, the dynamic response of the vehicle structure has been an active area of research which would affect the vehicle suspension system. This system is an important component which would alleviate the influence of forces excited by the road conditions [1]. On the other hand, the aerodynamic forces and moments are the key role of causing the vibration of any moving vehicle such as a road car, train and a lift car, see [2], [3] and [4].

In their work, [5] were first proposed the Ahmed body which is a generic car model considered as a bluff body in the community of aerodynamics as it has proximity to the ground. Ahmed body has three essential parts, namely a flat top and bottom panels, a rounded front section and the back slant is angled with a vertical back base at the rear. This model is widely studied in literature for better understanding the flow dynamics in the wake region of the body as well as the fluid forces are experienced in a wide range of configurations [6]. It has been also a reliable experimental database for several of aerodynamic investigations of CFD models [7], [8]. The flow over this type of vehicles is fully three-dimensional flow.

Therefore, the aerodynamic effects can play a significant role on cars vibration which could lead to instability. The objective of this paper is to studying and fully understanding the linear dynamic response which can be the reference of designing a more aerodynamically efficient vehicles. The above CFD common case study can be coupled to the finite element techniques to test the impact of the aerodynamic three forces and moments on the dynamic behaviour of the body. In particular, the overtaking manoeuvre causes additional forces and moments because of the sudden change of the fluid effects.

In the last two decades, an open-source code has been a common tool in academic research [9], [10] and it is attracting the attention of industrial users for multi-physics simulations and adjoined optimization as well [11], [12] and [13]. It has been a research platform and a black-box CFD and structural analysis solver. The CAD model has been generated by using solid modelling commercial software [14]. Another commercial package has been implemented to solve for the finite element analysis. This software is a CAE specific direct modeller which features a robust meshing solution and CAD clean-up workflow. This software provides a direct modelling for FEA to calculate a range of disciplines such as structural deformations, stress and vibrations. It uses matrix methods to process the finite elements which leads to the required results [15].

2. Ahmed Body geometry

The Ahmed body geometry used in [16] and scaled by 7/10 as shown in figure (1).

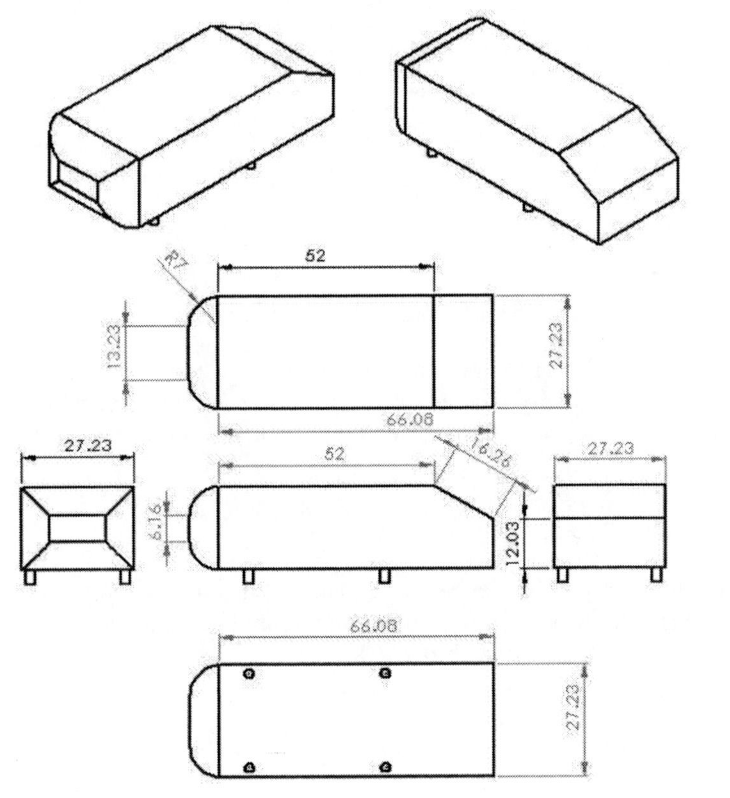

Figure 1. Ahmed bluff body, all dimensions are in cm.

As shown in the figure, the slant angle is 30° and the vehicle dimensions are (73.8, 27.3 and 20.16 cm) as the length, width and height, respectively. The ground clearance is set to be 55 mm. In addition, the Reynolds number based on the vehicle's height is equal to 3.9×10^5 as the velocity is 30.32 $^{m}/_{s}$.

3. The wind tunnel

The numerical domain is shown in figure (2). The virtual wind tunnel has been generated by using the BlockMesh dictionary which is available within the open-source software.

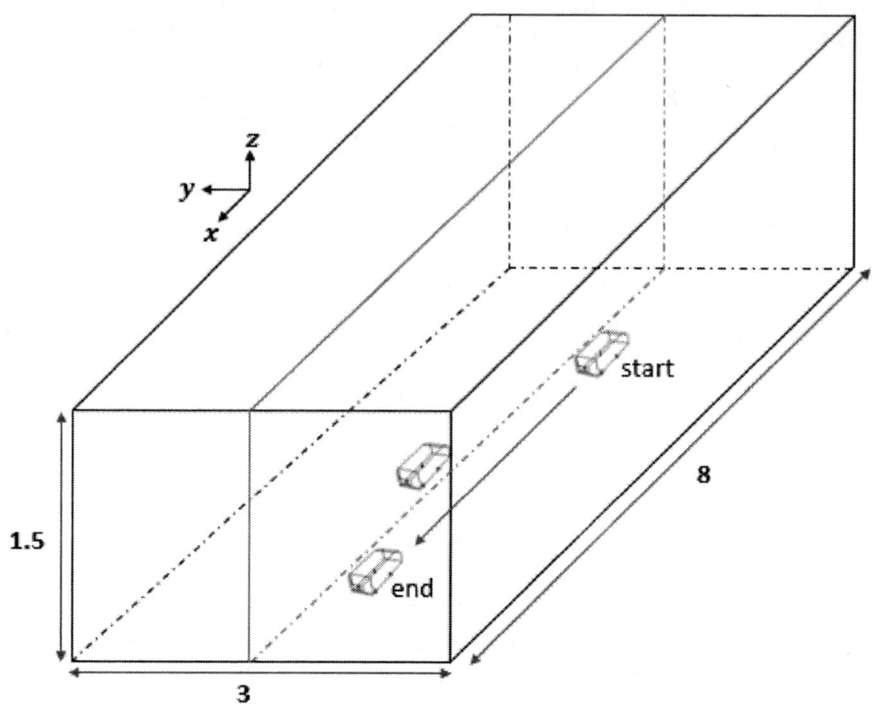

Figure 2. The wind tunnel layout

The length is set to be 8 m in order to have enough time for the simulation progress in terms of the mesh deformation. The uniform velocity was equal to 30.32 m/s at the inlet boundary. The pressure was applied to be static at the outlet boundary. The Ahmed body and the lower wall was set to be No-slip boundary conditions.

4. The fluid flow modelling

The RANS (Reynolds Averaged Navier Stokes equations) has been used to predict the flow behaviour around the vehicles. The conservation of momentum and mass equations are shown below:

$$\frac{\partial(\rho\tilde{u}_i)}{\partial x_j} = 0 \quad (1)$$

$$\rho\frac{\partial\tilde{u}_i}{\partial t} + \rho\tilde{u}_i\frac{\partial\tilde{u}_i}{\partial x_j} = -\frac{\partial\tilde{p}}{\partial x_j} + \frac{\partial}{\partial x_j}\left(\rho v\frac{\partial\tilde{u}_i}{\partial x_j} - \rho\widetilde{u_i'u_j'}\right) \quad (2)$$

where: $\widetilde{u_i}$, $\tilde{p}$ denote the mean velocity and pressure components, respectively. The Reynolds stresses are shown as $\rho\widetilde{u_i' u_j'}$; where: u_i', u_j' refer to the turbulent fluctuation values. Also, ρ denotes the air density and ν is the kinematic viscosity. The parameters also t, u_i denote the time and velocity components in the x_i direction.

In the transient state, the hybrid $\boldsymbol{SST\ k-\omega}$ turbulence model is used to solve the aerodynamic coefficients during an overtaking event of two bluff bodies [17]. The modelled equations for the specific turbulence kinetic energy (k) and the specific dissipation rate (ω) (turbulence frequency) are shown as follows:

$$\frac{\partial(\rho k)}{\partial t}+\frac{\partial(\rho k u_j)}{\partial x_j}=\widetilde{P_k}-\beta^*\rho\omega k+\frac{\partial}{\partial x_j}\left[\left(\mu+\sigma_k\frac{\rho k}{\omega}\right)\frac{\partial k}{\partial x_j}\right] \quad (3)$$

$$\frac{\partial(\rho\omega)}{\partial t}+\frac{\partial(\rho\omega u_j)}{\partial x_j}=\alpha\rho S^2-\beta\rho\omega^2+2(1-F1)\sigma_d\frac{\rho}{\omega}\frac{\partial k}{\partial x_j}\frac{\partial\omega}{\partial x_j}+\frac{\partial}{\partial x_j}\left[\left(\mu+\sigma_\omega\frac{\rho k}{\omega}\right)\frac{\partial\omega}{\partial x_j}\right] \quad (4)$$

where: P_k denotes the production limiter; $\beta, \beta^*, \alpha, \sigma_\omega, \sigma_d, \sigma_k$ denote the closure coefficients in the turbulence-rate equation to be 0.075, 0.09, 0.55, 0.5, 0.856 and 0.85, respectively [17]; μ denotes the dynamic viscosity; S denotes the strain rate; $F1$ is the first blending function.

Two blending functions are used in this numerical model to keep the k-ω model active at the region near the wall. Then, it will switch to the k-ε model outside the boundary layer region.

An open source package has been used to solve the system which is applicable for simulating a wide range of CFD applications. It is based on a cell-centred Finite Volume Method (FVM).

The SIMPLE algorithm explained in [18] has been used in the steady state to run a potential solution to initialise the flow. The simulation stability has been approached by running 500 iterations with a limited pressure gradient. The solution has been converged by running another 500 iterations without any limitation of the pressure gradient. The results from the steady state can be used as initial conditions for the transient solution.

Eventually, a transient solver has been set with a new turbulence variables for the Delayed Detached Eddy Simulation (DDES) turbulence model. The DDES model is a combination between both the speed of RANS modelling at the boundary region and the accuracy of Large Eddy Simulation (LES) modelling at the core region and so is called a hybrid solver as published in [19]. The Delayed model uses two blended functions which work very well with the high Reynolds number. Then, the hybrid PIMPLE algorithm which combines the SIMPLE and PISO algorithms which has been used for transient calculations, see [20].

5. The structure modelling

The above system can be considered as a multi-body system (MBS) with various distributed parameters such as discrete and/or continuous components. The multi-body dynamic consists of solid bodies connected to links by joints which can restrict their relative motion. The study of the MBD is to analyse the effects of excitations which can move mechanism systems and it can also be called forward dynamics. In contrast, the inverse dynamics address the necessary excitations which could move a mechanical system. The Ahmed body car suspension system can be shown in figure (3).

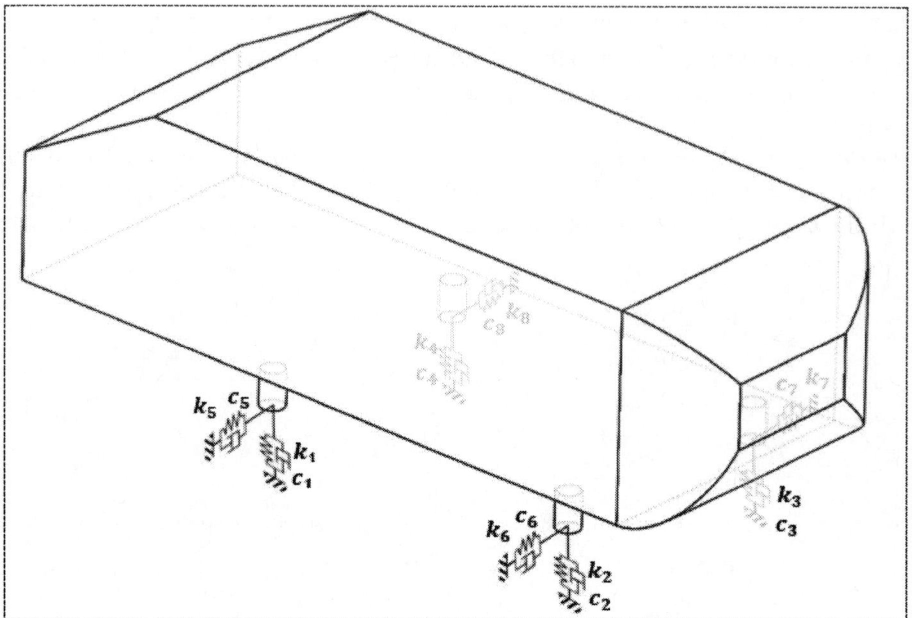

Figure 3. Ahmed body model shows its suspension system

A proper level of accuracy can be achieved and a mathematical and physical description of the mechanism can be constructed. It seems that the FEA analysis is readily applicable to predict the dynamic behaviour of this kind of systems. Ahmed body dynamics due to the fluid forces will be studied in this work, and the simulation tool has been developed to capture specific phenomena or instability.
The fluid-structure interaction is considered to be a one-way coupling method due to the small deformation. This means that only the fluid pressure acting on the body calculated from the fluid solver will be transferred to the structure solver [21].
The impulse conservation equation below has been developed by using the Lagrange's approach [22]. It can be solved by using one of the FEA approaches.

$$[M]\{\ddot{x}(t)\} + [C]\{\dot{x}(t)\} + [K]\{x(t)\} = \{F_i(t)\} \quad (5)$$

where: $[M]$ denotes the mass matrices for the body, $[K]$ and $[C]$ are the system stiffness and damping matrices, $\{\ddot{x}(t)\}, \{\dot{x}(t)\}$ and $\{x(t)\}$ represent the system acceleration and displacement vectors at time t. The fluid forces are denoted by $\{F_i(t)\}$.
The body mass is estimated to be equal to 9.5 kg and the aluminium alloys were used as a body material. The stiffness and damping of each of the springs can be obtained from the following two equations.

$$k_e = 4\pi^2 m f^2 \quad (6)$$

$$c_e = 2\zeta\sqrt{k_e m} \quad (7)$$

where: k_e and c_e denote the effective stiffness and damping coefficients, respectively; m represents the Ahmed body mass; ζ denotes the damping ratio. The natural frequency f has been chosen from one of the natural rigid modes listed in table 1. Therefore, the stiffness and damping coefficients can be equal to 1202.85 $^{N}/_{m}$ and 76.57 $^{(N\ s)}/_{m}$, respectively.

6. CFD mesh generation

Two meshing strategies are available within the OpenFOAM package. BlockMesh utility has been utilised to generate a simple block consists of 81000 cells which are fully structured as a hexahedral mesh. Then, the CAD model has been imported to OpenFOAM from SolidWorks as STL file. It was checked to be as smooth as possible and the correct orientation and position have been accomplished.

Eventually, the snappyHexMesh utility was implemented to generate a high quality hex-dominant mesh. The total mesh volume after using the snappyHexMesh technique was 17,119,330 cells. This mesh has been accomplished in fully parallel execution. The wall layers have been added on both car structures and all boundary conditions to have a better prediction to the flow near wall. The refined mesh near each car surface is shown in figure (4).

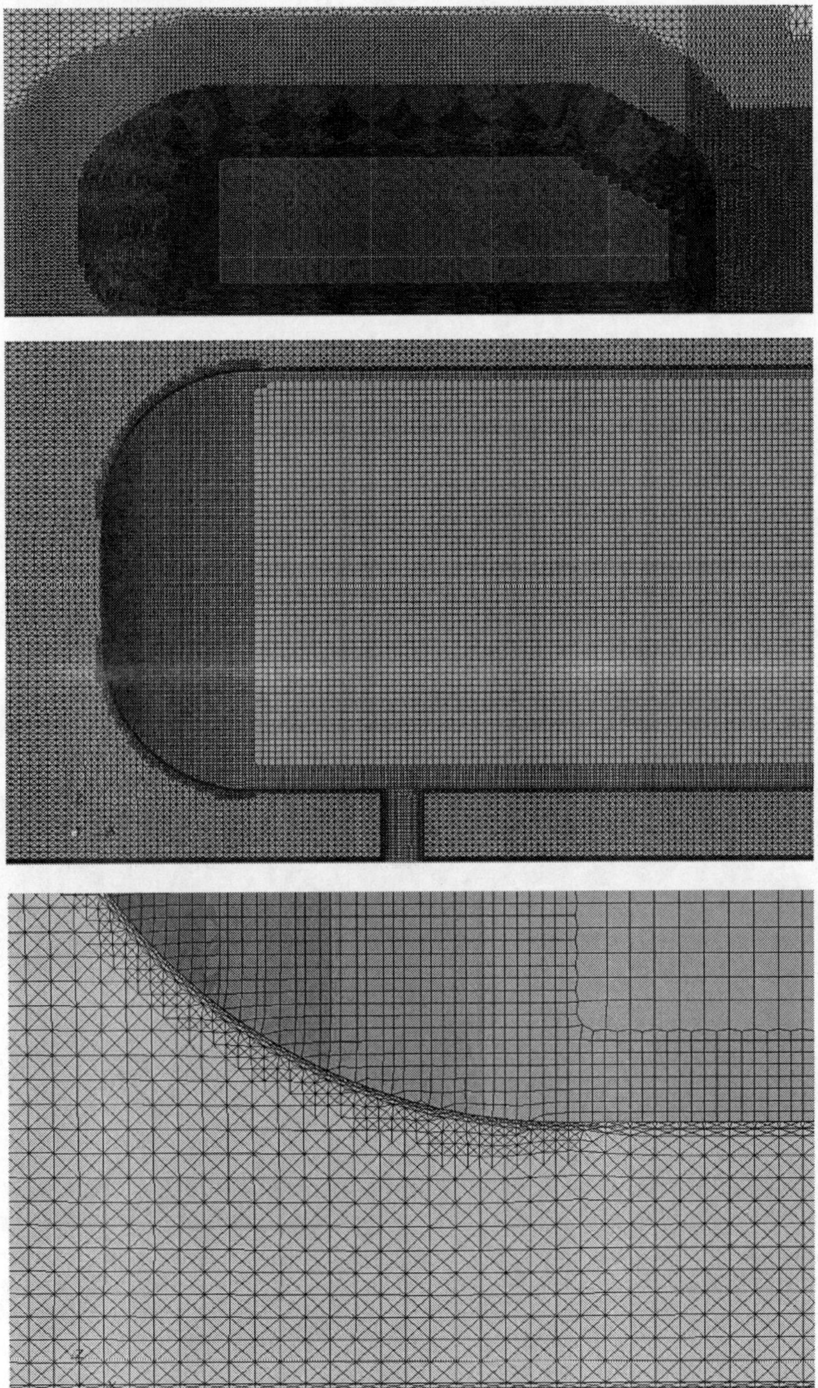

Figure 4. The surface and volume mesh of one car

The cell next to the wall thickness is equal to 0.3 mm and the wall function approach has been implemented which allows the dimensionless wall distance y^+ to be between 30 and 300 as mentioned in [23]. The y^+ average value in this simulation at the log-law region was approximately 75. The y+ is a non-dimensional distance which is used in CFD to describe how coarse of fine the mesh is for a particular fluid flow.

(a)

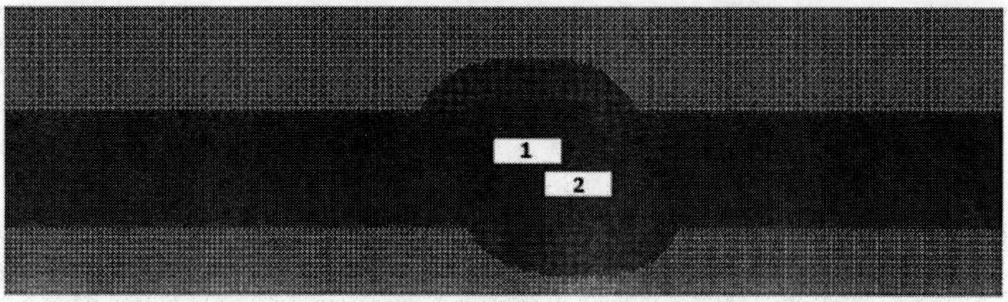

(b)

(c)

Figure 5. The computational mesh deformation for the overtaking manoeuvre. (a) Start, (b) at the overtaking event, (c) end.

The dynamic mesh technique was used to achieve the rectilinear displacement of the body which allows to control the mesh deformation and sliding. The movement of the mesh is illustrated in figure (5) at three different times of the scenario. Two subdomains were used, the stationary (overtaken) car 1 is located in the top subdomain, while, the moving (overtaking) car 2 is located in the bottom subdomain as shown in the figure above. The scenario was successfully used for two trains pass each other as illustrated in [24].

7. The FEA dynamic scenario

Figure (6) shows the body mesh as well as the suspension system at the pre-possessing stage for FEA study. The fluid forces/moments calculated during the CFD simulation have been applied through the centre of pressure of the Ahmed body car to study its dynamic behaviour. This fluid excitation was a time dependent loading which can be used to capture the transient modal dynamic response. A mixed mesh includes quadrilateral and triangular elements has been used to generate the surface mesh.
The mesh size range was set to be 10 mm to generate 20038 cells. Six sensors were located to capture the acceleration time history at particular locations.

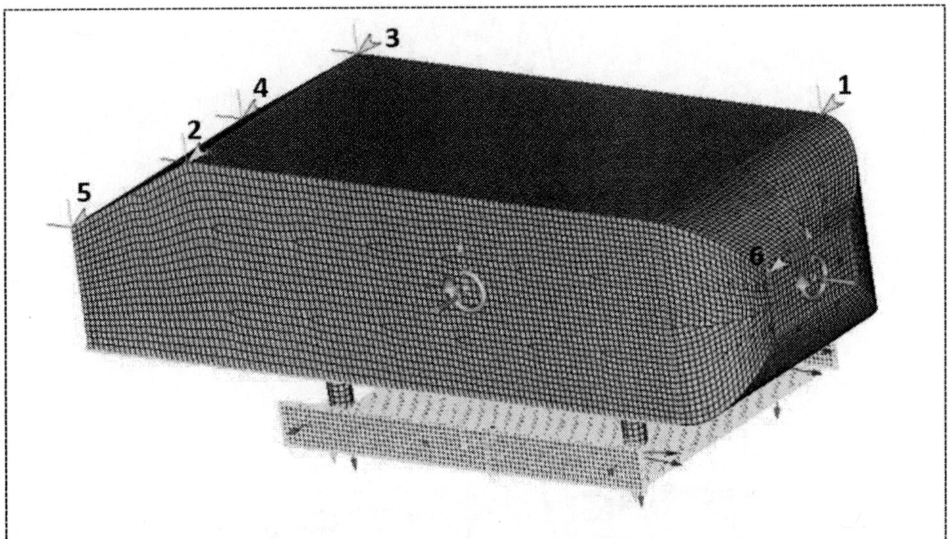

Figure 6. The location of dynamic loads and six sensors.

The simulation was accomplished in three stages, first, static pre-stiffening analysis, then, Modal extraction procedure and Dynamic analysis.
The eigen-frequency extraction procedure was conducted in order to extract the modes which includes the eigen-modes. This subset of modes is used to predict the structure's response. The analysis procedure is shown as follows:

- Processing the input data.
- Preparing Model topology data.
- Generating the model matrices.
- Assembling system matrices.
- Computing the normal modes and residual vectors has already being done in the modal analysis.
- Processing external loadings (the fluid excitation).
- Applying rigid-body constraints.
- Solving for unknowns.
- Generating model results.

8. Results and Discussion

8.1. CFD results

The aerodynamic coefficients have been obtained numerically and compared with the experimental data [25]. As shown in figure the numerical results are close to the experimental data with maximum

difference 5% between them. The focus will be on the aerodynamic effects on the overtaken car (stationary).
The side force is consistent with the experimental data between the positions from X/L=5 to X/L=1 without the influence of the overtaken car. However, between X/L=1 to X/L=0.5 the numerical side force is slightly larger than the experimental curve. The resultant fluid forces were obtained by multiplying the fluid pressure by the cross-sectional area of the body.

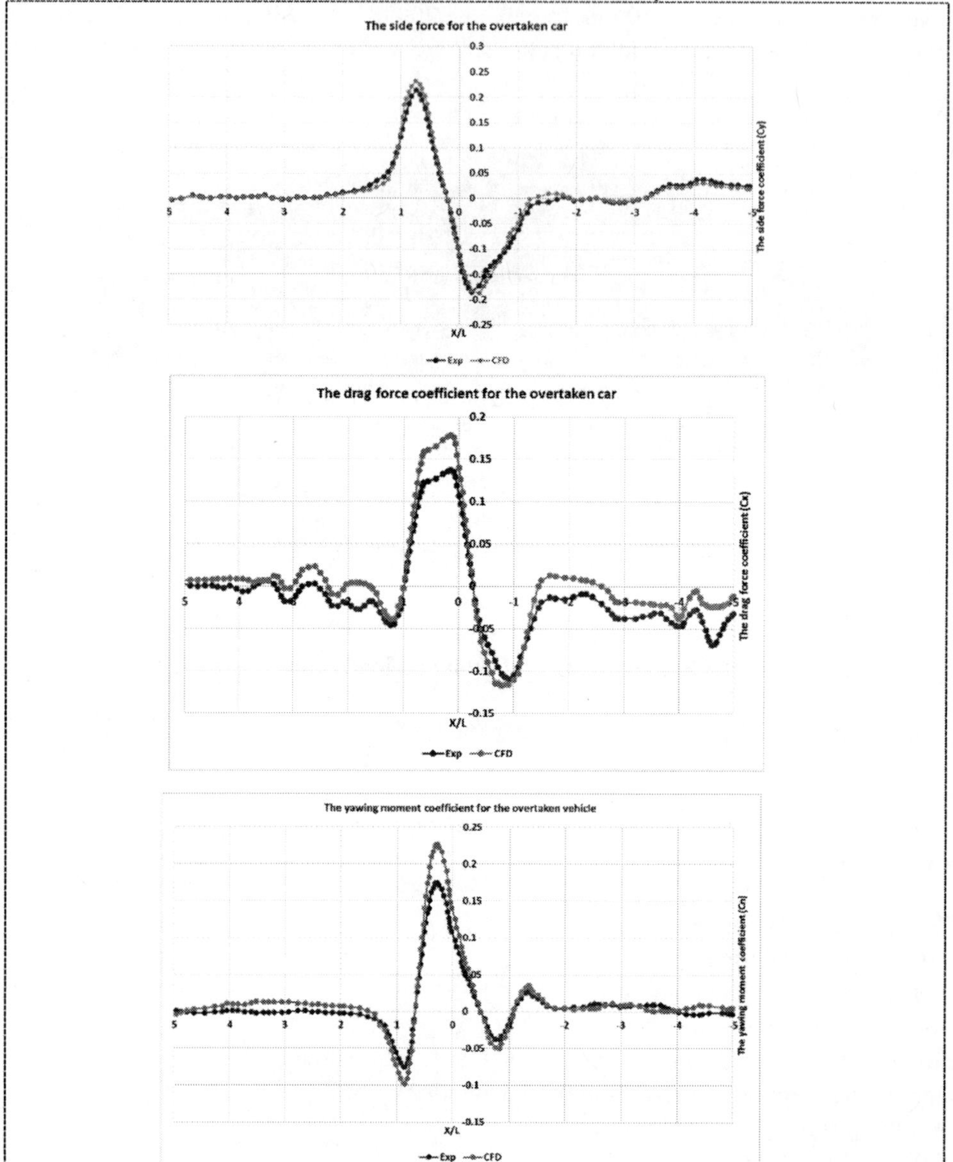

Figure 7. The aerodynamic coefficients comparison between the numerical and experimental data.

Figure (7) shows the aerodynamic coefficients acting on the overtaken (stationary) car1 during the computational analysis. The horizontal axis represents the dimensionless car location and the vertical axis represents the values of the aerodynamic force/moment coefficients. At the position between X/L=2 to X/L=1, it is clear that the drag is increasing when the overtaking event starts. This sudden impulse in the forces/moments is due to the high pressure in front of the moving car which leads to increase in pressure on the back panel of the stationary car. The pressure on the back panel of the stationary car will then be reduced because of the negative pressure on the front panel of the moving car. This reduction will continue as the cars get closer and the flow will accelerate more. On the other hand, the side force increased because of the moving pressure which causes the pressure reduction on the back side of the stationary car. The yawing moment decreases as the pressure on the inner side of the stationary car is decreasing as well.

At the position between X/L=1 to X/L=0, the drag force keep increasing before it decreases due to the low pressure affects at the back of the stationary car reduces as the front edge of the moving car overtakes the back of the stationary car. The side force behaves slightly similar to the drag force as it increases then decreases because of the pressure reduction at the inner side of the stationary car. The back side of the stationary car will is pulled into the path of the moving car which will cause the yawing moment to increase. Then, the low pressure effect also on the front half of the stationary car which causes to nose to be pulled into the path of the moving car causing the yawing moment reduction.

At the position between X/L=0 to X/L=1, the front of the moving car will overtake the front of the stationary car, therefore, the low pressure in front of the moving car will certainly decrease the pressure at the region in front of the stationary car causing the drag force to decrease. Then, the back of the moving car will be alongside the front side of the stationary car and the overtaking manoeuvre is finished at this location which will return the drag force to the normal values. The side force increases as the low pressure on the inner side of the stationary car decreases which cause the two cars to repel each other. The yawing moment increases because of the flow separation behind the two cars and then decreases again. The decrease then come from the low pressure acting on the front half of the stationary car which will try to pull this half towards the path of the moving car. The surface pressure coefficient on the two cars is shown in figure (8) for a better understanding of the aerodynamic forces/moments explained above. Figure (8) shows also a comparison between the CFD data obtained from OpenFOAM and the CFD data published in [26]. A good agreement between the data can be shown clearly.

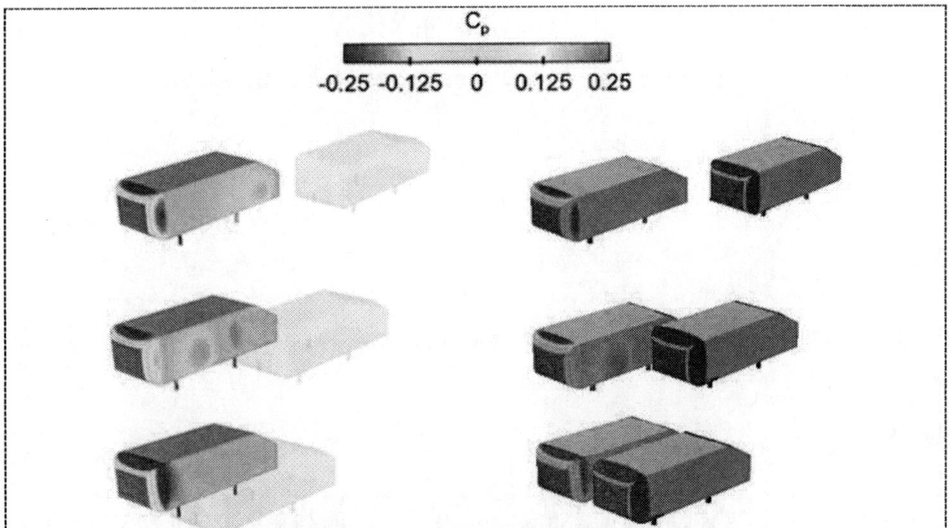

Figure 8. The comparison between two CFD data; CFD data from the open-source code used in this work (right); CFD published data (left)

The pressure coefficients were obtained by using the equation below:

$$C_P = \frac{P - P_\infty}{0.5\rho U^2} \tag{8}$$

Where: C_P denotes the pressure coefficient, $P - P_\infty$ is the relative pressure.

Note that the pressure coefficient is utilised in aerodynamic analysis to describe the pressure throughout the fluid field. It can be defined as a dimensionless number used to evaluate the airflow performance due to the presence of bluff bodies such as two moving cars inside the same tunnel. A complete aerodynamic analysis has been studied and performed to understand the effects acting on both cars. The above explanation showed that the OpenFOAM computational code is applicable to be trusted and its results are reliable. Figure (9) illustrates the all other aerodynamics effects which will be used to predict the dynamic response of the stationary car such as the lift force, the pitching and rolling moments.

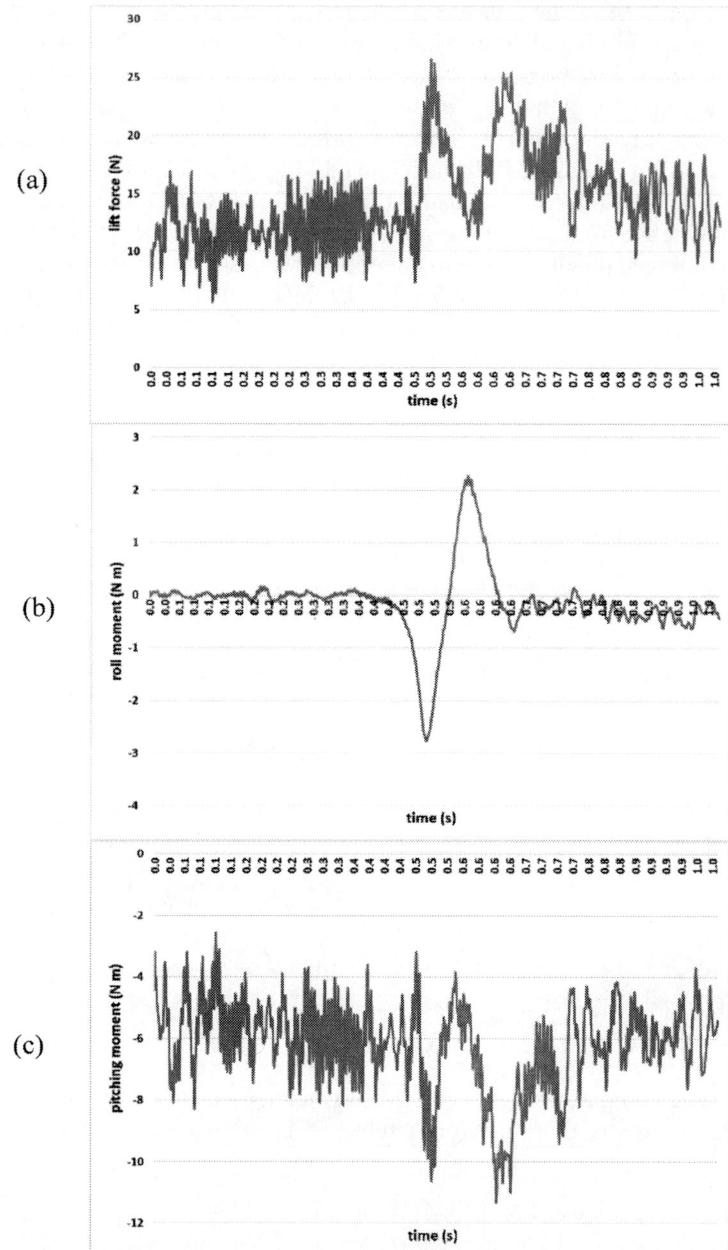

Figure 9. The time history of (a) lift force; (b) roll moment; (c) pitching moment acting on the overtaken car

8.2. FEA results

A finite element model was built by using a commercial software which is a powerful tool for simulating linear vibration analysis. The FEA software was also used for results visualisation. The stationary was modelled as a fully deformable body made of aluminium and the thickness is set to be 5mm.
Table 1 illustrates the natural modes of the Ahmed body oscillating system. The natural frequencies are calculated depending on the structure, material and boundary conditions.

Table 1. The natural modes and their natural frequencies

Mode	Natural frequency (Hz)
1	0.276
2	0.759
3	1.16
4	1.21
5	1.96
6	2.62
7	132
8	202
9	216
10	258
11	292
12	312
13	374
14	391
15	421
16	469
17	527
18	788
19	1100
20	1.5×10^6
21	2.64×10^6

The first six modes represent the rigid body modes and the modes from 7-21 represent the elastic body modes. In addition, the dynamic analysis have been accomplished by incremental method of frequency selection. The translational displacement response for the modes (3-6) can be shown in figures (10), (11), (12) and (13).

IOP Conf. Series: Journal of Physics: Conf. Series **1106** (2018) 012009 doi:10.1088/1742-6596/1106/1/012009

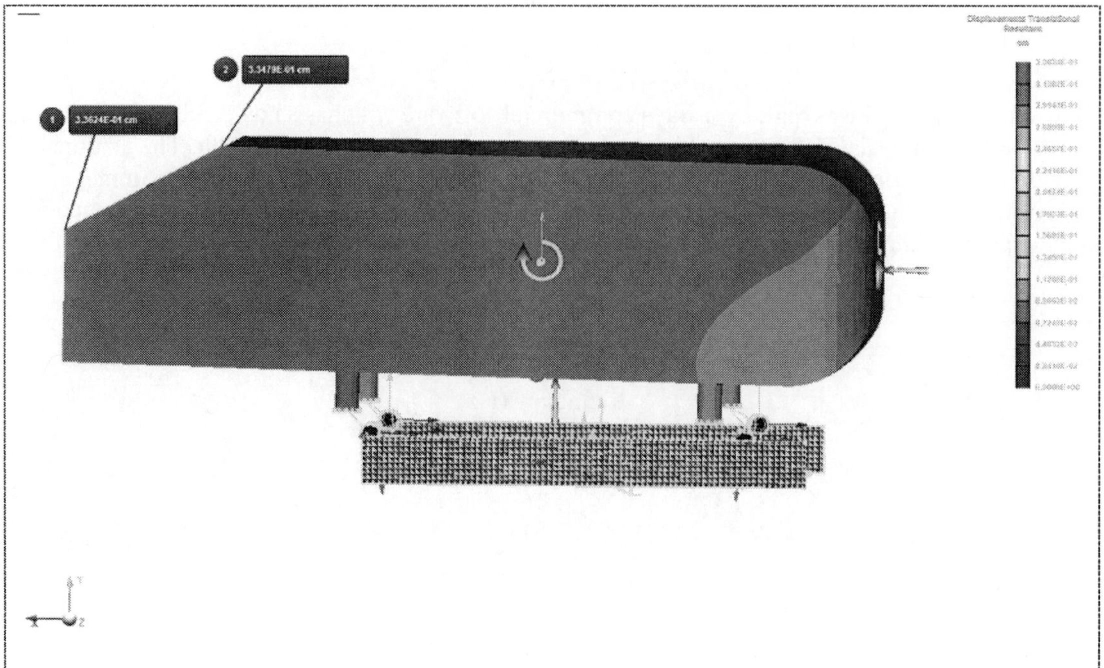

Figure 10. The fully dynamic response at the 3rd mode

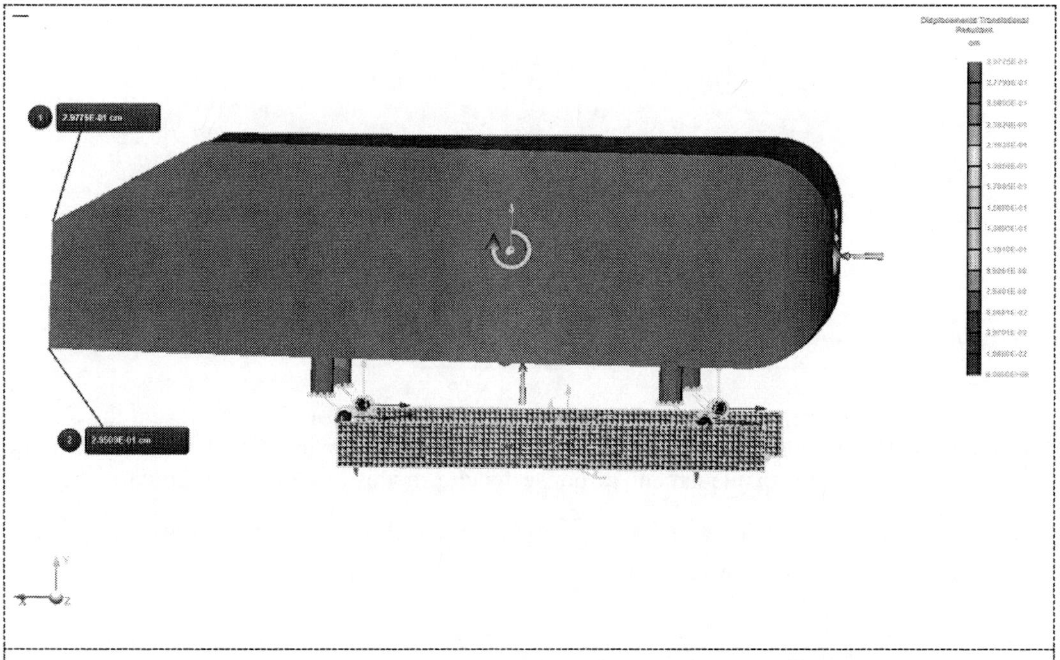

Figure 11. The fully dynamic response at the 4th mode

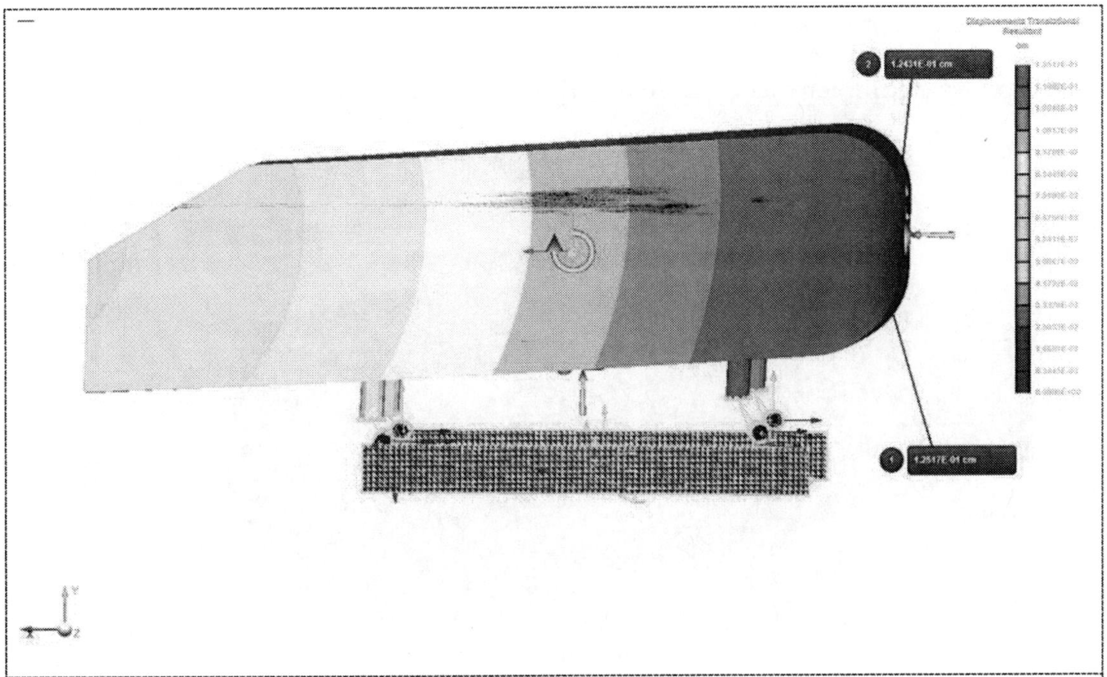

Figure 12. The fully dynamic response at the 5th mode

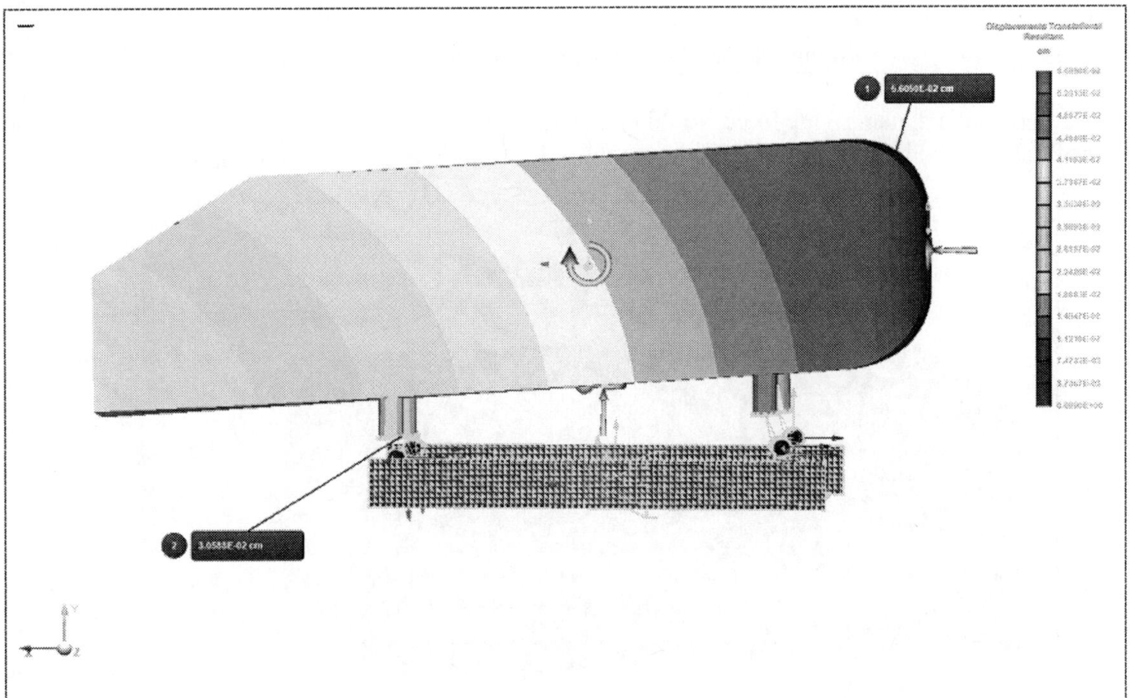

Figure 13. The fully dynamic response at the 6th mode

The maximum displacements for the rigid body frequencies are for 3rd, 4th, 5th and 6th are 0.34, 0.23, 0.19 and 0.056 centimetres, respectively.

Furthermore, the flexible body frequencies which cause the body deformation would be explained to identify the fluid forces contribution. As shown in figure (14), the top and bottom panels are influenced more by the dynamic lift force and the resultant translational displacement is equal to 7.9×10^{-5}cm.

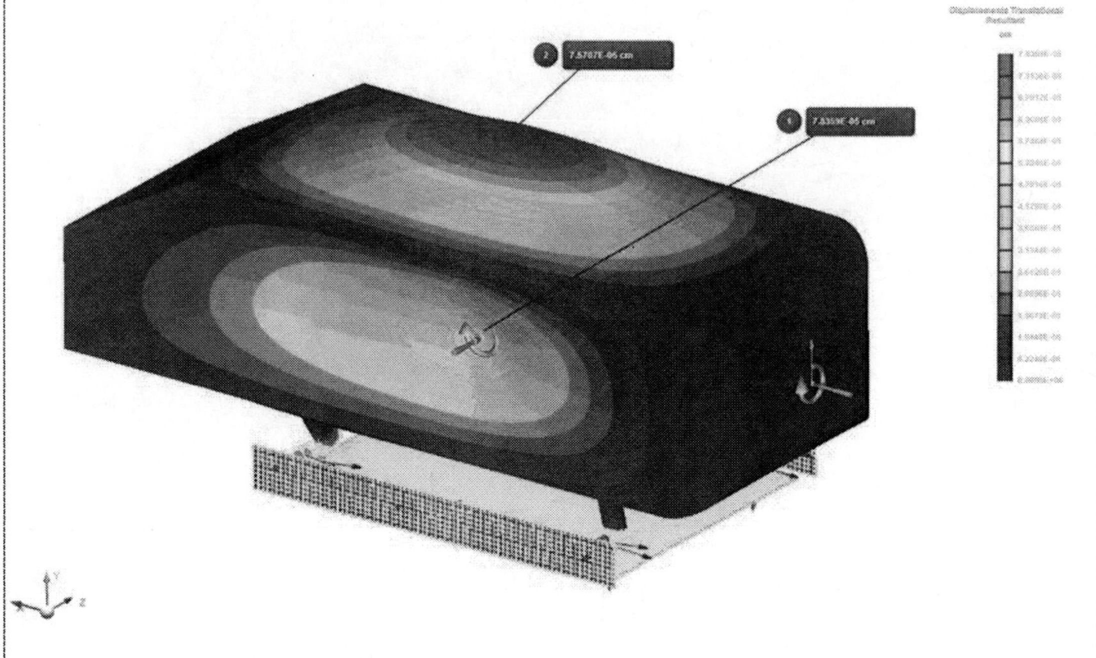

Figure 14. The fully dynamic response at the 7th mode

Nevertheless, the dynamic side force would be the main contributor in the 8th mode to deform the back half of the body bottom panel. The displacement is 1.2×10^{-4}cm as shown in figure (15).

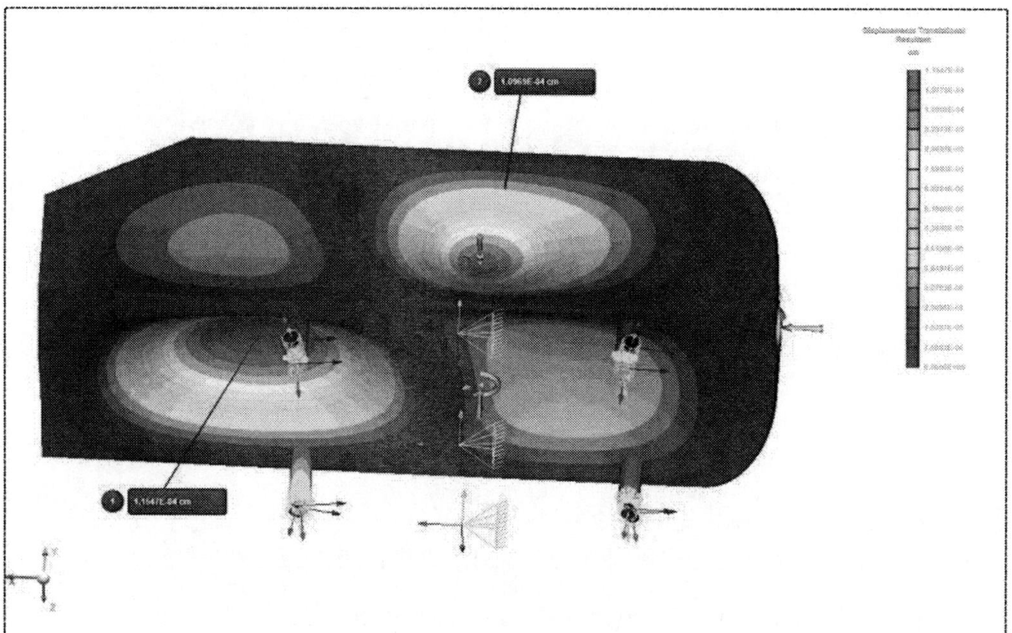

Figure 15. The fully dynamic response at the 8th mode

IOP Conf. Series: Journal of Physics: Conf. Series **1106** (2018) 012009 doi:10.1088/1742-6596/1106/1/012009

The frequency response has been calculated by the data conducted by the six sensors located on the body as shown in figure (6). The translational response shows the possible body oscillation as acceleration in a frequency domain. Figure (16) shows the body response to the lift force and yawing moment at the sensor (1) location. It is clear that the acceleration values increase at the natural frequencies. The maximum modal frequency is set to be at the 10th mode.

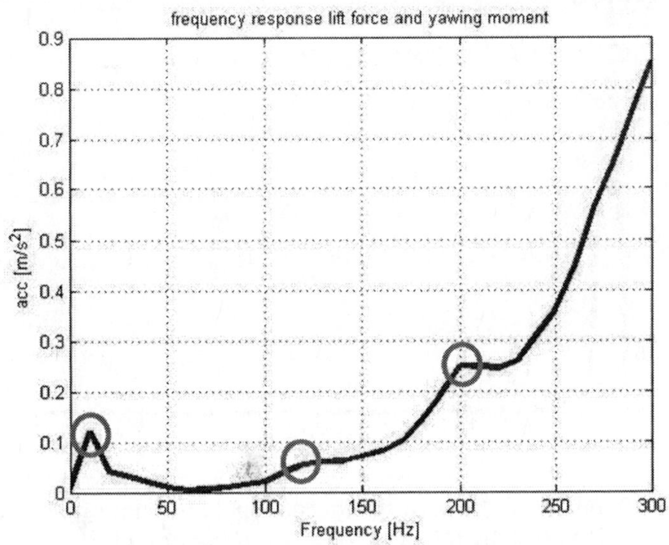

Figure 16. The frequency domain of the translational response of the body at sensor (1) cause by the lift force and yawing moment

The body response due to the dynamic side force and pitching moment is shown in figure (17). It shows also the magnitude spikes at the natural frequencies.

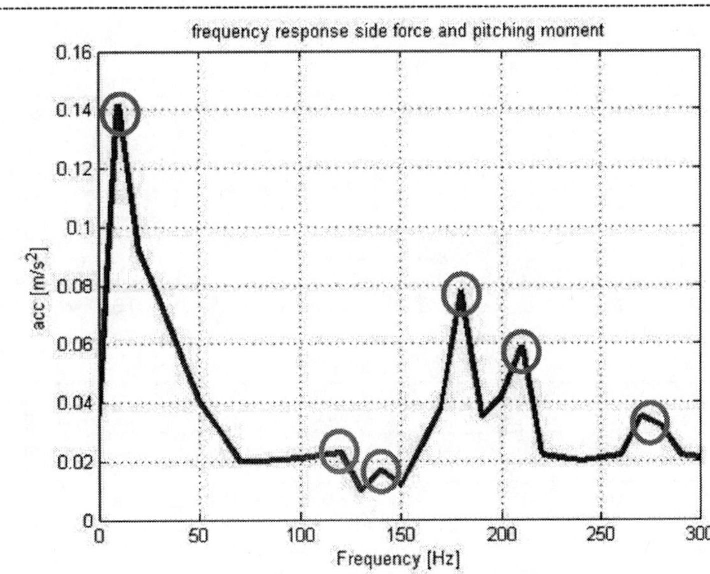

Figure 17. the frequency domain of the translational response of the body at sensor (1) cause by the side force and pitching moment

Interestingly, the drag force and rolling moment show a one high spike at the rigid body modes. While, the modes which cause deformation are not contributing as shown in figure (18).

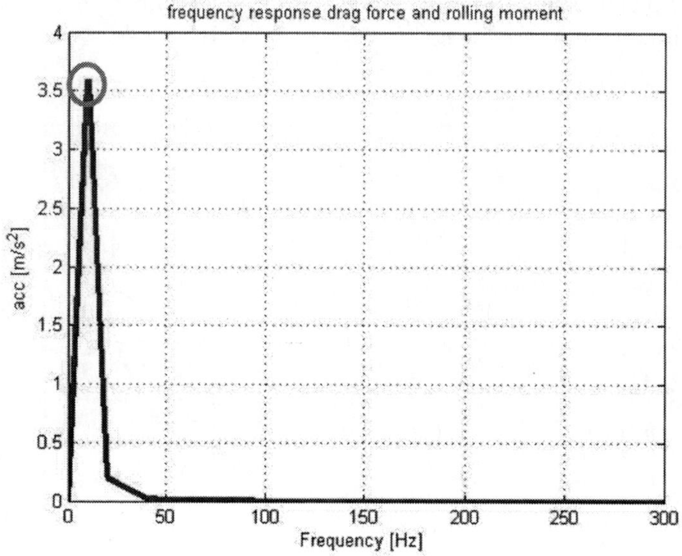

Figure 18. The frequency domain of the translational response of the body at sensor (1) cause by the drag force and rolling moment

By using the Inverse Fast Fourier Transformation (IFFT) operation [27], the frequency domain can be converted to time domain to show the body vibration as shown in figure (19).

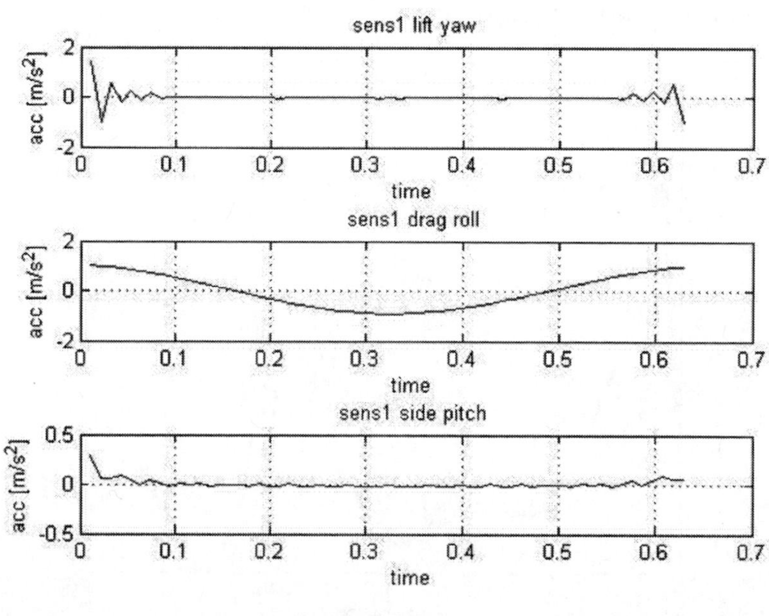

Figure 19. The time domain of the translational response of the body at sensor (1) location for the dynamic fluid excitations.

9. Conclusion

This paper examined the transient effects of fluid structure interaction of a freely vibrating Ahmed body during an overtaking manoeuvre. The Computational Fluid dynamics analysis has been achieved by the utilising an open-source package and some modifications to the code have been done. The dynamic fluid forces and moments have been constructed by running a steady and unsteady analysis.

Those aerodynamic loadings have been applied to the body as the only excitation to test their effect. A Finite Element Analysis has been accomplished by using a commercial software and the mode shapes, dynamic and frequency response have been studied. It is clear from this work that the fluid dynamics has a significant contribution to both the stationary and the moving cars. It was concluded that the transient phenomena is very important in terms of structural dynamics. This work has been done for only 1 second due to the computational cost, however, future work is required for longer CFD simulation in order to capture more data of aerodynamic effects. Overall, this study has proved that designers should consider the aerodynamic effects and make useful improvements to mitigate their effects.

References

[1] Zhang, B., Zhang, J., Yi, J., Zhang, N., Jin, Q. (2016) Modal and dynamic analysis of a vehicle with kinetic dynamic suspension system. Shock and Vibration. 2016.

[2] Sosnowski, M. (2018) The influence of computational domain discretization on CFD results concerning aerodynamics of a vehicle. Journal of Applied Mathematics and Computational Mechanics. 17(1), 79–88.

[3] Gallagher, M., Morden, J., Baker, C., Soper, D., Quinn, A., Hemida, H., Sterling, M. (2018) Trains in crosswinds – Comparison of full-scale on-train measurements, physical model tests and CFD calculations. Journal of Wind Engineering and Industrial Aerodynamics. 175, 428–444.

[4] Singh, N., Kaczmarczyk, S., Ehrl, T. (2017) An analysis of airflow effects in lift systems. In Lift and Escalator Symposium. The University of Northampton, CIBSE Lifts Group, LEIA, p. 24.

[5] Ahmed, S.R., Ramm, G., Faltin, G. (1984) Some Salient Features Of The Time-Averaged Ground Vehicle Wake.

[6] Huminic, A., Huminic, G. (2012) Numerical Flow Simulation for a Generic Vehicle Body on Wheels with Variable Underbody Diffuser.

[7] Ashton, N., Revell, A. (2015) Comparison of RANS and DES methods for the DrivAer automotive body. SAE Technical Paper.

[8] Tian, J., Zhang, Y., Zhu, H., Xiao, H. (2017) Aerodynamic drag reduction and flow control of Ahmed body with flaps. Advances in Mechanical Engineering. 9(7), 1687814017711390.

[9] Passalacqua, A., Laurent, F., Madadi-Kandjani, E., Heylmun, J.C., Fox, R.O. (2018) An open-source quadrature-based population balance solver for OpenFOAM. Chemical Engineering Science. 176, 306–318.

[10] Islam, H., Soares, C.G. (2018) Prediction of ship resistance in head waves using OpenFOAM.

[11] TotalSim Ltd. (2018) About TotalSim Experts in Computational Fluid Dynamics. [online]. Available from: https://www.totalsimulation.co.uk/about-us/ [Accessed March 14, 2018].

[12] IconCFD Ltd. (2018) ICON :: OpenFOAM®. [online]. Available from: http://www.iconcfd.com/services/consultancy/openfoam [Accessed March 14, 2018].

[13] OpenCFD Ltd (2018) Acquisition of OpenCFD Ltd., The leader in Open Source software in Computational Fluid Dynamics | ESI Group. [online]. Available from: https://www.esi-group.com/company/investors/news/acquisition-opencfd-ltd-leader-open-source-software-computational-fluid-dynamics [Accessed March 14, 2018].

[14] Solidworks, 2018. 3D CAD Design Software. Available at: https://www.solidworks.com/ [Accessed May 24, 2018].

[15] Apex, M. (2018) MSC Apex Modeler - Direct Modeling & Meshing Solution. [online]. Available from: http://www.mscsoftware.com/product/msc-apex-modeler [Accessed April 25, 2018].

[16] Noger, C., Regardin, C., Szechenyi, E. (2005) Investigation of the transient aerodynamic phenomena associated with passing manoeuvres. Journal of fluids and structures. 21(3), 231–241.

[17] Menter, F., Ferreira, J.C., Esch, T., Konno, B., Germany, A.C. (2003) The SST turbulence model with improved wall treatment for heat transfer predictions in gas turbines. In Proceedings of the international gas turbine congress. pp. 2–7.

[18] Ferziger, J.H., Peric, M. (2012) Computational methods for fluid dynamics. Springer Science & Business Media.

[19] Spalart, Pr., Allmaras, S. (1992) A one-equation turbulence model for aerodynamic flows. In 30th aerospace sciences meeting and exhibit. p. 439.

[20] Mohammad, A.F., Zaki, S.A., Ali, M.S.M., Yakub, F., Hagishima, A. (2018) Numerical Investigation of Wind-Induced Interference in a Random Urban Array.

[21] Friedrich-Karl Benra, Hans Josef Dohmen, Ji Pei, Sebastian Schuster, and Bo Wan, "A Comparison of One-Way and Two-Way Coupling Methods for Numerical Analysis of Fluid-Structure Interactions," Journal of Applied Mathematics, vol. 2011, Article ID 853560, 16 pages, 2011. https://doi.org/10.1155/2011/853560.

[22] WANG, F. and LI, S., (2018). Linearly Stable Periodic Solution for Lagrangian Equation. Journal of Computational Analysis and Applications, p.1454.

[23] Ariff, M., Salim, S.M., Cheah, S.C. (2009) Wall y+ approach for dealing with turbulent flow over a surface mounted cube: Part 2—High Reynolds number. In Proceedings of 7th International Conference on CFD in the Minerals and Process Industries CSIRO, Melbourne, Australia.

[24] KrajnoviÄ, S., Bengtsson, A., Basara, B. (2011) Large eddy simulation investigation of the hysteresis effects in the flow around an oscillating ground vehicle. Journal of Fluids Engineering. 133(12), 121103.

[25] Noger, C., Széchényi, E. (2004) Experimental study of the transient aerodynamic phenomena generated by vehicle overtaking. Flow Induced Vibration.

[26] Uystepruyst, D., Krajnović, S. (2013) Numerical simulation of the transient aerodynamic phenomena induced by passing manoeuvres. Journal of Wind Engineering and Industrial Aerodynamics. 114, 62–71.

[27] Ida, Y., Matsumoto, T., Matsufuji, S. (2018) Packet splitting and adaptive modulation based on CSI of time domain for decode-and-forward cooperative OFDM systems in different channel model. International Journal of Communication Systems. 31(2).

IOP Conf. Series: Journal of Physics: Conf. Series **1106** (2018) 012010 doi:10.1088/1742-6596/1106/1/012010

Investigation into Squeeze-Film Induced Levitation of Light Objects

A. Almurshedi[1], **T. Stolarski**[1], **M. Atherton**[1], **C. Mares**[1]
[1]Brunel University London, College of Engineering, Design and Physical Sciences UK

The corresponding author's e-mail: Ahmed.Almurshedi@brunel.ac.uk

Abstract. The performance of a system consisting of a thin rectangular plate under plane stress due to the action of PZT actuators, thus subjected to the Poisson's contraction effect, and a light circular object freely floating over it was studied both numerically and experimentally. The results show that performance of the system is influenced by a few factors, including the magnitude of elastic deformation of the plate and the electrostatic action of the PZTs attached to the plate, which combine to form oscillating dimples created by the Poisson's effect to separate the object from the plate with a film of compressible fluid. The importance of operating frequency of PZTs, shape and size of the dimple, offset and amplitude voltage under which PZTs operate, and distribution of PZTs on the plate's surface were all examined theoretically and verified experimentally. The plate was made of aluminium and stainless steel in order to find out the importance of the material's energy absorption rate for the effectiveness of squeeze-film induced levitation. The agreement between theoretical models and experimental measurements proved to be quite satisfactory.

1. Introduction

Frictionless or non-contact transportation is essential for manufacturing ultra-precision products in micro-fabrication and nanotechnology, for example, silicon wafers and integrated circuits, where tiny flaws, infinitesimal scratches and contamination must be avoided. Techniques such as air–bed flotation and magnetic levitation, have been utilised for frictionless, non-contact production–line transportation. However, air–beds require huge amounts of clean air at high cost due to exterior equipment, e.g. piping and pumps and magnetic levitation only works for ferrous material [1–3].

The literature review reveals that levitation comprises three main categories, namely: Standing Wave Levitation (SWL), Near-Field Acoustic Levitation (NFAL) and Squeeze film levitation (SFL). SWL lifts tiny particle substances placed between two structures (reflector and radiator) due to the reflection of a standing pressure wave. In a near-field acoustic levitation (NFAL) system, there is no need for the reflector as it is represented by the floating item itself.

Squeeze-film levitation (SFL) relies on a similar working principle to NFAL, occurring when the object to be floated is located adjacent to a vibrating structure. However, it is the viscosity and compressibility of a thin film of air (squeeze-film) that is trapped between the two surfaces (vibrating surface and floating object) that is the basis of SFL rather than sound pressure.

There is some confusion in the associated literature with the SFL term being used interchangeably with NFAL. While there is some similarity between NFAL and SFL, they differ as mentioned above and this paper is concerned with the SFL effect.

1.1 Squeeze-Film Levitation (SFL)

The squeeze-film levitation (SFL) occurs when a flat item is located above an oscillating structure. A basic physical realisation of SFL comprises an elastic plate, piezoelectric actuators and a levitated object with air gap. Throughout a compression stroke, a comparatively high overpressure will be generated that attempts to thrust the air out of the boundaries (Figure1). On the other hand, the air-film will turn out to be very small, and only a little air flow takes place. Conversely, throughout decompression cycle, the air-film will come to be much bigger than before, so that a comparatively small under pressure will already be adequate to pull sufficient air back into the system to recompense for the losses throughout the compression stroke. Salbu [4] used the nonlinear equation of Reynolds to explain the steady-state sinusoidal squeeze film action. This equation was used in different papers [5,6] to analyse the SFL dynamics. SWAL has a different design to the NFAL method, whereas the former is only utilised for tiny particles, the latter can be used for a heaver floating object. However, the principle underpinning both methods (SWAL and NFAL) is based on sound radiation. In contrast, the squeeze film levitation method is fundamentally based on compressing an air. It should be noted that in some studies it is assumed that there is a similarity between NFAL and SFL [7,8] but in reality there are significant differences regarding the two as aforementioned. Stolarski and Woolliscroft [9] examined levitation features for lightweight objects placed on a vibrating plate which was excited by actuators bounded underneath the plate. The study was based on the SFL phenomenon and it was found that the actuators position governed its effectiveness. Wang and Wei [10] studied the SFL generated by a mixed-modal disk excitation utilising a horn transducer design. The physical studies identified the modal shapes of the disk that produced levitation. The current investigations will focus on the conditions that the design can create a non-contact pressure capable to carry lightweight and delicate items.

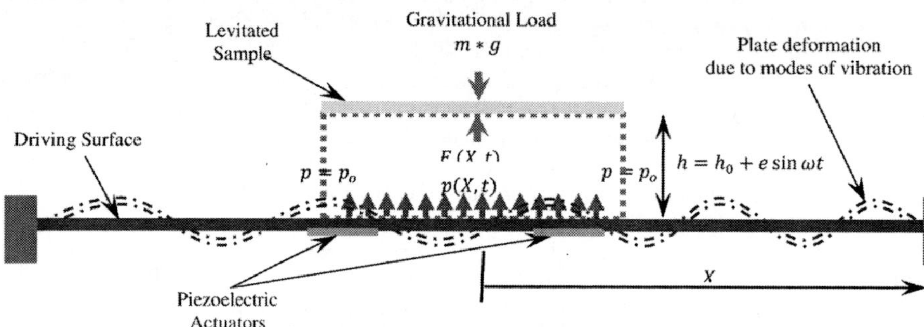

Figure 1: SFL schematic design (side view).

1.2 *Statement of the problem*

Figure 1 presents a freely moving flat item being raised as a result of the boosted average pressure formed by a vibrating surface, which is a thin flat rectangular plate clamped at both ends. A group of single layer hard piezoelectric actuators are mounted onto the lower surface. The plate can oscillate excited by the piezo actuators and a number of parameters affect the load capability: the excitation frequency, the plate boundary conditions and the materials and plate dimensions. Driving the actuators at similar resonance frequencies to the plate can lead to effective levitation [11,12]. Moreover, the attached piezoelectric actuators on the underside of the vibrating plate at a suitable location can further enhance the design's performance [13].

1.3 *Numerical simulation*

The coupled system: plate and fluid film, was modelled using ANSYS Workbench. The design consists of a thin flat plate and four rounded piezoelectric actuators bonded onto the plate, exciting the plate at various vibrational frequency. The dynamic response of the plate was determined when driving piezoelectric actuators by an AC voltage with a sine - waveform signal [11,13,14]. The piezoelectric actuator was simulated with the SOLID227 element with ten nodes with up to five degrees of freedom per node (UX, UY, UZ, VOLT and TEMP), which can be used for elastic, plastic and large strain deformations. The piezoelectric properties matrices have been included in the simulation throughout using the Piezo ACT (ANSYS Customization Toolkit). The SOLID187 element has been used to model the plate finite element. This is a higher order 3-D with 10-nodes and three degrees of freedom (3-DOF) at each node (UX, UY, and UZ). The material properties used for two different designs of the plate are presented in Table 1.

Table 1: Plates design configurations [18,19]

Design No.	Plate size [mm]	Material	Modulus of Elasticity [GPa]	Density [kg/m^3]	Poisson's Ratio
A	200*100*2	Aluminium	69	2676	0.31
B	200*100*2	Stainless Steel	193	7781	0.33

The plate dynamics must be optimised in order to maximize the vibration amplitude and minimisation of pressure leakage of air trapped between the two surfaces of the vibrating surface and floating object, which will lead to increased levitating loads. The squeeze film action takes place in an air film with a floating height at micrometre scale and the dynamics is governed by the Reynold's equation, in this complex case, a numerical solution of a 2D coupled model being determined with the air film modelled using CFX. The following conditions are assumed: air considered as an ideal gas, the flow being laminar and isothermal, the plate and load have smooth surfaces and the air layer thickness is much smaller than the size of the plate surface. The velocity gradients in the normal direction are more significant than those in the radial and tangential path, so the latter two can be ignored [11,15]. The studies revealed that the effect of the boundaries on the flow cycle show that, throughout the compression step, the growth in pressure creates an outward flow and at the same time, the air-gap between the planes reduces, permitting only little airflow to take place. On the contrary, throughout the decompression the pressure drops and the flow alters its path to being an inward flow. At this step, the air-gap rises, permitting a greater flow amount to pass in at a fairly slow velocity [16,17]. The levitated object (freely moving rigid mass) has only one degree of freedom in the vertical direction. The instantaneous location of the floating object is calculated by balancing of its inertia, gravity and the fluid flow force, which is induced by the air pressure. An air film that is squeezed between a lightweight object (a disc of mass 5 g) and an aluminium vibrating plate (Design A) is considered. In order to study the levitation performance, first, an air film of 20 μm thickness, which represents the clearance between the floating object and the vibrating plate is simulated. The area of the air film modelling is same as the floating object area. Three boundary conditions introduced are: at the top related to the levitating load surface, at the bottom modelled as a deformed boundary with an amplitude obtained from the dynamic deformation of the plate and the lateral boundary considered an open boundary allowing unrestricted flow.

2. Experimental analysis

A plate with dimensions 200×100×2 mm, clamped at both ends was excited by four foil type PZTs of circular shape (dia. 28 mm and thickness 0.5 mm) attached to the bottom of the plate's surface. PZTs were driven by a power supply unit together with a sine wave signal generator. Floating objects, in the form of a light disc, were monitored utilising a Digital Image Correlation (DIC) system (Figure 2). The experimental apparatus enabled direct observation of the localised elastic deformations of the plate for comparing them with that generated by the plate's finite element computer model. Moreover, it was also possible to measure the floating height of the object and corroborate it with the prediction of the fluid-solid interaction (FSI) model presented in the previous section. For example, for a given plate design, frequency of PZTs operation, and object mass of 5 g, the floating height was around 70 μm.

IOP Conf. Series: Journal of Physics: Conf. Series **1106** (2018) 012010 doi:10.1088/1742-6596/1106/1/012010

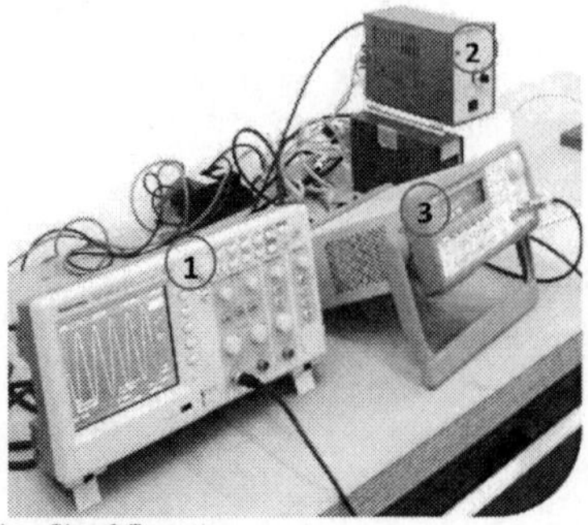

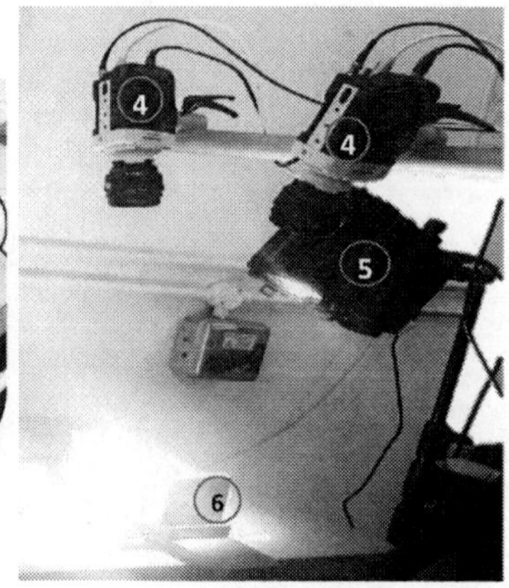

(1) Signal Generator
(2) Voltage Amplifier
(3) Oscilloscope
(4) High speed camera (DIC system)
(5) Light (DIC system)
(6) Vibrating plate design

Figure 2: Experimental rig.

3. Results and Discussion

A correlation analysis of the plate modal shapes was initially carried out between the finite element model and the experimental rig. Three types of modal shapes were observed and considered for model correlation: flexural modes in the X and Y directions and mixed wave shapes with diverse wave numbers alongside the X and Y directions coordinates (Figure 3). For load capability analysis, the plate was excited at 1.4 KHz close to a resonance frequency and the corresponding dynamic response in the X, Y directions showing a good correlation (Figures 4 and 5).

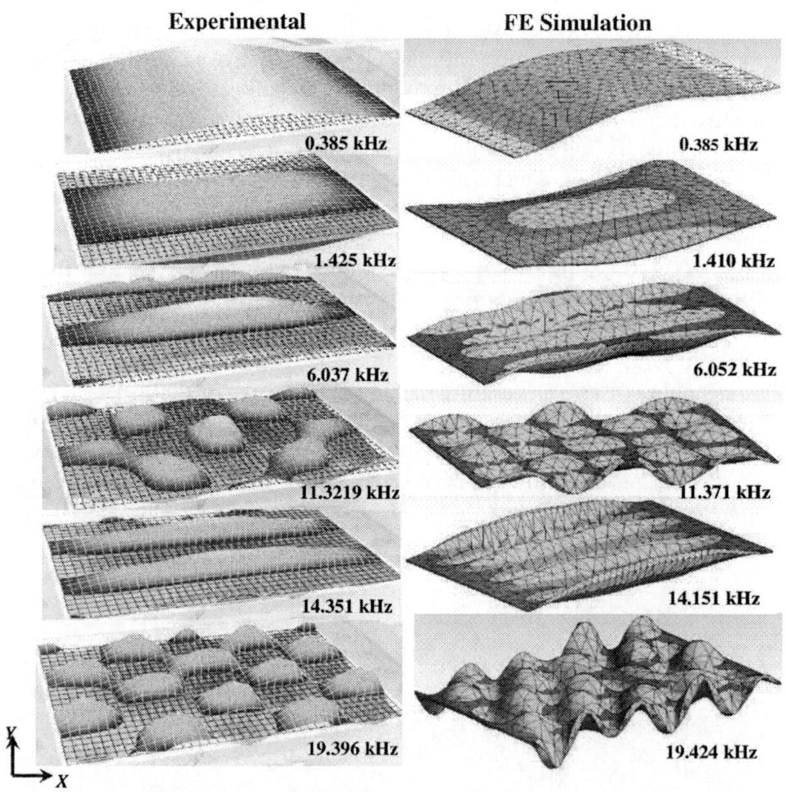

Figure 3: Experimental and simulation of modal shapes.

Figure 4: Experimental deformation.

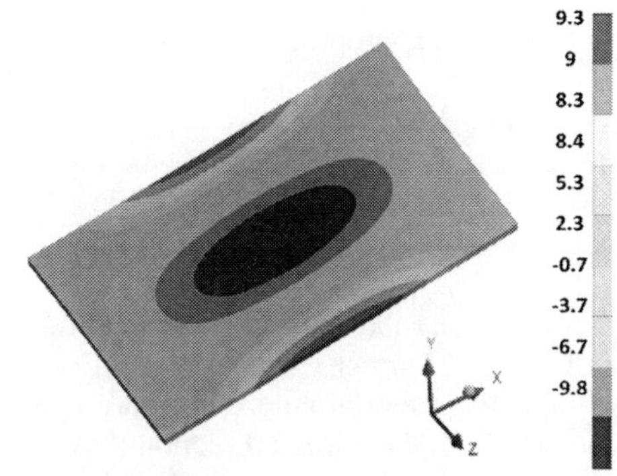

Figure 5: FE deformation.

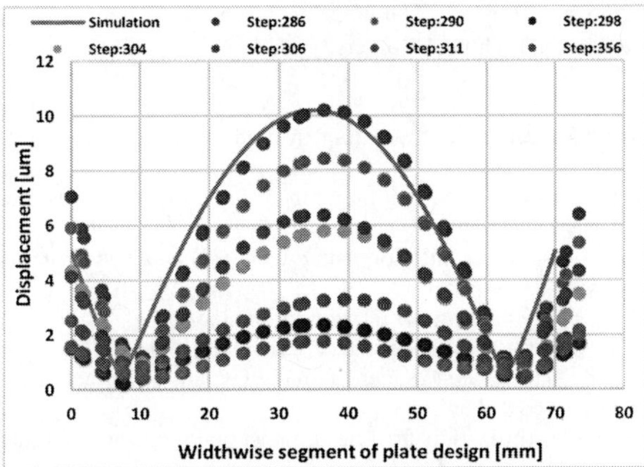

Figure 6: Theoretical and experimental plate deformation (width-wise segment).

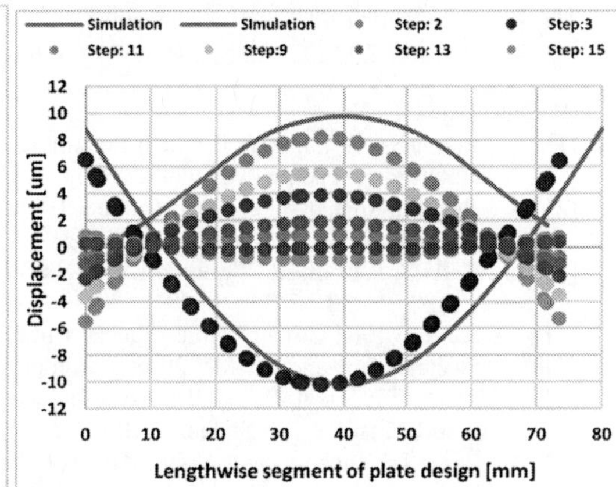

Figure 7: Theoretical and experimental plate deformation (Lengthwise segment).

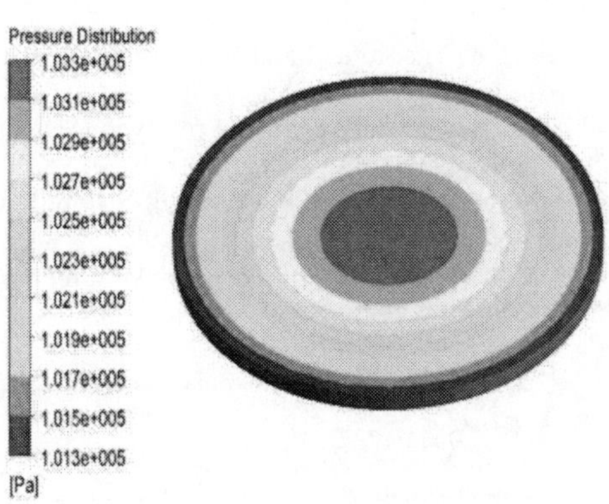

Figure 8: Squeeze air film pressure distribution.

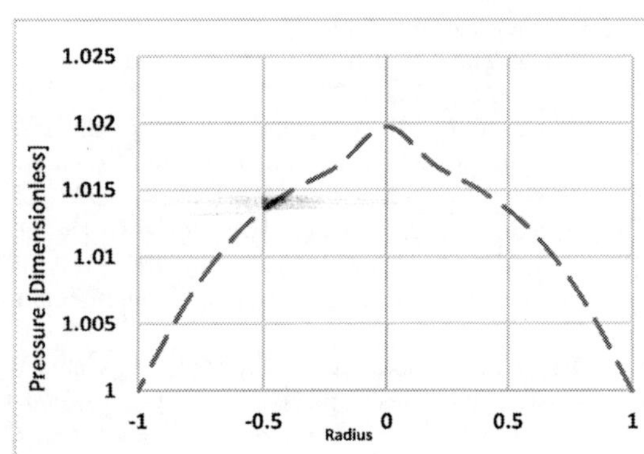

Figure 9: pressure distribution of the squeeze air film.

Furthermore, the simulated pressure profile and the resulting dimensionless profile of the squeezed air film shown in Figures 8 and 9, confirm that the average pressure is higher than the ambient pressure, leading to the levitating load movement. A significant increase in pressure can be observed in the central area, due to the appropriate vibrational excitation frequency and operational modal shape.

4. Conclusion

A coupled fluid-structure interaction model capable of simulating a squeeze film configuration was developed. The levitating capabilities of the proposed system were studied theoretically and experimentally on test rig for different configurations of vibrating plates with an optimization of the excitation piezo actuators location and operational frequency. The results of this investigation clearly demonstrate that the squeeze-film levitation can represent a practical approach for load levitation and transportation. Further studies are required concerning the levitating mechanisms and refinement of the load capability determination. The current assumption that the levitating object (disc) could be assumed to move only in the normal direction (having 1 DOF) should be extended by including the possibility of a rigid rotation while the load levitates vertically.

Acknowledgment
The authors acknowledge the financial support from the Iraqi Ministry of Higher Education and Scientific Research (http://mohesr.gov.iq/en/) and Al-Furat Al-Awsat Technical University / NAJAF.

REFERENCES

[1] Vandaele V, Lambert P and Delchambre A 2005 Non-contact handling in microassembly: Acoustical levitation *Precis. Eng.* **29** 491–505
[2] Hashimoto Y, Koike Y and Ueha S 1998 Transporting objects without contact using flexural traveling waves *J. Acoust. Soc. Am.* **103** 3230–3
[3] Ueha S, Hashimoto Y and Koike Y 2000 Non-contact transportation using near-field acoustic levitation *Ultrasonics* **38** 26–32
[4] Salbu E 1964 Compressible squeeze films and squeeze bearings *J. Basic Eng.* **86** 355–64
[5] Stolarski T A and Chai W 2008 Inertia effect in squeeze film air contact *Tribol. Int.* **41** 716–23
[6] Ma X, Xie S and Zhang W 2016 Influence of the Disturbance Frequency on the Dynamic Characteristics of an Ultrasonic Gas Squeeze Film *Tribol. Trans.* **59** 690–7
[7] Ilssar D, Bucher I and Cohen N 2014 Structural optimization for one dimensional acoustic levitation devices - Numerical and experimental study *Proceedings of ISMA 2014 - International Conference on Noise and Vibration Engineering and USD 2014 - International Conference on Uncertainty in Structural Dynamics* pp 317–29
[8] Wei B, Shaham R and Bucher I 2018 Theoretical investigation and prototype design for non-parallel squeeze film movement platform driven by standing waves *Tribol. Int.* **119** 539–48
[9] Stolarski T and Woolliscroft C I 2007 Use of Near-Field Acoustic Levitation in Experimental Sliding Contact *J. Appl. Mech.* **74** 816
[10] Wang Y and Wei B 2013 Mixed-Modal Disk Gas Squeeze Film Theoretical and Experimental Analysis *Int. J. Mod. Phys. B* **27** 1–22
[11] Chang X, Wei B, Atherton M, Mares C, Stolarski T and Almurshedi A 2016 NFAL Prototype Design and Feasibility Analysis for Self-Levitated Conveying *Tribol. Trans.* **59** 957–68
[12] Hashimoto Y, Koike Y and Ueha S 1995 Acoustic levitation of planar objects using a longitudinal vibration mode. *J. Acoust. Soc. Japan* **16** 189–92
[13] Atherton M A, Mares C and Stolarski T A 2014 Some fundamental aspects of self-levitating sliding contact bearings and their practical implementations *Proc. Inst. Mech. Eng. Part J J. Eng. Tribol.* **228** 916–27
[14] Perry M A, Bates R A, Atherton M A and Wynn H P 2008 A finite-element-based formulation for sensitivity studies of piezoelectric systems *Smart Mater. Struct.* **17**
[15] Langlois W E 1961 Isothermal squeeze films 50
[16] Stolarski T A and Chai W 2006 Self-levitating sliding air contact *Int. J. Mech. Sci.* **48** 601–20
[17] Zhao S 2010 Investigation of Non-contact Bearing Systems Based on Ultrasonic Levitation 107
[18] William D. Callister J and David G. Rethwisch 2010 *Materials science and engineering: An introduction*
[19] Theodor Krauthammer E V 2001 *Thin plates and shells, theory, analysis, and applications*

IOP Conf. Series: Journal of Physics: Conf. Series **1106** (2018) 012011 doi:10.1088/1742-6596/1106/1/012011

Neutron scattering study of strain behaviour of porous rocks subjected to heating and unconfined uniaxial compression

Nadimul Haque Faisal[a,1], Reza Sanaee[a], Rao Martand Singh[b], John Murray[a], Abbie McLaughlin[c], David Healy[d], Tung-Lik Lee[e]

[a] School of Engineering, Robert Gordon University, Garthdee Road, Aberdeen, AB10 7GJ, UK
[b] Department of Civil and Environmental Engineering, University of Surrey, Guildford, Surrey, GU2 7XH, UK
[c] Department of Chemistry, University of Aberdeen, Aberdeen AB24 3UE, UK
[d] School of Geosciences, University of Aberdeen, Aberdeen, AB24 3UE, UK,
[e] ISIS Neutron and Muon Source, Science and Technology Facilities Council (STFC), Rutherford Appleton Laboratory, Didcot, OX11 0QX, UK

[1]N.H.Faisal@rgu.ac.uk

Abstract. Neutron scattering (at ENGIN-X) was used to examine strain distributions in two hydrocarbon reservoir rock types (i.e. sandstone and chalk) under uniaxial stress (up to 35 MPa) at temperature as high as 70 °C to utilise the advantages of neutron scattering to draw a detailed picture of the structural evolution in the porous rock core plugs subjected to heating. Each sample was cylindrical in shape (i.e. 38.1 mm diameter and 48 mm length). The corresponding strain free lattice parameter (d_0) for each sample at each measurement point at 5 MPa compressive stress condition were obtained by testing at 25°C. Various microstructural and material characterisation were carried out using X-ray diffraction, energy dispersive spectroscopy and scanning electron microscopy. Comparison of Rietveld refinement for multiple peak of crystalline phases in both samples were made. Results are discussed in terms of the influence of temperature and compressive stress on the residual strain profile along radial direction of cylindrical rock sample. It was found that the sandstone sample has significantly high strain bearing capability when compared with the chalk, however, the overall strain profile from the central axis in the radial direction looked very similar.

1. Introduction

Quantifying the relationship between stress-strain state and rock failure (i.e. fracture initiation and propagation) is important in characterisation of fracture and design of a hydraulic fracturing job as a means of well stimulation. The development of predictive fracture growth monitoring tools (currently a challenge and industry interest) is critical in oil and gas, carbon sequestration, shale gas and the geothermal power industry particularly with regards to the high costs associated with field trials in extreme conditions (e.g. North Sea, Gulf of Mexico or Arctic hydrocarbon fields). There is numerous data associated with porous rock fracture investigations at macro-scale, but the value of this information is limited by inability to experimentally measure micro-scale stress-strain concentrations (leading to failure) within the sample mass near critical potential fracture planes.

Understanding and quantifying the micro-scale stress-strain relationship and porous rock material failure behaviour and having the ability to predict these quantities for identified load conditions, is crucial to all geo-mechanical studies and flow characterisation applications [1-3]. The constitutive equations governing the deformation of various rock types are typically adequate for bulk or large-scale deformation and stress analyses. However, they fail to make accurate predictions in physical and mechanical processes where highly localised heterogeneity exists or where the presence of geometric irregularities such as nano/micro-cracks are prevalent. Therefore, fracture initiation and propagation models at micro-scale are needed to understand hydraulic behaviour of naturally fractured reservoirs and design efficient hydraulic fracturing treatments particularly in presence of localised in-situ stress concentrations. Conventional rock mechanical tests subject samples to controlled load conditions and measure bulk deformations of the sample or more localized deformations only on exposed surfaces of the sample. They are currently not able to map the deformation state within the sample at high resolution.

In this research we used a neutron scattering facility (ENGIN-X) which gives high resolution and deep penetration to study localised micro-scale strain behaviour of porous rock materials. For stimulation treatments such as hydraulic fracturing, the in-situ stress profiles of a region within the rock mass can significantly influence the overall failure behaviour [4]. Therefore, the investigators believe that developing a means to measure stress-strains within samples subjected to fracture loading conditions (at high temperatures) will provide a useful tool for understanding the localised effects not captured by conventional techniques and may serve as a method for improving fracture evolution models and fracturing operations design. Neutron diffraction, unlike other measurement techniques, offers the opportunity to non-destructively measure the properties of polycrystalline materials, including porous geological materials [e.g. 1-3, 5-14]. Such properties include microstructure (grain size, grain shape, defect densities), texture (crystallographic preferred orientation) and stress.

The overall goal of this study is to better understand the micro-scale localized strain behavior of rocks and relate it to the residual stress distribution. It is important to note that time-of-flight (TOF) technique have been particularly important because they record continuous spectra that can be analysed for peak positions, peak intensities and peak shapes with the Rietveld method or multiple-peak analysis [2]. The test methodologies developed through this work can have a significant effect on developing future strategies in execution of hydraulic fracturing jobs and exploitation of naturally fractured reservoirs.

2. Methodology

The present study examines the strain distribution in two main different hydrocarbon reservoir rock types (i.e. sandstone/Clashach: SiO_2; chalk/limestone: $CaCO_3$). Neutron scattering facility (ENGIN-X) was utilised to exploit the advantages of neutron scattering to measure strain in the porous rock samples subjected to unconfined uniaxial stress at moderate to high temperature. The rock samples tested were cylindrical in shape and had 38.1 mm diameter and 48 mm length. As listed in **Table 1**, the sandstone samples were subjected to uniaxial compression stress of 20 MPa and 35 MPa (both at 70 °C), whereas, the chalk samples were subjected to uniaxial compression stress of 8 MPa (at 25 °C) and 12 MPa (at 25 °C and 50 °C), example shown in **Fig. 1b**. The compressive stress applied to the samples are different (i.e. chalk samples subjected to lower applied stress because of the lower ultimate strength). It is important to note that as per the original test plan, the chalk samples were also subjected to uniaxial compression stress of 20 MPa (at 70 °C), however, the samples failed while loading.

One set of three measurement points was evaluated across the radial direction per stress level (**Fig. 1b**). To apply temperature, the samples were mounted in a furnace and were heated at 2°C/min rate to achieve the final test temperature. Strain measurements were performed using a gauge volume of 4 × 4 × 4 mm^3 (fully submerged). Neutron diffraction data was collected at each measurement points (*a:* sample center, *b:* 10 mm and *c:* 15 mm away from the central axis in the radial direction) in the sample at each stress level, as shown in **Fig. 1b**. The corresponding strain free lattice parameter (d_0) for each sample at each measurement point at 5 MPa compressive stress condition were obtained by testing at 25 °C. The direct elastic strain in the material in the measured direction was calculated from the following expression: $\varepsilon = (d-d_0)/d_0$, where d and d_0 are the lattice parameter for the sample under stress and unstressed sample (at 5 MPa compressive loading and 25 °C in this analysis), respectively.

Sample	Pressure	Temperature	Measure
Sandstone	5 MPa	25 °C	d_0
	20 MPa	70 °C	d
	35 MPa	70 °C	d
Chalk	5 MPa	25 °C	d_0
	8 MPa	25 °C	d
	12 MPa	25 °C	d
	12 MPa	50 °C	d

Table 1. Uniaxial compression test matrix.

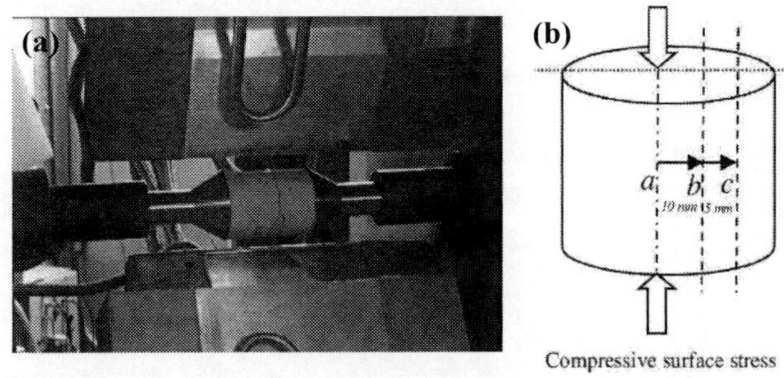

Fig. 1. (a) Uniaxial compression test (at ENGIN-X) of chalk at 5 MPa, and (b) measurement scheme.

Various microstructural and material characterisation were carried out using X-ray diffraction (XRD), energy dispersive spectroscopy (EDS) and scanning electron microscopy (SEM). Rietveld refinement for multiple peak of crystalline phases in both samples were made. The sandstone sample (Clashach) investigated previously [15] for its micro-scale fractured poro-elastic flow dynamics had the typical petrophysical characteristics as follows: porosity: 0.154, permeability: 315 mD, Young's modulus: 40 GPa, Poisson's ratio: 0.14, and compressibility: 3e-10 Pa^{-1}.

3. Results

3.1. Microstructure and X-ray diffraction

The determination of pore size distribution of a rock is often complicated, however, as shown in **Fig. 2**, the surface morphology of fractured surface reveals granular structure (average size about 100 µm to 250 µm in sandstone and 1 µm to 5 µm in chalk). The pore size ranges in nanometric-to-micrometric scale and all the spatial distribution of pores is roughly uniform.

Fig. 2. Scanning electron microscopy: (a) sandstone, and (b) chalk.

X-ray powder diffraction patterns were collected at room temperature on a PANalytical X'Pert Powder diffractometer with Cu K_α radiation. Data were recorded in the range $5° < 2\theta < 80°$, with a step size of 0.0131°. From the X-ray diffraction pattern (**Fig. 3a**), sandstone was mainly indexed as SiO_2 (quartz) with hexagonal crystal system. There was no evidence of any other polymorphs of quartz. There are some other small peaks that would tentatively be index as potassium aluminium silicate ($KAlSi_3O_8$). From the X-ray diffraction spectrum (**Fig. 3b**), chalk was indexed as calcium carbonate ($CaCO_3$) only, and there was no evidence of any un-indexed peaks. **Figure 3** also displays the crystallographic planes identified through X-ray diffraction analysis.

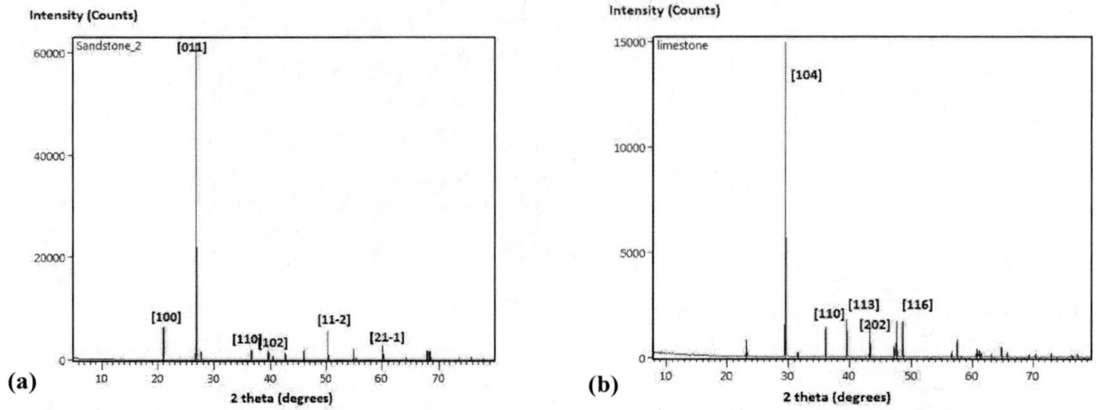

Fig. 3. X-ray diffraction spectrum: (a) sandstone, and (b) chalk.

3.2. Neutron scattering residual strain analysis

Figure 4 shows the *d*-spacing and crystallographic planes (example shown for 5 MPa compressive loading) identified through neutron diffraction analysis of the sandstone and chalk samples. From the neutron diffraction pattern (**Fig. 4**), 25 and 28 individual peaks (or crystallographic planes) were identified in sandstone and chalk, respectively. The obtained neutron diffraction pattern (**Fig. 4**) confirms the phase composition, as determined by quantitative X-ray analysis (**Fig. 3**).

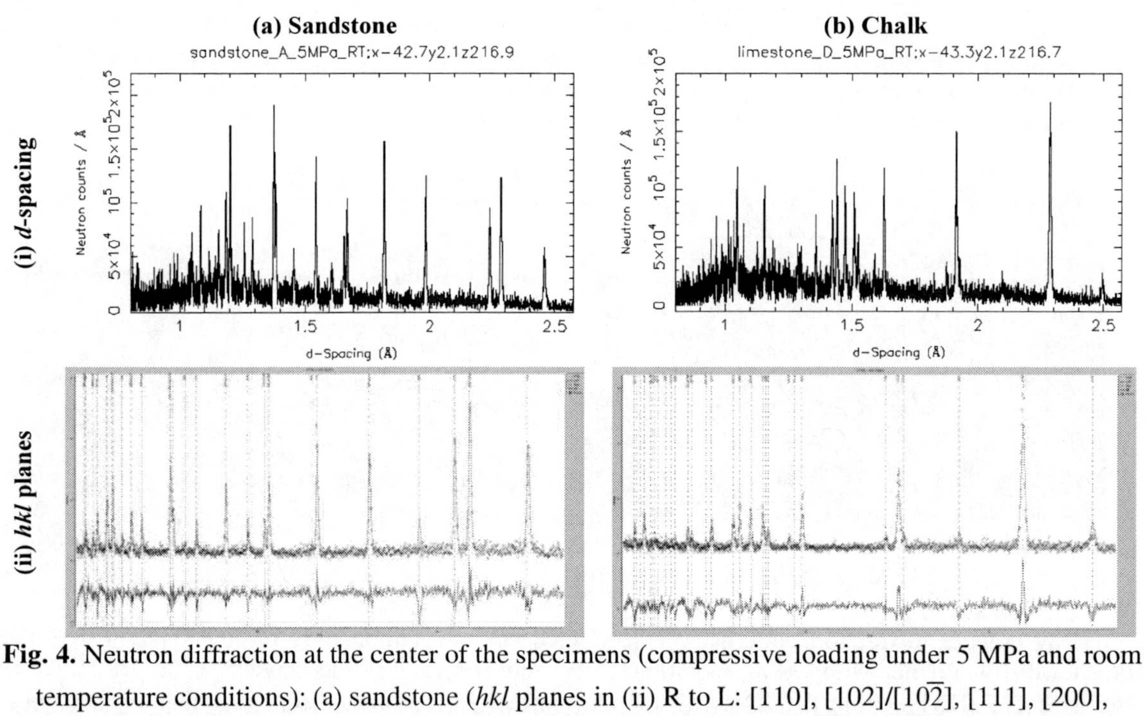

Fig. 4. Neutron diffraction at the center of the specimens (compressive loading under 5 MPa and room temperature conditions): (a) sandstone (*hkl* planes in (ii) R to L: [110], $[102]/[10\bar{2}]$, [111], [200], $[201]/[20\bar{1}]$, [112], [003], $[202]/[20\bar{2}]$, $[103]/[10\bar{3}]$, [210], $[211]/[21\bar{1}]$, [113], [300], $[301]/[30\bar{1}]$, $[203]/[20\bar{3}]$, $[104]/[10\bar{4}]$, $[302]/[30\bar{2}]$, [220], [221], $[213]/[21\bar{3}]$, [114], $[311]/[31\bar{1}]$, $[204]/[20\bar{4}]$, [222],

IOP Conf. Series: Journal of Physics: Conf. Series **1106** (2018) 012011 doi:10.1088/1742-6596/1106/1/012011

[303]/[$30\bar{3}$]), and (b) chalk (*hkl* planes in (ii) R to L: [110], [113], [202], [$20\bar{4}$], [10-8], [116], [211], [21-2], [1010], [214], [208], [119], [$21\bar{5}$], [300], [0012], [217], [$201\bar{0}$], [218], [$30\bar{6}$]/[306], [220], [1112], [223], [$31\bar{1}$], [312], [2110], [$101\bar{4}$], [314], [$211\bar{1}$]).

Comparison of Rietveld refinement (using GSAS code for ENGIN-X) for multiple peak of crystalline phases in both samples were made. Results shown in **Fig. 5** are discussed in terms of the influence of temperature and compressive stress on the residual strain profile along radial direction of cylindrical rock sample. As mentioned above and shown in **Fig. 6**, the chalk sample subjected to uniaxial compression stress of 20 MPa (at 70 °C) failed or crumbled while loading, whereas, the sandstone sample survived (e.g. **Fig. 5a**). Therefore, it can be estimated that the sandstone sample has significantly high strain bearing capability when compared with chalk sample, however, the overall strain profile from the central axis in the radial direction looked very similar.

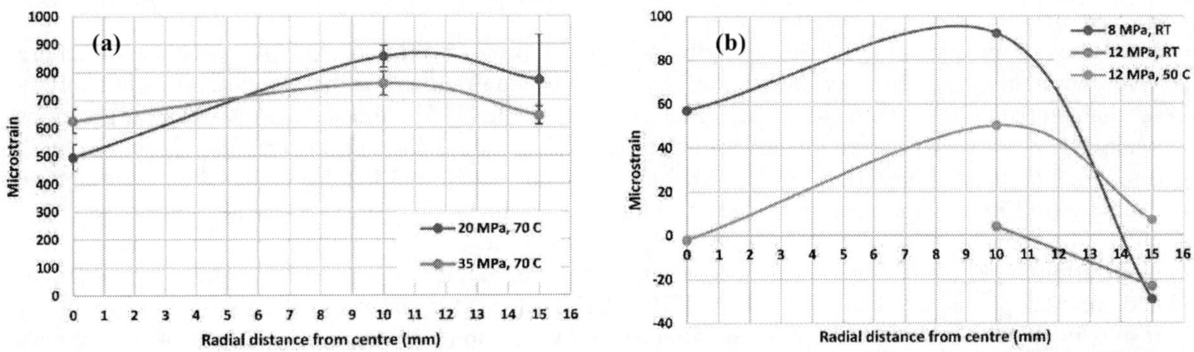

Fig. 5. Neutron diffraction residual strain (based on Rietveld analysis) comparison during uniaxial compression stress, showing strain variation along the radial direction from the centre of the sample: (a) sandstone, and (b) chalk.

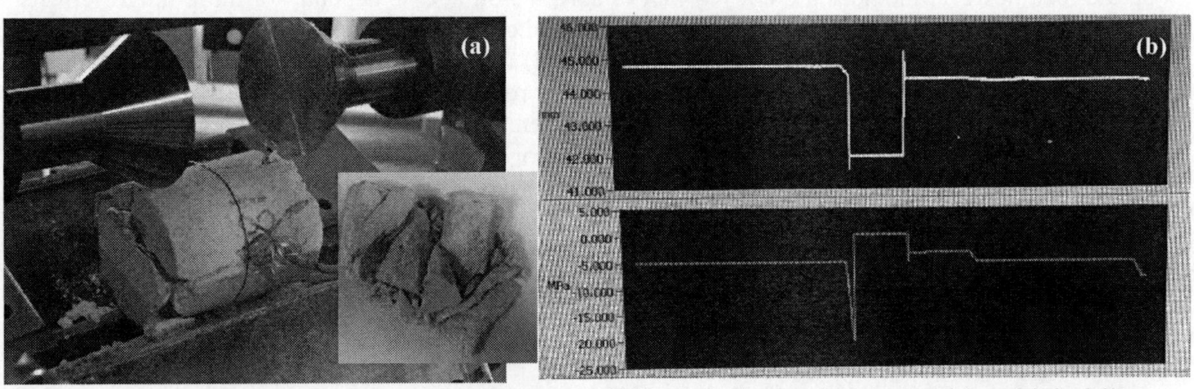

Fig. 6. Uniaxial compression stress of 20 MPa (at 70°C) of chalk sample fracture: (a) multiple failure during compression which includes diagonal shear planes, vertical fractures, vertical splitting, shear, conical and spalling (see inbox), and (b) displacement and stress jump leading to failure.

4. Discussion

Through the enhanced strain analysis of rock core samples both in this study and in previous and future investigations, it can be better understood how rocks may behave (i.e. deformation and fracture) under the different pressures induced through drilling and other production operations.

4.1. Deformation

Laboratory based uniaxial compression tests are generally conducted to determine the unconfined compressive strength of a rock (i.e. measuring whole-rock properties [5]). However, the difference between lattice strain data using neutron scattering and mechanically measured values during compression loading (e.g. **Fig. 5**) can reflect the influence of specific changes within the sample such as changes in the orientation of grains as well as the reconstruction of the pore volume or the anisotropy of the elastic constants within the involved crystals [8], but these can be part of further detailed analysis. The most precise way to analyse this phenomenon would be to track the distribution of stresses and strains inside the rocks (e.g. **Fig. 1b**) as they undergo mechanical loading but the scientific community in not near to succeeding in this task as of yet [18], however, *in-situ* neutron diffraction strain analysis may be useful in characterising deformation in rock samples (e.g. **Fig. 5**). To check the discrepancy of macro- and micro-strain for rocks of different structural characteristics, additional investigations of the microstructural change after deformation are needed [8].

4.2. Fracture

Fundamental understanding of the strength and fracture process of rocks such as sandstone which are very commonly found in reservoir formations and in mining requires realistic modelling and analysis of characteristics from both the single-grain and bulk scales [16]. Fracture takes place in locations where there are weaker stress planes in the rock; displacement of the grains usually appears to be in directions perpendicular to the lowest principal stress under external loading. Also shown through an example (in **Fig. 6a**), the common failure mechanisms for this kind of test are; axial splitting, spalling, single shear fracture or multiple fractures [17]. Although these failure modes represent a range where by a certain rock failure may be categorized, predicting the failure mode exactly is still a very difficult and complex procedure. The standard (ATSM D7012-04 [18]) for uniaxial compression tests generally results in axial splitting in a direction parallel to the maximum principal stress because of wing cracks that cannot be tolerated. These wing cracks are generated from tensile stresses exceeding the maximum tensile strength at the tip of any defect present in the rock.

Crack propagation is largely down to the relative displacement of the individual grains that make up the rock however methods are not yet well established on how to account for these 'micro-effects' in predicting bulk strength, and neutron diffraction could be beneficial. Methods like the finite element method (FEM) and the discrete element method (DEM) are advanced computational approaches that require realistic inputs to more easily accurately predict rock characteristics. To further investigate the influence of temperature on the residual strain a mathematical model approach can be applied. This approach can be based on contact mechanics (confined or unconfined) of cylindrical sample to predict the average strain due to the compressive loading and temperature. Computational numerical simulations as part of further work can allow the visual assessment of stress acting inside a rock which can provide insight into where and when the rock is likely to fatigue and fail.

Future application of ENGIN-X could be used to link pore orientation, pore volume, and mineral orientation for systematic studies on flow and mechanical properties of the microstructure, including testing at cryogenic conditions (currently under investigation by the authors [19]).

5. Concluding remarks

Compressive stress behaviour and general understanding of rock mechanics are essential to predict the wellbore stability, choose the appropriate drilling fluid and the type of hydraulic fracture. Rock failure is dependent on the ultimate strength of the material; a factor that is practically hard to determine when preparing to drill into a formation beneath the surface or assessing the wellbore stability. We have therefore presented a preliminary description of porous geological samples in micro-scale (sandstone and chalk), where the analysis gives new insights into the strain (deformation) distribution in radial direction of cylindrical sample during uniaxial compression loading, with following key conclusions:

a. From comparison of Rietveld refinement for multiple peak of crystalline phases in both samples, it was found that the sandstone has significantly high strain bearing capability when compared with chalk, however, the overall strain profile from the central axis in the radial direction looked very similar.

b. Comparison of the residual strain profile between the sandstone and chalk indicate that the average residual strain in both samples are largely tensile with some compressive residual strain in the chalk (near the outer periphery of the cylindrical sample).
c. This application of neutron scattering also makes it possible to target microscopic volumes and at the same time collect statistically relevant data for large sample volumes.

References

[1] Voigtlaender A, Leith K, Krautblatter M, Walter J M and Scheffzük C 2015 *Geophysical Research Abstracts* **17** EGU2015-11269-1
[2] Wenk H R 2005 *Geophysical Research Abstracts* **7** 03911
[3] Darling T W, TenCate J A, Brown D W, Clausen B and Vogel S C 2004 *Geophysical Research Letters* **31** L16604
[4] Sanaee R, Shadizadeh S R and Riahi M A 2010 *International Journal of Rock Mechanics and Mining Sciences*, **47(4)** p 599-605
[5] Schofield P F, Covey-Crump S J, Stretton I C, Daymond M R, Knight K S and Holloway R F 2003 *Mineralogical Magazine* **67(5)** p 967-987
[6] Wenk H R 2006 *Reviews in Mineralogy and Geochemistry* **63(1)** p 399-426
[7] Wenk H R, Lutterotti L and Vogel S C 2010 *Powder Diffraction* **25(3)** p 283-296
[8] Frischbutter A, Neov D, Scheffzuk Ch, Vrana M and Walther K 2000 *Journal of Structural Geology* **22** p 1587-1600
[9] Wenk H R, Lutterotti L and Vogel S C 2003 *Nuclear Instruments and Methods in Physics Research A* **515** p 575-588
[10] Drobek T, Strobel J, Park S, Phuong K L, Altermann W, Lemmel H, Lindner P and Stark R W 2011 http://jgmaas.com/SCA/2011/SCA2011-48.pdf [SCA2011-48] p 1-6
[11] Radlinski A, AGSO Research Newsletter 1999 http://www.agso.gov.au/information/publicationslresnews/ p 29-30
[12] Radlinski A P, Ioannidis M A, Hinde A L, Hainbuchner M, Baron M, Rauch H and Kline S R 2002 www.jgmaas.com/SCA/2002/SCA2002-35.pdf [SCA2002-35] p 1-12
[13] Kahlel A, Winkler B, Radulescu A and Schreuer J 2004 *European Journal of Mineralogy* **16** p 407-417
[14] Wenk H R 2012 *JOM* **64(1)** p 127-137
[15] Sanaee R, Oluyemi, G F, Hossain M and Oyeneyin M B 2013 In: *Proceedings of the 47th U.S. rock mechanics/geomechanics ARMA symposium*, 23 - 26 June 2013, San Francisco, US. Alexandira, VA: ARMA, document ID 13-442
[16] Antony S, Olubenga A and Ozerkan N 2017 *Rock Mechanics and Rock Engineering* **51(2)**, p 451-464
[17] Mardalizad A, Scazzosi R, Manes A, Giglio M 2018 *Journal of Rock Mechnaics and Geotechnical Engineering* **10** p 197-211
[18] ASTM D7012-04. Standard test method for compressive strength and elastic moduli of intact rock core specimens under varying states of stress and temperatures. West Conshohocken, USA: ASTM International; 2004
[19] Faisal N H, Sanaee R, Singh R M, Mclaughlin A, Lee T-L, Kelleher J 2018 Neutron scattering of porous rocks subjected to uniaxial compression in cryogenic conditions for offshore drilling operations in Arctic, Neutron beamtime by STFC/Engin-X (RB1810235); https://data.isis.stfc.ac.uk/doi/INVESTIGATION/92924528/

Acknowledgements

The authors acknowledge the award of ENGIN-X beam time at the STFC ISIS Facility (experiment number RB1720331 (September 2017) and RB1810235 (April 2018) for the neutron diffraction measurements of geological rock samples.

IOP Conf. Series: Journal of Physics: Conf. Series **1106** (2018) 012012 doi:10.1088/1742-6596/1106/1/012012

Lattice strain and texture development in coarse-grained uranium – a neutron diffraction study

P Earp[1,4], S Kabra[2], J Askew[3] and T J Marrow[1]

[1] Department of Materials, University of Oxford, Parks Road, Oxford, OX1 3PH, UK
[2] ISIS Neutron and Muon Source, Science and Technology Facilities Council, Rutherford Appleton Laboratory Didcot, OX11 0QX, UK
[3] AWE plc, Aldermaston, Reading, Berkshire, RG7 4PR, UK

philip.earp@materials.ox.ac.uk

Abstract. A neutron diffraction study of deformation in a cast uranium has been conducted for the first time. Lattice-scale plasticity in this coarse-grained material initiates at a lower stress than in a previous study of fine-grained material in the literature. This is attributed to a combination of larger thermal residual stresses in the coarse-grained material and the Hall-Petch effect making twinning easier in large grains. Asymmetry between the tensile and compressive response shows that twinning is the dominant plastic deformation mechanism at low strains. Axial texture changes for the cast uranium were calculated by post processing of the full diffraction spectra, which shows that lattice rotations associated with twinning occurred at yield. This lattice rotation was observed to disappear after unloading, which indicates that de-twinning can occur in uranium. ©British Crown Owned Copyright 2018/AWE

1. Introduction

Understanding fracture processes in materials such as cast alpha-uranium is a complex task; the combined effects of large grains [1], residual stresses and anisotropic mechanical properties creates unusual responses to deformation, and variable mechanical properties [2,3].

Uranium metal is highly elastically anisotropic, both elastically and plastically. In the room temperature alpha phase it has an orthorhombic crystal structure [4,5], an anisotropic coefficient of thermal expansion [6], and each grain has a limited number of slip modes; in particular, there are no active slip modes at room temperature to accommodate deformation perpendicular to the (001) plane [7]. Consequently, it deforms extensively by twinning, the cooperative shear of sections of the crystalline lattice, with (130) and '(172)' twins commonly observed [8]. Incompatibilities between deforming grains can result in stress concentrations that may initiate fracture. To fully understand local processes such as fracture, there is a need to develop models, such as the crystal plasticity finite element (CPFE) method, which can be used to predict the strength of engineering components whilst incorporating the internal microstructure of the material. A number of visco-plastic self-consistent (VPSC) modelling schemes have been developed for fine-grained wrought uranium grains [9] which has then been incorporated into a finite element framework [10] which was then used to model compressive and rolling deformation. Uranium presents unique challenges to the CPFE modelling

[4] Author to whom any correspondence should be addressed

IOP Conf. Series: Journal of Physics: Conf. Series **1106** (2018) 012012 doi:10.1088/1742-6596/1106/1/012012

community, specifically how to incorporate the effects of twins into a dislocation-based model [11], and how to experimentally validate these models in coarse-grained microstructures.

This paper presents results from a neutron diffraction experiment with measurements performed on coarse-grained cast uranium under tension and compression. Analysis is performed using both single-peak fitting and Rietveld refinement, with a focus on lattice strain accumulation and texture changes due to the applied deformation.

2. Experimental methods

2.1. The ENGIN-X Diffractometer

The neutron diffraction experiment was performed at the ENGIN-X beamline [12], ISIS Neutron Source, Rutherford Appleton Laboratory, UK under experiment number 1710227. The specimen is held between the grips of a horizontally mounted 100 kN Instron servo-hydraulic stress rig at an angle of 45° to the incoming beam direction, and two neutron detector banks are located as shown in figure 1. Bank 1 can measure changes in lattice spacing in the family of grains that have diffracting plane normals parallel to the loading axis and detector bank 2 measures in the orthogonal direction. The diffracting gauge volume is defined by slits and radial collimators positioned in front of the detectors.

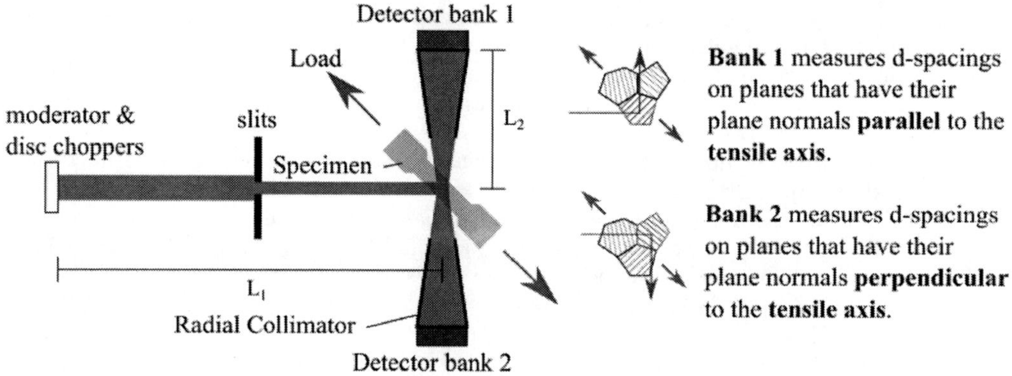

Figure 1. Schematic plan view of the ENGIN-X beamline at the ISIS neutron source.

2.2. Specimen design and microstructural characterisation

The depleted uranium was provided by AWE plc (UK) as a cast plate; the material contained many carbides.

The tensile specimens had a gauge diameter of 8.00 mm and a parallel gauge length of 31.42 mm, and the compression specimens a 14.00 mm diameter and 28.00 mm.

Grip extensions (HSLA steel) reduced the potential risk of material transfer to the ENGIN-X stress rig grips. In order to satisfy the radiation protection requirements, a conventional strain gauge could not be attached to the specimens, and the gauge region of each specimen was wrapped in aluminium foil to reduce the risk of contamination if the sample were to fracture. Digital image correlation was therefore used to measure the displacement of the sample grips at each stage of the test. The total strain was calculated from the relative displacement in the horizontal x-direction (parallel to the loading axis) of the moving grip with respect to the fixed grip using a gauge length equal to the specimen length between the grips.

The material's microstructure was investigated using scanning electron microscopy (SEM) and electron backscatter diffraction (EBSD). Cuboid specimens (6 mm × 6 mm × 1.5 mm) from the same cast plate were prepared by grinding and polishing to 3μm diamond, then electropolished at 12V, 0.4A for 5 minutes using a 10:6:6 volumetric ratio of ethanol, orthophosphoric acid and ethylene glycol as the electropolishing solution [13]. The EBSD analysis obtained 8 scans, each approximately 3 mm × 1.5 mm, that together form a large area map (figure 2). The grains are clustered into areas of similar orientation with approximately 5° misorientations between the grains in a cluster and a larger misorientation of greater than 30° between the clusters. The average grain size within a cluster is approximately 200 μm whilst the clusters are > 4 mm in size.

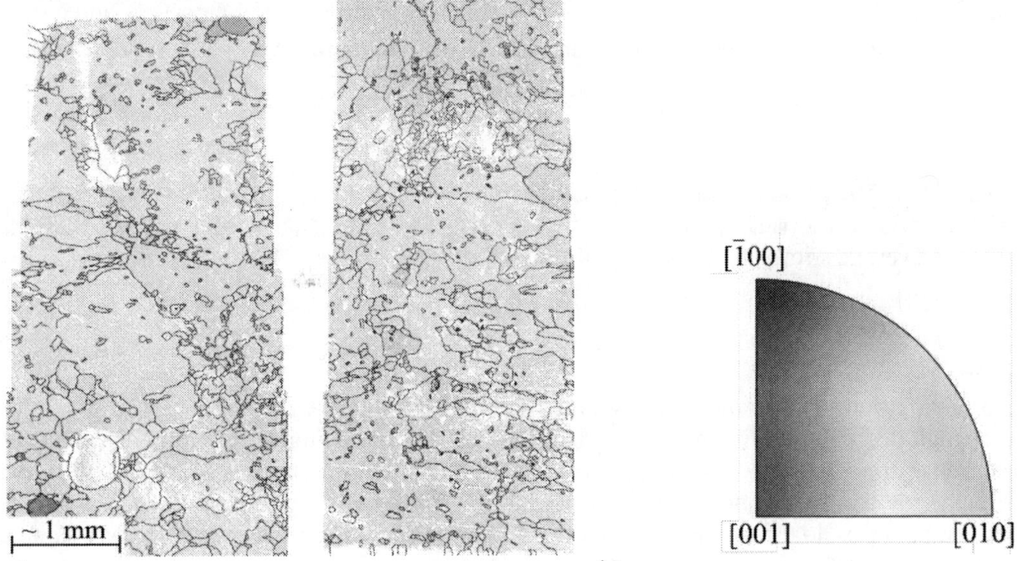

a) b)

Figure 2. Stitched EBSD orientation maps (a) showing a representative microstructure of the uranium metal used in this experiment. Each grain is coloured according to its crystal orientation with respect to the out-of-plane direction, shown in the inverse pole figure (b). Thin black lines are drawn along boundaries with misorientation greater than 3°.

2.3. Neutron diffraction experiment

Sample T1 was strained in load-control in steps of ~10 MPa, from a pre-load of 2 MPa up to a maximum load of 310 MPa with a neutron diffraction measurement taken at each step (each requiring ~25 minutes). The gauge volume of 4 mm × 4 mm × 8 mm sampled approximately 16000 grains.

Sample T2 was tested in the same manner, but the load steps were adjusted to better capture the elastic-plastic transition, after performing the measurement at maximum load, the specimen was to 10 MPa for a final measurement.

From a compressive pre-load of -4 MPa, sample C1 was loaded in steps of -10 MPa to -310 MPa, followed by an unloaded measurement at -10 MPa. The gauge volume of 4 mm × 4 mm × 12 mm sampled approximately 24000 grains.

2.4. Data processing methodology

Initial processing of the experimental data used the ENGIN-X OpenGenie software [14] to perform single-peak fits. Lattice strains for a reflection $\{hkl\}$, ε^{hkl}, were calculated from the measured d-spacing using equation 1 where the reference lattice, d_0^{hkl}, is the value in the initial unloaded state.

$$\varepsilon^{\mathrm{hkl}} = \frac{d^{hkl} - d_0^{hkl}}{d_0^{hkl}} \qquad [1]$$

A Rietveld refinement method in the GSAS software package [15] was used to fit multi-peak profiles to the measured spectra. This fits parameters including the lattice parameters, peak shape, and peak intensity (sample texture) to the entire spectrum at once.

Quantifying the ratio between the measured intensity in a $\{hkl\}$ reflection, and the $\{hkl\}$ peak intensity expected from a random distribution of grains, gives a measure of texture along the diffraction vector, in units of multiples of random distribution (MRD). The preferred orientation correction factor quantifies this, and is calculated from the coefficients of the 10th order spherical harmonic function fitted as part of the Rietveld refinement. The values of this factor as a function of crystal direction were plotted on an inverse pole figure using the 'SMARTSinvpol' script [16].

Lattice parameter strains, ε_a, ε_b, and ε_c were calculated from the lattice parameters a, b, and c using equation 2, where a_0, b_0, and c_0 are the lattice parameters measured at pre-load.

$$\varepsilon_a = \frac{a - a_0}{a_0} \qquad \varepsilon_b = \frac{b - b_0}{b_0} \qquad \varepsilon_c = \frac{c - c_0}{c_0} \qquad [2]$$

Each diffraction peak {*hkl*} contains contributions from the family of grains that have their {hkl} planes oriented in the diffraction condition (that is their {*hkl*} plane-normals parallel or perpendicular to the loading axis for detector banks 1 and 2 respectively). The measured {*hkl*} lattice strain is therefore proportional to the average stress on that family of grains [17].

3. Results

3.1. Macroscopic flow curve

The macroscopic flow curve for the coarse-grained samples (CG T1, CG T2, and CG C1) are plotted in figure 3 along with data from a fine-grained rolled uranium deformed along the rolling direction (FG RD) and transverse direction (FG TD) from Brown et al. [17]. The random error in the strain was estimated from the standard deviation in the DIC-measured displacements to be on the order of 0.05 %. This assumes that the grips do not themselves deform, so the total strain for the coarse-grained samples is only indicative and should not be used for detailed comparison with the data from the fine-grained uranium. The elastic modulus of the material varies with orientation and is therefore not known exactly in these coarse-grained specimens, however its value must lie between the maximum and minimum single crystal moduli of 288 GPa and 149 GPa, indicated on figure 3 as a pair of straight dotted lines. The flow curves all have a very small linear elastic region, with non-linearity observed from approximately 70 MPa in tension and 30 MPa in compression.

3.2. Lattice strain development

Figure 4 shows the development of lattice strains measured with the bank 1 detector (diffraction vector parallel to the loading direction), as a function of applied stress. Lattice strains from fine-grained uranium deformed in tension along the rolling direction [17] is shown for comparison. There is a clear similarity in the lattice strain response between the 2 specimens deformed in tension. A linear behaviour was observed during loading from the pre-loaded state until a critical stress at which the {200} and a-strains begin to saturate with further increases in stress resulting in no additional lattice strain on these planes. This critical stress occurs at 90 MPa for specimen T1 and 110 MPa for specimen T2. The gradients of the ε_a, ε_b, and ε_c in the elastic region are (244 ± 8) GPa, (236 ± 13) GPa, and (224 ± 8) GPa for specimen T1 and (276 ± 8) GPa, (198 ± 3) GPa, and (220 ± 7) GPa for specimen T2. In compression, the lattice parameter and single-peak strains respond linearly with applied stress up to 130 MPa, a higher threshold stress than in tension. The lattice parameter moduli for specimen C1 are (191 ± 3) GPa, (201 ± 2) GPa, and (192 ± 4) GPa for a, b and c respectively.

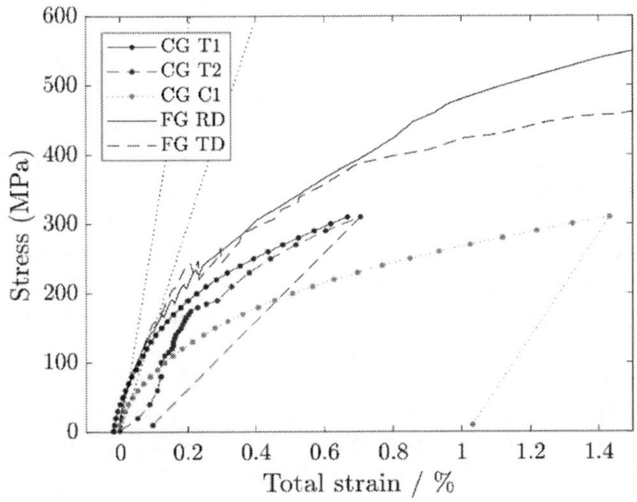

Figure 3. Plot of applied stress against total strain. Lines labelled CG are from the coarse-grained material tested in this experiment, with strain measured using DIC. Lines labelled FG are from fine-grained rolled material tested by Brown et al. [17] using a conventional extensometer for strain measurement. The two straight dotted lines represent the maximum and minimum possible young's moduli for an alpha-uranium polycrystal of 288 GPa and 149 GPa.

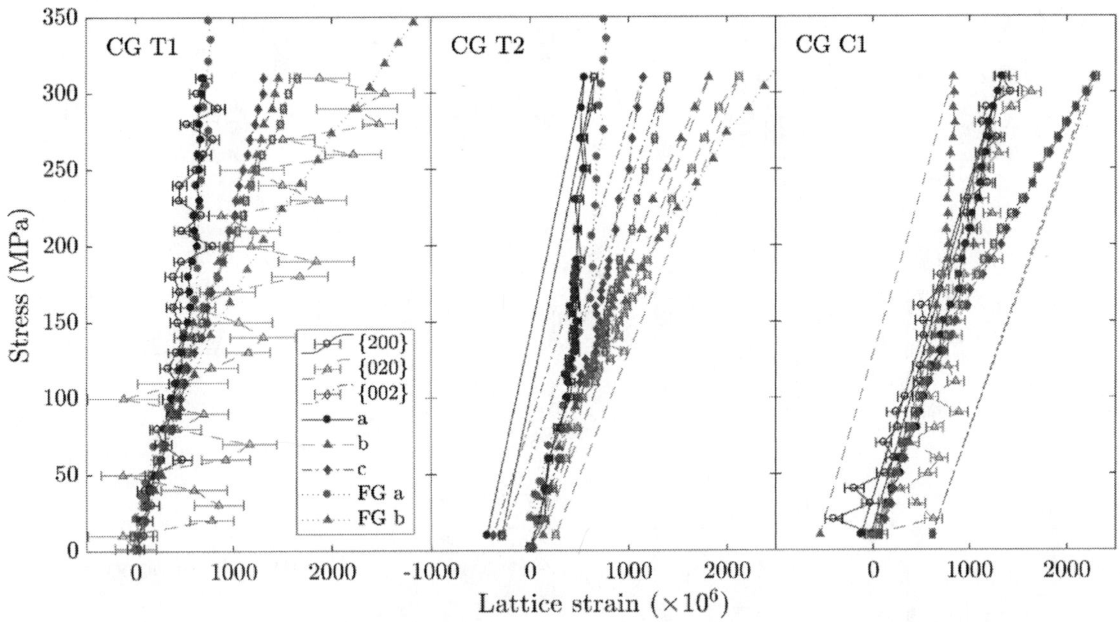

Figure 4. Lattice strain development as a function of applied stress. Open markers represent single-peak lattice strains whilst filled markers represent lattice parameter strains measured from the Rietveld refinement.

3.3. Texture changes

3.3.1. Inverse pole figures

Figures 5a, 5b, 5c, and 5d are inverse pole figure representations of the sample texture as calculated from the Rietveld refinement with the samples in their pre-load state. There are two preferred orientations in sample T1; the predominant orientation is where the loading direction is parallel to {115} plane normals ($n_{\{115\}}$), and the secondary orientation is where it is parallel to $n_{\{312\}}$. Sample T2 has a uniaxial texture, with most grains in the gauge volume having $n_{\{020\}}$ parallel to the loading direction. Sample C1 has two predominant families of planes in the diffraction condition; {190}, the peak closest to {010} in the inverse pole figure, and {350}, the peak situated halfway between {010} and {100}.

IOP Conf. Series: Journal of Physics: Conf. Series **1106** (2018) 012012 doi:10.1088/1742-6596/1106/1/012012

Figures 5e, 5f, and 5g show the absolute change in texture, in MRD, between the initial pre-load state, and the state at maximum tensile or compressive load. A positive value in a particular crystal direction on the inverse pole figure corresponds to an increase in the volume of grains oriented with the diffraction vector (the loading direction) parallel to that particular crystal direction. Figure 5h shows the error in the values presented in figure 5e (sample T1 at maximum tensile load). The inverse pole figure representation of the error was calculated by propagation of the errors in the spherical harmonic coefficients. Sample T1 shows an increase in texture in the upper sector of the inverse pole figure, with a maximum increase of around 0.3 MRD close to {171}, and a decrease of similar magnitude close to {115}. In sample T2, an increase in texture close to {100} is accompanied with slight decreases near {024}. The decrease visible near to {010} is of a similar magnitude to the error shown in figure **5**h. Sample C1 shows an increase in intensity of 0.4 MRD parallel to the {310} plane normal and an equivalent decrease parallel to {350}.

Figures 5i and 5j show the absolute change in texture between the initial pre-load state and the state at unload at the end of the experiment for samples T2 and C1. No unload was performed for specimen T1. The texture of sample T2 after unload is almost identical to its initial state; the texture change that was present at maximum load has disappeared. Sample C1 however shows very similar textures at maximum load and after unloading, with no significant change during unloading.

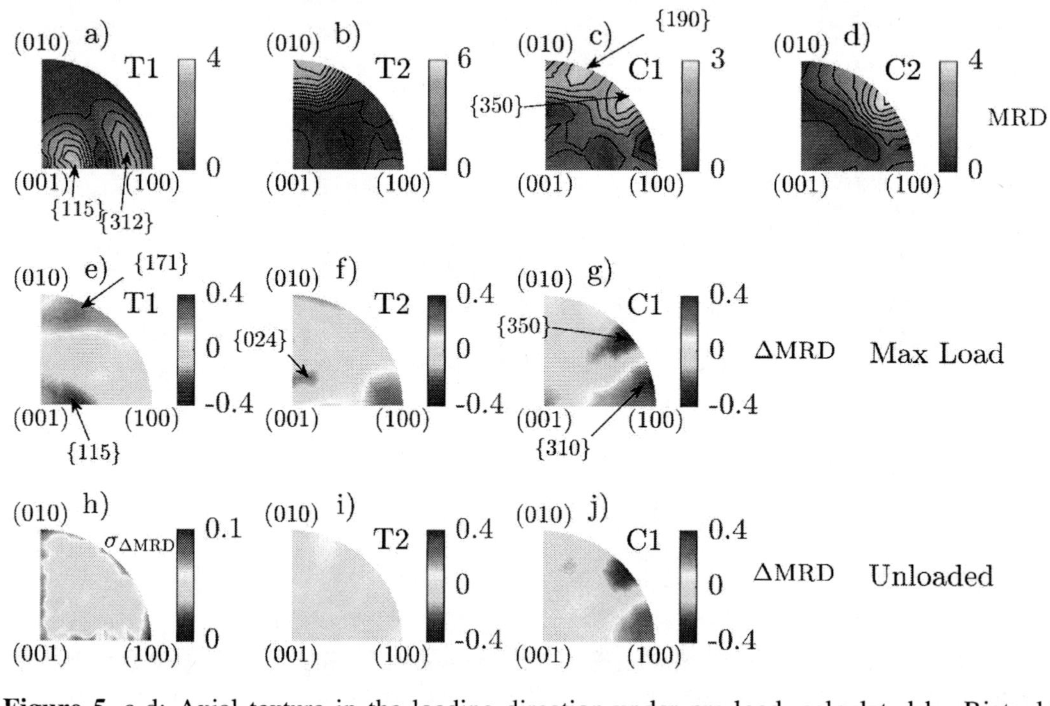

Figure 5. a-d: Axial texture in the loading direction under pre-load, calculated by Rietveld refinement, for specimens T1, T2, C1 and C2. Plotted as inverse pole figures (IPFs) in multiples of random distribution (MRD). Higher values represent a strong texture, i.e. a preferred orientation. e-g: Absolute change in texture, in MRD, between pre-load and maximum load state. i-j: Absolute change in texture, in MRD, between pre-load and final unloaded state. h: Error in values presented in figure 5e, calculated from errors in the fitted spherical harmonic coefficients. Representative of errors in all differential texture measurements e-g and i-j.

4. Discussion

4.1. Macroscopic response and lattice strains

The gauge volume for neutron measurement is considerably smaller than the total sample size, and deformation in a region outside of the gauge volume will contribute to the total strain even if the gauge region is still behaving elastically. That there is still an elastic region to the stress vs. lattice strain plot in figure 4 shows that plasticity is not initiating in all grains at the start of the test, therefore there must be residual stresses such that a subset of grains are already at their yield point to give the non-linear macroscopic flow curve.

Such high residual stresses are indeed possible when we consider the processing route of the coarse-grained material. During the casting process, after solidification, the sample undergoes a solid-state transformation from the beta phase to the orthorhombic alpha phase [18] and large thermal stresses are built up due to the thermal expansion between neighbouring grains [6]. It is well known that these stresses are large enough to cause plastic deformation; thermal cycling of textured uranium causes significant shape change known as 'thermal cycling growth' [19]. In this material, the small grains comprising the larger clusters are all oriented in similar directions, so large contiguous volumes could be at the yield stress in the initial unloaded state. The stress at which the lattice parameter strains in figure 4 deviate from linearity is the lattice yield stress. In both tensile specimens it is the a-parameter that saturates, and the lattice yield stress is lower than in the fine-grained material. This effect is most pronounced in the specimen T2 where yield occurs approximately 30 MPa earlier than for the fine-grained material. This would be expected from a Hall-Petch-type grain size effect [20].

4.2. Texture changes

The measured texture in the gauge volume of each sample is very strong, with only one or two clusters of orientations present in each case. An increase in texture in one direction is seen alongside a decrease in another direction. Since the production of a twin reduces volume fraction of parent material and increases the volume fraction of material in the twin orientation, the directions of positive and negative texture changes can be assigned to the parent and twin orientations respectively.

At the small applied strains in this test, lattice rotation due to dislocation slip is negligible. The observed texture changes are therefore related to the formation of deformation twins. The twin is effectively a new grain in a different orientation, so the formation of twins will be associated with a texture decrease in the parent orientation and an increase in the twin orientation. The twin has a fixed misorientation with the parent grain that depends on the active twinning variant; the variant is dependent on the local stress condition via the Schmid factor [21].

5. Conclusion.

The macroscopic yield point in the deformation of cast uranium in tension and compression is ill-defined, due to large residual stresses.

Despite significant texture differences, very similar macroscopic and lattice-strain response was seen in deformed in tension, with saturation of the a/{200} strains. The specimen deformed in compression responded differently, with saturation of the b-parameter strains only. The asymmetry between tensile and compressive response suggests twinning is the dominant deformation mechanism.

The texture affects the active twinning systems in a particular volume, however the exact identification of the active twin variant was not possible from axial texture measurement only. Large grains make the interpretation of the data difficult and any future experiment would have to include a larger neutron gauge volume in order to provide a more representative measurement of the whole-sample response.

Acknowledgments

The authors thank the Science and Technology Facilities Council for beamtime at the ISIS Neutron and Muon Source under experiment 1710227, as well as the facility staff, in particular Dr Joe Kelleher for experiment planning and Dr Marek Jura for radiation protection advice. Thanks are extended to Dr Keith

Hallam and Dr Chris Jones at the Interface Analysis Centre, University of Bristol, for help with the EBSD mapping, and to Bjorn Clausen and Don Brown at Los Alamos National Laboratory, USA, for assistance with Rietveld refinement. We acknowledge the assistance of Dr Dong Liu, Talha Pirzada and Shixiang Zhao from the University of Oxford during the neutron diffraction experiment.

References

[1] Sariel J, Pelleg J and Kimmel G, 1986 Evaluation of Anisotropy in Cast and Heat Treated "Adjusted-Uranium." *Journal of Nuclear Materials* **140** 288–92

[2] Olofson CT, Meyer GE and Hoffmanner AL, 1976 *Processing and Applications of Depleted Uranium Alloy Products*. (Columbus, OH, USA: Battelle Columbus Laboraties)

[3] Caskey Jr GR, 1959 *Mechanical Properties of Uranium Plate*. (Savannah, GA: E. I. Du Pont de Nemours & Co.)

[4] Jacob CW and Warren BE, 1937 The Crystalline Structure of Uranium. *Journal of the American Chemical Society* **59** 2588–91

[5] Cahn RW, 1953 Plastic Deformation of Alpha-Uranium ; Twinning and Slip. *Acta Metallurgica* **1** 49–70

[6] Frost BRT, 1994 *Nuclear Materials - Materials Science and Technology Volume 10A*

[7] Daniel JS, Lesage B and Lacombe P, 1971 The influence of temperature on slip and twinning in uranium. *Acta Metallurgica* **19** 163–73

[8] McCabe R, Field R, Brown D, Alexander D and Cady C, 2008 Electron Backscatter Diffraction (EBSD) Characterization of Twinning Related Deformation and Fracture in α-Uranium. *Microscopy and Microanalysis* **14** 638–9

[9] Zecevic M, Knezevic M, Beyerlein IJ and McCabe RJ, 2016 Origin of texture development in orthorhombic uranium. *Materials Science and Engineering: A* **665** 108–24

[10] Zecevic M, Knezevic M, Beyerlein IJ and McCabe RJ, 2016 Texture formation in orthorhombic alpha-uranium under simple compression and rolling to high strains. *Journal of Nuclear Materials* **473** 143–56

[11] Ardeljan M, McCabe RJ, Beyerlein IJ and Knezevic M, 2015 Explicit incorporation of deformation twins into crystal plasticity finite element models. *Computer Methods in Applied Mechanics and Engineering* **295** 396–413

[12] Santisteban JR, Daymond MR, James JA and Edwards L, 2006 ENGIN-X: A third-generation neutron strain scanner. *Journal of Applied Crystallography* **39** 812–25

[13] Jones CP, Scott TB and Petherbridge JR, 2015 Structural deformation of metallic uranium surrounding hydride growth sites. *Corrosion Science* **96** 144–51

[14] Akeroyd FA, Ashworth RL, Campbell S, Johnston SD, M Moreton-Smith C, Sergeant RG, et al., 2018 Open Genie Reference Manual - Version 2.0

[15] Larson AC and Von Dreele RB, 2004 General Structure Analysis System (GSAS). *Los Alamos National Laboratory Report LAUR* **86** 748

[16] Clausen B, 2003 *SMARTSware manual*. (Los Alamos National Laboratory, NM, USA)

[17] Brown DW, Bourke M a M, Clausen B, Korzekwa DR, Korzekwa RC, McCabe RJ, et al., 2009 Temperature and direction dependence of internal strain and texture evolution during deformation of uranium. *Materials Science and Engineering A* **512** 67–75

[18] McCabe RJ, Capolungo L, Marshall PE, Cady CM and Tomé CN, 2010 Deformation of wrought

uranium: Experiments and modeling. *Acta Materialia* **58** 5447–59

[19] Kelman LR, 1949 Dimensional changes in Uranium Under Thermal Cycling. *Journal of Metallurgy and Ceramics* **4**

[20] Meyers MA, Vöhringer O and Lubarda VA, 2001 The onset of twinning in metals: A constitutive description. *Acta Materialia* **49** 4025–39

[21] Godet S, Jiang L, Luo AA and Jonas JJ, 2006 Use of Schmid factors to select extension twin variants in extruded magnesium alloy tubes. *Scripta Materialia* **55** 1055–8

IOP Conf. Series: Journal of Physics: Conf. Series **1106** (2018) 012013 doi:10.1088/1742-6596/1106/1/012013

Numerical Inspection based on Quasi-Static Analysis using Rousselier Damage Model for Aluminium Wingbox Aircraft Structure

M.I.M. Ahmad[1,2], J.L. Curiel-Sosa[1], M. Akbar[1], N.A. Abdullah[1]

[1]Computer-Aided Aerospace and Mechanical Engineering Research Group (CA^{2}M), The University of Sheffield, Sheffield, UK

[2]Department of Mechanical and Materials Engineering, Faculty of Engineering and Build Environments, University Kebangsaan Malaysia, 43600 UKM Bangi, Selangor, Malaysia

E-mail: mimeorahmad1@sheffield.ac.uk

Abstract. The present paper illustrates a modelling technique of void damage prediction using constitutive Rousselier damage model in wingbox aircraft structure. The high possibility of aluminium wingbox structure to deform during operating become a driving force in this study. The Rousselier model is implemented by the user-defined material subroutine UMAT in the ABAQUS/ Standard. The wingbox is a part of a commercially-made commuter aircraft prototype, in which steady aerodynamic pressure is applied as the quasi-static, ranging from cruise load factor until the ultimate load factor. This analysis is divided into three assessments: load factor assessment, damage detection and implementation of Rousselier-UMAT-XFEM (RuX) model solution. The results indicate the maximum stress concentration occurs at the bottom-root part of the skin. The void volume fraction and the crack grow as the load factor increases to the ultimate load factor. A comparison with the literature is shown in some details.

1. Introduction

Aircraft structural members are designed to carry a load and resist the stress during operation. The significant effect on the aerodynamic performance of the aircraft existed due to material deform during the extreme condition on the flight. Several studies have been done to investigate the behaviour of damage occur in the aircraft structure in order to improve and to make sure it is safe during flight. In experimental approach, Sundaresan et al. 2003 [1] discussed the damage detection on a wing panel inside the wingbox using vibration detection shapes. Following with mathematical investigation by Grondel et al. 2004 [2], which has developed a health monitoring system to detect the damaging impact of the wingbox structure. Then, Chen et al. 2007 [3] proposed a dynamic-based damage detection method in composite wingbox structure using Hillbert-Huang transform (HHT).

Another method that can be applied is by using continuum damage mechanics (CDM) model [4, 5]. Previously, the CDM model based on Hashin failure criterion was used to model the damage assessment of rotary wing aircraft cabin door [6]. In this study, the numerical inspection based on Rousselier damage criteria of CDM model is investigated in wingbox structure where most of the wings's load is carried [7]. The Rousselier model, which has been introduced over last two decades, has become a powerful tool for modelling the damage criteria in structural analysis.

For example of the model application such as in predicting the fracture resistant behaviour of Zircaloy fuel pin specimens [8], simulation damage behaviour in eletron beam welded joints [9] and welded Esshete 1250 pipe [10]. This model can predict the behaviour of materials from the micromechanics processes leading to ductile fracture [11, 12]. From the authors' knowledge, this is a first attempt that the structural analysis is proposed for an aircraft wingbox by using the Rousselier damage model, in order to represent the damage behaviour of the material. Moreover, further investigation is carried out by implementing the Rousselier model with the XFEM solution, so called RuX model, to predict the void damage with the crack development in the material.

2. Numerical frameworks

2.1. Rousselier's Damage Model

Rousselier damage model was implemented using an Abaqus UMAT subroutine to describe the material damage. In the subroutine, the integration scheme utilizing the radial return algorithm and implicit solution method are formulated in the finite element method. The Rousselier plastic potential is used to estimate the onset of yielding of the material point which shows the presence of the void volume fraction, considered as an internal state variable in the material constitutive model. Rousselier[13] proposed the classical plastic potential as:

$$\phi = \frac{q}{\rho} - R(\varepsilon_{eq}) + B(\beta)\, De^{\left(\frac{-p}{\rho\sigma_1}\right)} \tag{1}$$

where q and p are the von Mises equivalent stress and hydrostatic pressure respectively, D and σ_1 are Rousselier material parameters, ρ is a relative density, β is an internal variable describing damage and ε_{eq} is the equivalent plastic strain which represents hardening of the material. With loading, the void volume fraction evolves from the initial void volume fraction, f_0, of the material. The void growth rate can be obtained as:

$$\dot{f} = (1-f)\,\dot{\epsilon}^{p}_{kk} = (1-f)\,\dot{\epsilon}_p \tag{2}$$

where $\dot{\varepsilon}^{p}_{kk}$ is the component of the plastic strain tensor.

By considering a material which contains a void, the relationship between the relative density, ρ, and f can be written as:

$$\rho = \left(\frac{1-f}{1-f_0}\right) \tag{3}$$

since the value of f_0 is very small compared to unity, Eq. (3) becomes as [14]:

$$\rho = 1 - f \tag{4}$$

However, f_0 is still be calculated in the formulation of this study, in order to see the influence of f_0 in the formation of void growth in the material. The function of $B(\beta)$ in Eq. (1) can be written in terms of the void volume fraction as:

$$B(\beta) = \sigma_1 f \tag{5}$$

By substituting Eq. (4) and Eq. (5) into Eq. (1), the Rousselier plastic potential can be written as:

$$\phi = \frac{q}{1-f} - R(\varepsilon_{eq}) + D\sigma_1 f e^{\left(\frac{-p}{(1-f)\sigma_1}\right)} \tag{6}$$

To establish the yield criterion, the values of yield function, ϕ are classified as follow:

$$\phi \leq 0 : \text{Elastic deformation regime} \tag{7}$$

$$\phi > 0 : \text{Plastic deformation regime} \tag{8}$$

3. Technical wingbox structure

The layout of the wingbox used in the present work is shown in Fig. 1. The wingbox prototype is developed under the joint program of Indonesian Aerospace, National Institute of Aeronautics and Space of Indonesia and Agency for Assessment and Application of Technology of Indonesia. Fig. 2 shows a wingbox structure which consists of spars, ribs, stringers and skin.

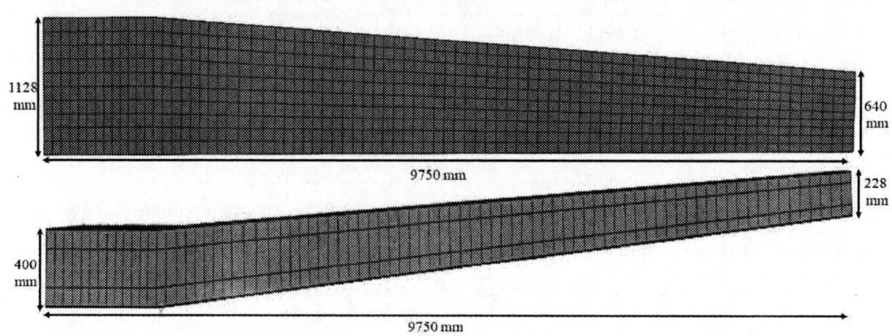

Figure 1. Wingbox layout (top and side view)

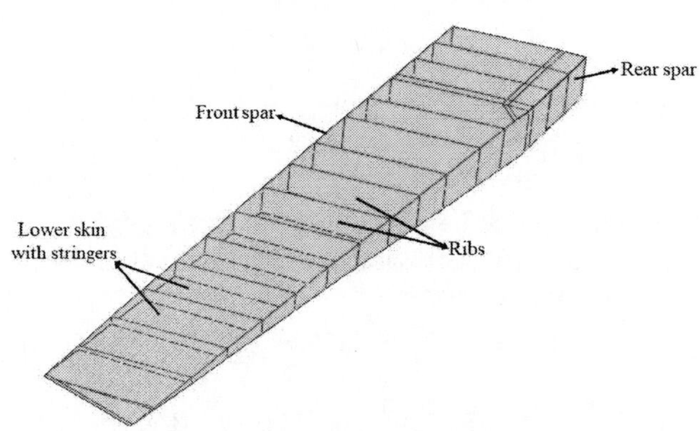

Figure 2. Wingbox structure

The wingbox model consists of three type of elements, which are B31: 2-node linear beam, S4R: 4 node thick shell with reduced integration and hourglass control and C3D81: 8-node linear brick with the incompatible mode. For the boundary condition, the quasi-static load obtained from aerodynamic analysis is applied and fixed support in all-direction at the root of the wing, as shown in Fig. 3. The load increment is considered in terms of load factor, N. Initially no load condition is increased until the appropriate cruise load condition (N=1). It is further increased until ultimate load condition, which is 2.5 times cruise load (N=2.5). The details mechanical properties of the wingbox materials and hardening data can be accessed in Table 1 and 2, respectively.

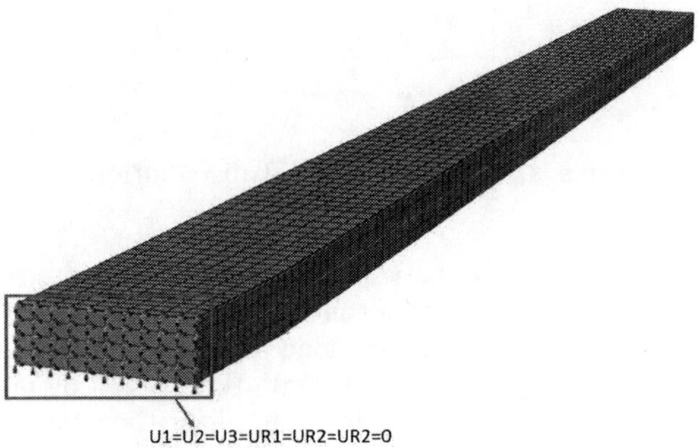

Figure 3. Boundary conditions applied to the wingbox structure

Table 1. Mechanical properties and Rousselier parameters of 6065-T6, aluminium alloy [15]

Young modulus, GPa, (E)	68.9
Poisson ratio, (v)	0.33
Yield strength, MPa, (σ_y)	270
Rousselier parameter, MPa, (σ_1)	180
Rousselier parameter, (D)	2
Initial void volume fraction, (f_o)	0.0001

Table 2. Hardening data [16]

True strain, (ϵ_T)	True stress, MPa, (σ_T)
0.012	296
0.0182	301
0.027	309
0.0396	320
0.0509	330
0.0641	338
0.079	348
0.0926	355
0.1028	359
0.1129	361

4. Load factor assessment

As explained in Section 3, aerodynamic pressure is applied as the quasi-static load. The steady aerodynamic pressure is calculated using Lifting-Line Theory (LLT)[17, 18]. The LLT provides an analytical treatment for the evaluation of the aerodynamics of a finite wing, which is in the form of an infinite sine series for the circulation distribution, as shown in Fig. 4.

IOP Conf. Series: Journal of Physics: Conf. Series **1106** (2018) 012013 doi:10.1088/1742-6596/1106/1/012013

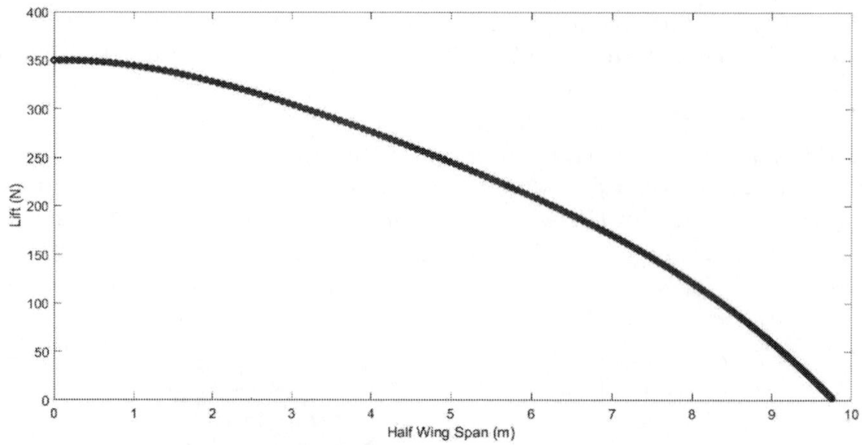

Figure 4. Aerodynamics lift distribution

Figs. 5 and 6 present the distribution of stresses at the top and bottom skin of the wingbox, where it can be seen that the maximum stresses concentration occurs in the region near the root section. As expected, the highest load factor influenced the formation of highest stress in the wingbox structure, which create 139 MPa in N=1, 208 MPa in N=1.5 and 305 MPa in N=2.5. Then, the distribution of the maximum principal strain along the wingbox structure is shown in Fig. 7. From Figs. 5, 6 and 7 (a and b), the contour plots do not show any significant difference in their patterns distribution due to deformation within elastic region during a load increment. Meanwhile, the wingbox structure deformed plastically during the ultimate load condition, where the void damage starts to grow. The formation of the void volume fraction along the wingbox structure can be seen as shown in Fig. 8.

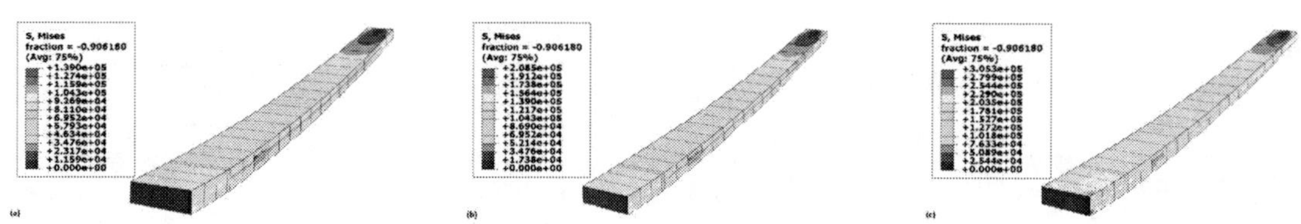

Figure 5. Von Mises stress at a)N=1, b)N=1.5 and c)N=2.5 along the upper skin of wingbox structure

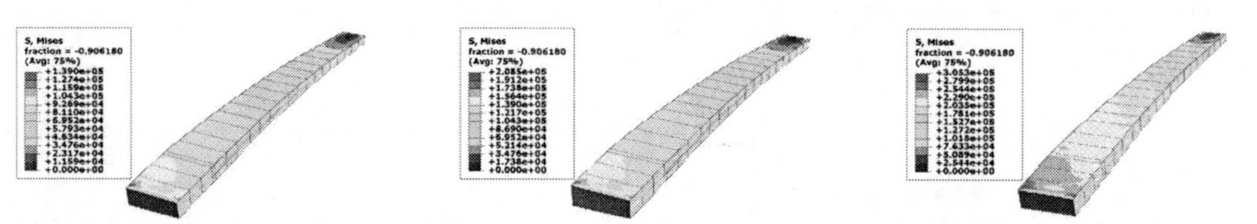

Figure 6. Von Mises stress at a)N=1, b)N=1.5 and c)N=2.5 along the bottom skin of wingbox structure

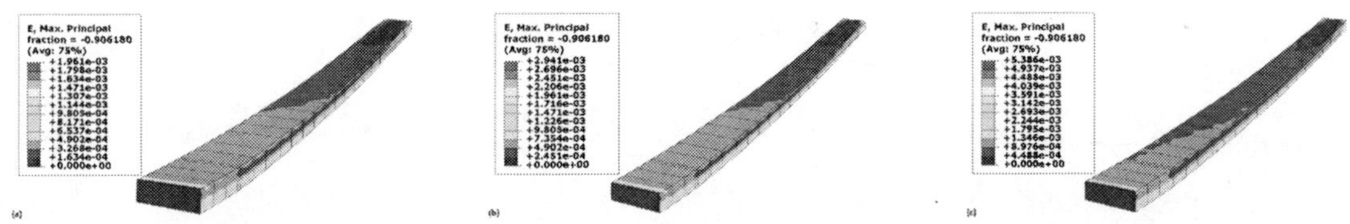

Figure 7. Maximum principal strain at a)N=1, b)N=1.5 and c)N=2.5 along the upper skin wingbox structure

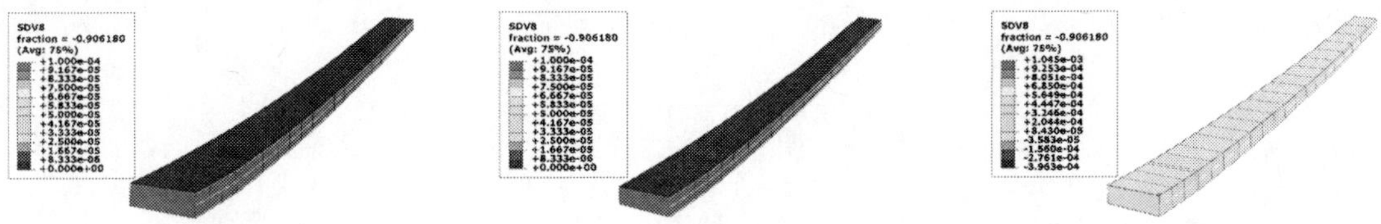

Figure 8. Void volume fraction at a)N=1, b)N=1.5 and c)N=2.5 along the upper skin wingbox structure

As can be seen clearly in Fig. 8(a and b), the contour plots also do not show any significant different, and the values of void damage are still below the value of the initial void volume fraction ($f_o = 0.0001$). However, the void damage emerged during the ultimate load condition as shown in Fig. 8(c), which represents the values of the void damage distribution along the wingbox structure. To determine the maximum displacement acting in the wingbox structure, Fig. 9 is depicted, which bring the maximum displacement of 724 mm in N=2.5, followed by 434 mm in N=1.5 and 289 mm in N=1. Therefore, it means that the wingbox structure design is safe from damage under the consideration of load factor range from 1-1.5. However, its need more attention concerning the safety of the wingbox structure with a situation of load factor in 2.5 or above.

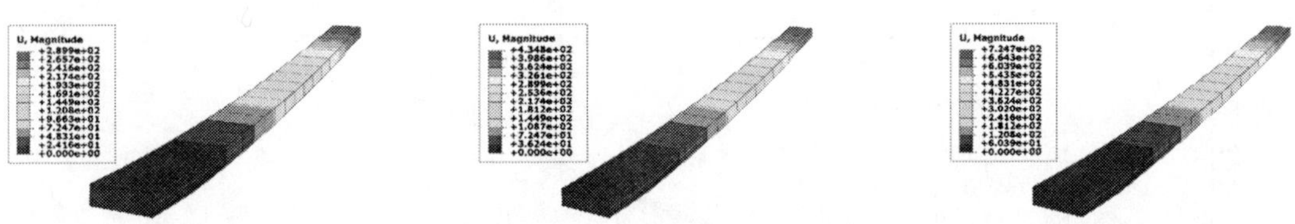

Figure 9. Displacement at a)N=1, b)N=1.5 and c)N=2.5 along the upper skin wingbox structure

Next, the comparison of stress-strain relation with the literature's result [19] is shown in Fig. 10. The benchmark results are considered under static load analysis, thus provides the stress-strain point for every increment of load factor. As a result, it shows that a good comparison has been achieved between both results in the load factor of 1 and 1.5 respectively. This finding can be verified by calculating the numerical value of similarity where it has revealed that 84%-86% similarity between a predicted value and benchmark value. Nevertheless, the comparison of stress-strain in the ultimate load factor gives small variances different, which admit 75% similarity. This is due to of different finite element formulation that used to represent the plasticity behaviour of the wingbox structure. In overall, the trend shows a reasonable agreement between both results and the validity of the result is successfully verified.

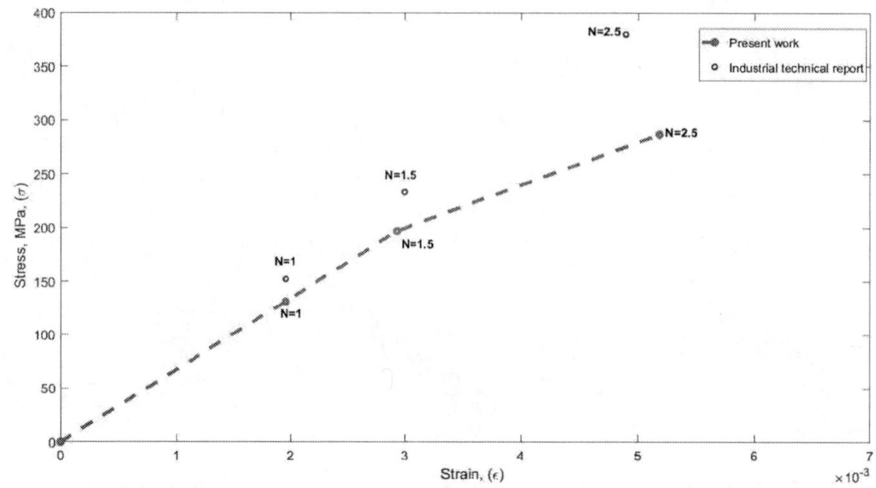

Figure 10. Stress strain relation with different load factors

5. Damage detection

This assessment is focused on the ultimate load factor, where the formation of void damage exists in the wingbox structure. The relationship between the void volume fraction and the strain is plotted as shown in Fig. 11. The initial void volume fraction, f_o, is increased with the increase of the strain, which revealed that the strain distribution had influenced the formation of the void damage, thus affect the material behaviour of wingbox structure to deform plastically.

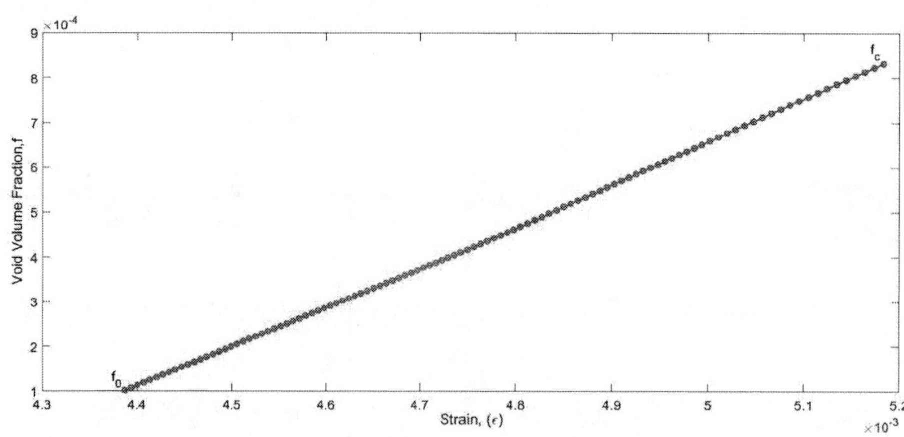

Figure 11. Void volume fraction with strain relation

To justify the results of damage model used in this analysis, the contour damage has been compared with the results from technical report [19]. Fig. 12 shows a damage parameter based on the failure indices criteria which is stated as more than one. In that case, the wingbox structure is suffering from fully damaged at the bottom skin of the root wing under the ultimate load condition. While from Fig. 13, the void damage based on the void volume fraction parameter is detected forms at the same location with the damaged region from the benchmark result and the value of critical void volume fraction, f_c is recorded approximately as 0.001. Thus, a good reasonable comparison has been achieved and verified.

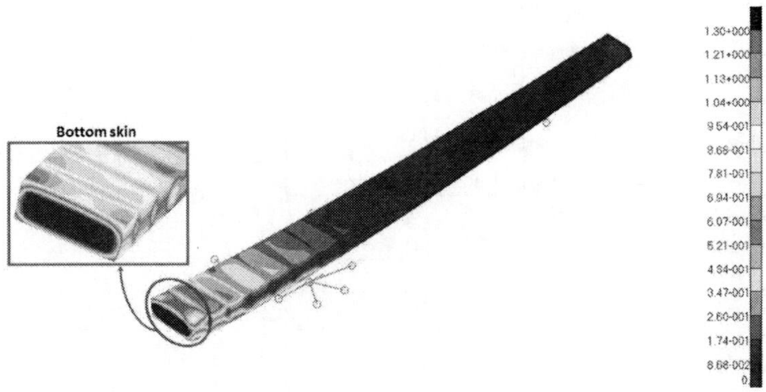

Figure 12. Failure indices in wingbox structure issued by industrial technical report [19] based on Hill theory

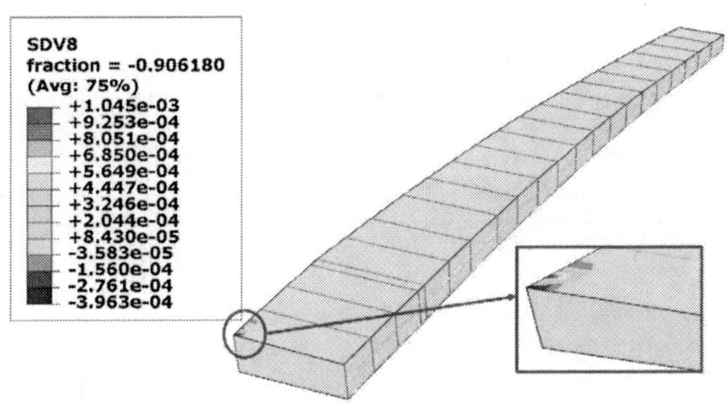

Figure 13. Void damage in wingbox structure based on the Rousselier damage model

6. Implementation of RuX model

RuX model solution has been proposed to predict the micromechanical damage of the material with the existing of the crack development. The RuX model is a combined term of the Rousselier model (constitutive damage model), UMAT subroutine (interface user-defined material) and XFEM solution (crack development). Fig. 14 illustrates the flow-diagram, displaying the steps involved in the model.

The XFEM provides a significant benefits in the numerical modelling of crack propagation where a discontinuous function is added to the finite element approximation of the displacement field [20, 21, 22, 23]. This solution is proposed where the crack initiation and evolution based on the damage of traction-separation law and displacement/ energy release rate criterion are selected for characterizing the crack behaviour, as shown in Table 3. Fig. 15 shows how the crack is propagated due to the void growth detected at the bottom skin of the root of the wingbox structure. As expected, the crack was initiated, and then it has propagated in the region

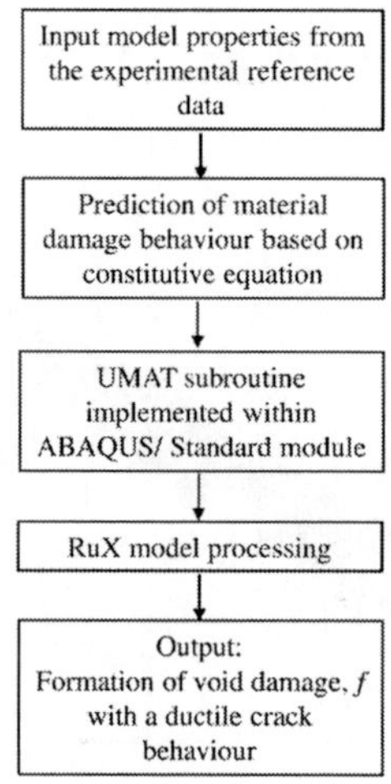

Figure 14. The flow diagram of RuX model

where the concentration of maximum stresses existed. It gives a good sign for RuX model as a constitutive damage model to predict the void-crack damage of the material behaviour problem.

Table 3. Material parameter and fracture toughness values of 6065-T6, aluminium alloy [24]

Max principal stress, MPa	310
Fracture energy, kJ/m^2, (G_{Ic})	121

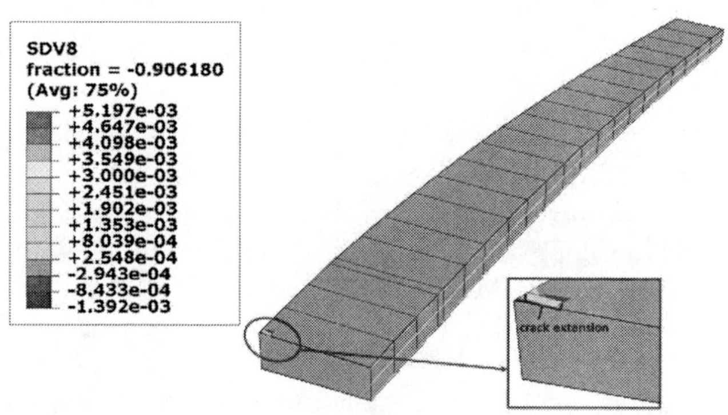

Figure 15. Formation of void damage and crack propagation at the root-bottom skin of the wingbox structure

IOP Conf. Series: Journal of Physics: Conf. Series **1106** (2018) 012013 doi:10.1088/1742-6596/1106/1/012013

6.1. Relative error

The evaluation of relative error is conducted to validate the stability and accuracy of the simulation results. In this case, the residual displays information determines either an iteration has produced an equilibrium solution [25]. It can be seen in Fig. 16, a solution achieved a stable condition and the data recorded are below the tolerance value, which sets to 0.005.

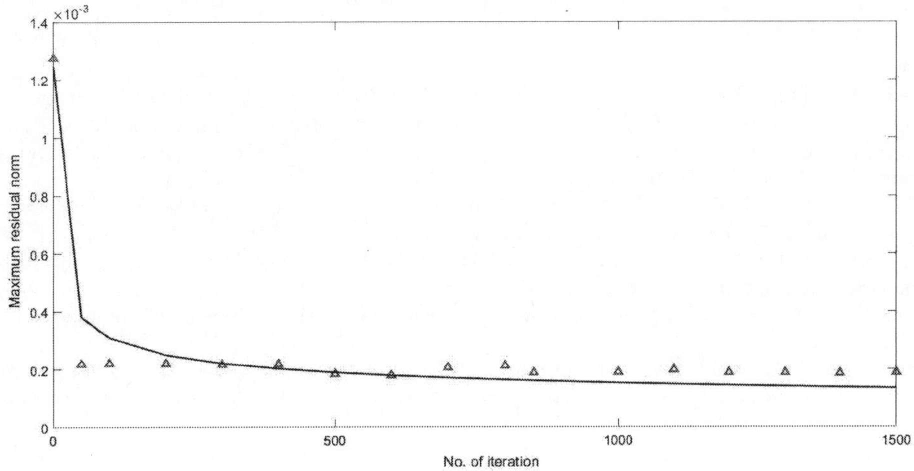

Figure 16. Maximum residual norm vs No. of iteration

7. Conclusion

This research has presented a structural analysis of aluminium wingbox aircraft structure based on quasi-static analysis using constitutive Rousselier damage model. The ability of this model to define the material damage behaviour of the wingbox structure become a novel achievement of this research finding. The investigation with the proposed numerical solution resulted in the following:

- The Rousselier damage model was successfully installed in the user-defined material subroutine UMAT in the ABAQUS/ Standard.
- The aerodynamic distributed forces were calculated using the Lifting-Line Theory (LLT). By referring to the indicator of load factor, N, the highest stresses created: 139 MPa in N=1, 208 MPa in N=1.5 and 305 MPa in N=2.5. It is well predicted when the results have been compared from previous literature and show a good agreement between both results.
- The damage was formed at the bottom skin of the root wing, and it is verified by comparing the result from previous literature. The results show a good agreement have been achieved between both results.
- The crack development has been introduced by the RuX model solution, which purposes to predict the void damage and the position of the crack in the wingbox structure. As been expected, the crack was initiated and extended in the area of the formation of the void damage, where the maximum stress concentrates at the same place.

8. Acknowledgments
The authors would like to express their gratitude to the University of Sheffield, the National University of Malaysia and Ministry of Higher Education Malaysia through the fund for supporting this research project. We also acknowledge the technical support provided by the National Institute of Aeronautics and Space of Indonesia, Indonesian Aerospace and Agency for Assessment and Application of Technology of Indonesia, especially to Mr. Nanda Wirawan (Team leader, Aero-structure division, Aeronautics Technology Center).

References
[1] Sundaresan M.J, Ghoshal A, Li J, Schulz M.J, Pai P.F and Chung J.H. Experimental damage detection on a wing panel using vibration deflection shapes, Structural Health Monitoring. 2003, 2(3): 243-256.
[2] Grondel S, Assad J, Delebarre C and Moulin E. Health monitoring of a composite wingbox structure, Ultrasonics. 2004, 42(1-9): 819-824.
[3] Chen H.G, Yan Y.J and Jiang J.S. Vibration-based damage detection in composite wingbox structures by HHT, Mechanical Systems and Signal Processing. 2007, 21(1): 307-321.
[4] Chaboche J.L, Boudifa M and Saanouni K. A CDM approach of ductile damage with plastic compressibility, International Journal of Fracture. 2006, 137: 51-75.
[5] Besson J. Continuum models of ductile fracture: a review, International Journal of Damage Mechanics. 2010, 19: 1-52.
[6] Gangadhara Rao T.B, Vijaya K.R and Subba R.V.V. Numerical modelling and damage assessment of rotary wing aircraft cabin door using continuum damage mechanics model, Applied Composite Materials. 2017, 24(1): 235-250.
[7] Akbar M and Curiel-Sosa J.L. Piezoelectric energy harvester composite under dynamic bending with implementation to aircraft wingbox structure, Composite Structures. 2016, 153: 193-203.
[8] Samal M.K and Shah P.K. On the application of Rousselier's damage model to predict fracture resistance behavior of zircaloy fuel pin specimens, Procedia Engineering. 2013, 55: 710-715.
[9] Tu H.Y, Schmauder S, Weber U, Rudnik Y and Ploshikhin. Simulation of the damage behaviour of electron beam welded joints with the Rousselier model, Engineering Fracture Mechanics. 2013, 103: 153-161.
[10] Arun S, Sherry A.H, Smith M.C and Sheikh M. Simulations of the large-scale four point bending test using Rousselier model, Engineering Fracture Mechanics. 2017, 178: 497-511.
[11] Rousselier G, Devaux J-C, Mottet G and Devesa G. A methodology for ductile fracture analysis based on damage mechanics: An illustration of a local approach of fracture. STP27716S Nonlinear Fracture Mechanics: Volume II Elastic-Plastic Fracture, ASTM International, West Conshohocken, PA. 1988, 332-354.
[12] Rousselier G. Dissipation in porous metal plasticity and ductile fracture, Journal of The Mechanics and Physics of Solids. 2011, 49: 1727-1746.
[13] Rousselier G. Finite deformation constitutive relations including ductile fracture damage. Three-dimensional constitutive relations and ductile fracture, Amsterdam, North-Holland Publishing Co. 1981, 331-355.
[14] Rousselier G. Ductile fracture models and their potential in local approach of fracture, Nuclear Engineering and Design. 1987, 105(1): 97-111.
[15] Guo J, Zhao S, Murakami R and Zang S. Experimental and numerical investigation for ductile fracture of Al-alloy 5052 using modified Rousselier model, Computational Materials Science. 2013, 71: 115-123.
[16] Xu F, Zhang H, Zhao S, Chen C, Cao M and Chen W. Optimized calibration procedure of the damage parameters of 6082-T6 sheet, Materials. 2018, 11(2): 248.
[17] Pakalnis E. Lift and drag force calculation methods using non-linear section data. History and recent research, Aviation. 2010, 8:2, 9-13.
[18] Phillips W.F and Snyder D.O. Modern adaptation of Prandtl's classic Lifting-Line Theory, Journal of Aircraft. 2000, 37:4, 662-670.
[19] Scaling and Analysis of N219 Half Wing for Wind Tunnel Flutter Model Design, Structure Analysis Department, The Indonesian Aerospace (IAe) Company. Private and confidential.
[20] Belytschko T, Liu W, Moran B and Elkhodary K. Nonlinear finite elements for continua and structures, 2nd Ed. Wiley. 2013.
[21] Giner E, Sukumar N, Tarancn J and Fuenmayor F. An abaqus implementation of the extended finite element method, Engineering Fracture Mechanics. 2009, 76(3): 347-368.
[22] Curiel-Sosa J.L and Karapurath N. Delamination modelling of GLARE using the extended finite element method, Composite Science and Technology, 2012, 72: 788-791.
[23] Curiel-Sosa J.L, Tafazzolimoghaddam B and Zhang C. Modelling fracture and delamination in composite laminates: Energy release rate and interface stress, Composite Structures. 2018, 189: 641-647.

[24] P.S, Singh K.K, Tripathi V.K, Sarkar P.K and Kumar P. Fracture toughness of thin aluminium sheets using modified single edge notch specimen, International Journal of Engineering and Innovative Technology (IJEIT). 2012, 1(5): 283-288.

[25] Curiel-Sosa J.L, Beg O.A and Murillo J.M.L. Finite element analysis of structural instability using an implicit/ explicit switching technique, International Journal for Computational Methods in Engineering Science and Mechanics. 2013, 14(5): 452-264.

IOP Conf. Series: Journal of Physics: Conf. Series **1106** (2018) 012014 doi:10.1088/1742-6596/1106/1/012014

Analysis on Cracked Commuter Aircraft Wing Under Dynamic Cruise Load By Means of XFEM

N Wirawan[1], N A Abdullah [2,3], Mahesa Akbar[2] and J L Curiel-Sosa[2,4]

[1] Aeronautics Technology Center, National Institute of Aeronautics and Space of Indonesia (LAPAN), Bogor, Indonesia
[2] Department of Mechanical Engineering, Computer-Aided Aerospace and Mechanical Engineering Research Group (CA^2M), The University of Sheffield, The Portobello Centre, Sheffield, S1 4ET, United Kingdom
[3] Department of Mechanical Engineering, International Islamic University Malaysia, Malaysia

E-mail: nanda.wirawan@lapan.go.id

E-mail: makbar2@sheffield.ac.uk

Abstract. In this paper, a new investigation of crack propagation on a commuter aircraft wing is discussed. The aircraft is a 19-seater multi-purpose commuter aircraft designed for a mission on the remote area. The dynamic load acting on the wing is assumed as harmonic load considering aerodynamic lift at a particular excitation frequency. A 3D wingbox CAD model is developed for the numerical simulation. Extended finite element (XFEM) is applied to analyse the crack propagation on the aircraft wing. The crack growth simulation is successfully done, and the results are shown in some details. These results provided a novel insight on the computational simulation of cracked wing structure under dynamic flight load.

1. Introduction

In the industrial aircraft design process, the structural damage analysis commonly evaluated through static/quasi-static load analyses [1], and to some extent, the dynamic analysis concerning fatigue-related issues [2–4]. Mathematical and numerical models related to those subjects have as well been developed since decades ago [5; 6]. However, despite this fact, structural damage analysis exerted by the dynamic load under the aeroelastic condition is still received little attention.

Two of the earliest numerical investigations on damaged lifting surface concerning aeroelastic condition were presented in [7; 8]. Both articles presented the studies on a cracked panel made of under supersonic condition. Similar schemes were also applied, finite element method and piston theory were utilised to model the structural and aerodynamic parts, respectively. However, Chen and Lin [7] used isotropic materials, while the latter, Lin et al. [8], used anisotropic materials. The results of both articles concluded that the existence of cracks might decrease the critical flutter speed. Despite this fact, it was found, interestingly, that for anisotropic panels at a particular configuration, the flutter boundary increases with the existence of cracks [8].

[4] Web address: http://www.jlcurielsosa.org/

The results of Lin et al. [8] were supported by the investigations done by Wang et al. [9] and Abdullah et al. [10]. Both observed that cracking could increase the flutter speed for a specific fiber orientation. However, different with Lin et al. [8], both investigations in [9; 10] were done for the subsonic condition. Other investigations on cracked composites lifting structure can be found in [11–14]. Pidaparti [11] observed the panel with microcracks; Kapania and Castel [12] observed cracks on the swept wing; Wang and Inman [13], Georgiou et al. [14] introduced a probabilistic aeroelastic model for the cracked panels via the Polynomial Chaos Expansion (PCE).

However, the investigations in [11–14] were all assumed a static crack condition, without crack growth. To the authors′ knowledge, to this date, only Strganac and Kim [15] considered the damage growth on a cracked panel subjected to an aeroelastic load. They utilised Internal State Variable (ISV) as the damage model to quantify the growth of crack density. It was observed, as the crack is grown, the flutter boundary initially dropped. However, up to a certain point, the boundary is almost unaffected by the crack growth due to the damage accumulation getting closer to a full damage evolution [15].

In the present work, an initial effort to observe the crack propagation of lifting structure under the aeroelastic condition is presented. In this effort, firstly, the aerodynamic load is calculated via steady model estimation, The Lifting Line Theory (LLT). Then, the aerodynamic loads are assumed oscillating harmonically within a range of excitation frequencies. Extended finite element method (XFEM) is utilised to model the cracked structure.

In the past decade, along with the growth of more advanced numerical methods, the structural testing has been moving towards virtual experiments [16]. In terms of damage modeling, various high-fidelity approaches, i.e., continuum damage mechanics [17], cohesive elements [18], have been applied to represent cracks. XFEM, first introduced by Belytschko and Black [19], has attracted large attention in the field of numerical damage mechanics due to the ease of modeling the crack propagation with minimum remeshing [20]. Several researchers have shown its robustness as the numerical results could closely represent the experiments for various cases, i.e., delamination [21], multiaxial loading [22], transverse crack and delamination [23; 24].

Therefore, the work presented in this article is aimed, for the first time, to establish the scheme to evaluate the crack propagation on a structure under aerodynamic loads. Within the continuous work of the authors, this scheme is expected, to be extended for a more realistic aeroelastic condition with the coupling of the unsteady aerodynamic loads. In addition, due to the ease of XFEM, it is possible to investigate crack growth on a more complicated and detailed structure, i.e., aircraft wingbox, as presented herein.

2. The Extended Finite Element Method (XFEM)

In this section, a brief introduction to the extended finite element method (XFEM) is presented. For a more detailed overview of XFEM, interested readers are referred to [20; 25].

In general, standard finite element formulation uses the singular element to model the crack tip. Thus, when the crack is growing, the domain around the crack tip should be remeshed to maintain the geometry of the propagating crack. Therefore, for a more complicated structure, the computational cost may increase significantly due to remeshing. XFEM added special enrichment functions into the finite element formulation. Belytschko and Black in [19] first proposed this concept of discontinuous enrichment functions. The enrichment functions used to construct additional displacement fields on the meshed domain. These additional displacement fields represent the discontinuities, i.e., cracks, within an element.

The displacement field of a domain discretised into n-node elements with the addition of the enrichment function can be written as follows [19]:

$$\mathbf{u^h}(x) = \mathbf{u^{FE}} + \mathbf{u^{Enr.}} = \sum_{i=1}^{n} N_i(x)\mathbf{u_i} + \sum_{j=1}^{m} N_j(x)H(x)\mathbf{a_j} \tag{1}$$

Where $\mathbf{u^{FE}}$, denotes the displacements of the standard degree of freedoms (DOFs), $\mathbf{u_i}$, of the n-node elements. The selected nodes around the discontinuity now have extra DOFs as these nodes are enriched. $\mathbf{u^{Enr.}}$ represents the additional displacement field of the enriched DOFs, $\mathbf{a_j}$. N is the regular shape functions of the finite element, while H represents the enrichment function, i.e., Heaviside function.

Concerning the displacement field in Eq.(1), the enriched element stiffness matrix is much larger than the standard FE stiffness matrix. It will contain the components associated with the regular DOFs, enriched DOFs and also the coupling components between the regular and enriched DOFs. For example, if the size of the standard FE stiffness matrix is n $\times$ n, then the enriched stiffness matrix is (n+m) $\times$ (n+m), where m denotes the number of enriched DOFs.

3. Computational scheme algorithm

In the present computational simulation, the XFEM module in ABAQUS software is used. Periodic load function is applied to model the excitation load on the structure. The workflow to apply the oscillating aerodynamic load on the structure is defined as follows:

(i) Perform modal analysis to obtain the mode shapes and the related natural frequencies.
(ii) Evaluate the aerodynamic pressure distribution on the structure for a certain flight condition, i.e., airspeed and density, via the Lifting-Line Theory (LLT).
(iii) Select a range of excitation frequencies in relation to the mode of interest.
(iv) Construct the excitation load by
    (a) applying the aerodynamic pressure as the force amplitude to the elements/nodes,
    (b) then select the periodic amplitude function, and input one excitation frequency from the previously selected range.

After the excitation load is formed, the XFEM module can be performed in the ABAQUS considering a particular loading time. However, it is important to note, the XFEM module in ABAQUS may or may not require a crack initiation as an input. In the case crack initiation is present; usually, it is applied to the area with large stress concentration to observe the criticality of the structure.

4. Numerical investigation on cracked wingbox structure

In the present work, a commuter aircraft wingbox in [26] is adopted. The layout of the wing is depicted in Figure 1. The mass configuration for maximum take-off weight (MTOW) at maximum fuel is used.

The wingbox is discretised using several types of elements. The skins, ribs and spars are modeled by the shell and solid elements. The stringers are represented by the beam elements. The engine, the control surfaces at the trailing edge and the fuel masses are modeled by lumped mass elements. The rigid bar elements are used to connect the lumped masses of engine and control surfaces to their respective hinge points. The wing components are all made of aluminium 6061-T6 with Young's modulus 68.9 GPa, density 2710 kg/m^3 and maximum principal stress of 290 MPa. Figure 2 shows the corresponding elements type of the structural components.

Validation of the wingbox finite element model of the wingbox is done by comparison with the numerical investigation done by the industry that developed the aircraft. The stick model, a simplification of the structure using only beam elements and lumped masses, is used in their

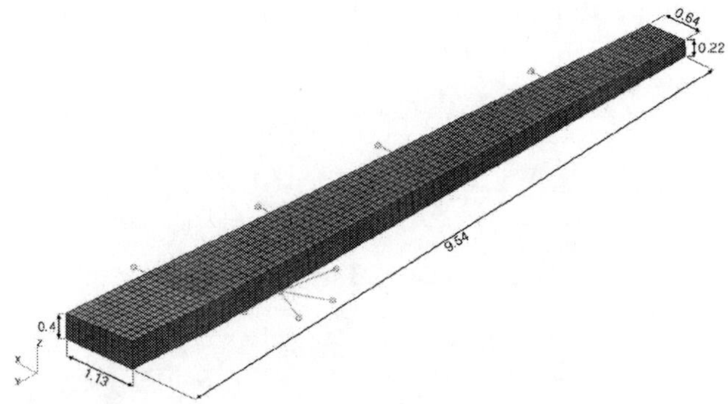

Figure 1: Wingbox layout (unit length in m)

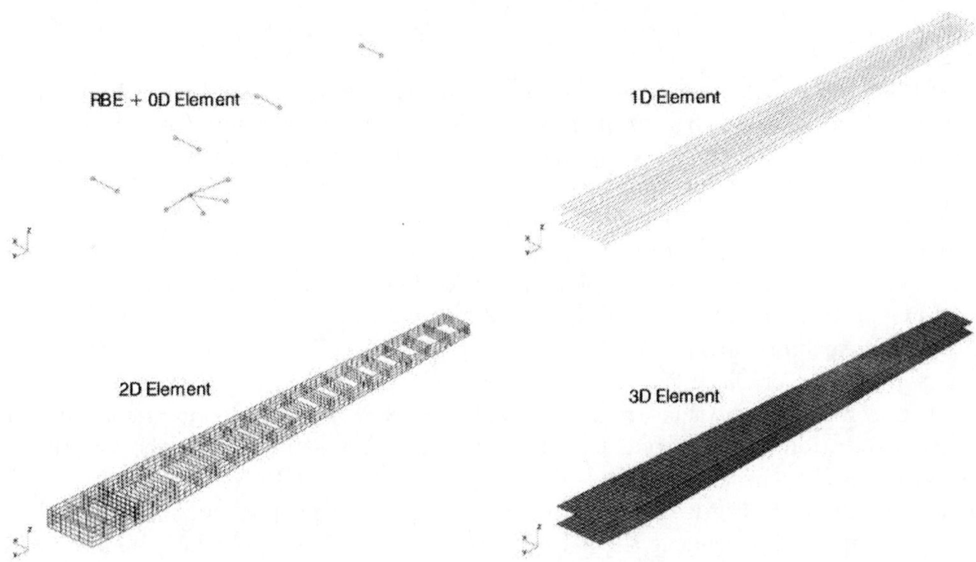

Figure 2: Wingbox components and their respective elements

Table 1: Natural Frequency Comparison of The Present Detailed Model with The Stick Model

	Natural frequency (Hz)			
Mode	Present	Stick Model [27]	Δ	Remarks
1	2.59	2.47	4.6 %	1st Vertical Bending
2	5.58	4.58	17.9 %	1st In-plane Bending
3	8.74	8.28	5.6 %	2nd Vertical Bending
4	12.76	11.56	9.4 %	1st Torsion

investigation. Their results were presented in [27]. The present model is in good comparison with their results as depicted in Table 1.

The cruise flight condition is observed. The aerodynamic lift distribution is depicted in Figure 3a. As aircraft structure commonly have a low-frequency vibration, the range of excitation frequencies around the 1st vertical bending mode is applied. Prior investigating the crack, a dynamic response analysis in the frequency domain is performed to the wingbox without the crack. The frequency response function (FRF) of the tip displacement is shown in Figure 3b.

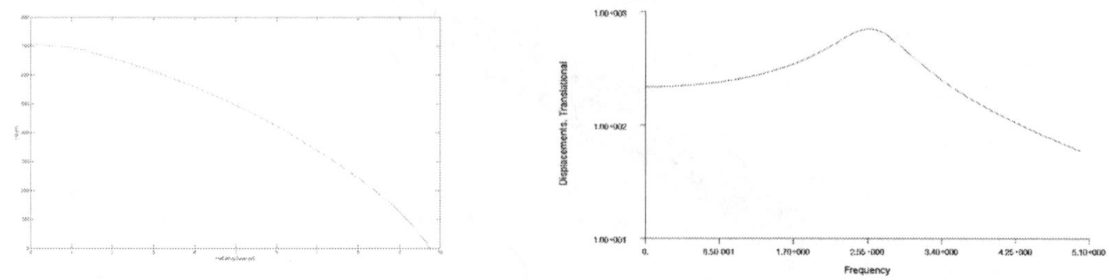

(a) (b)

Figure 3: (a) Aerodynamic lift distribution along the span (b) Tip displacement FRF due to aerodynamic load

It can be seen from Figure 3b, the response peak is when the excitation frequency equal to the 1st vertical bending natural frequency, or so-called resonance point. The critical aspect here is that until around 1.7 Hz, the response insignificantly changes from the static condition, around 220 mm. However, afterwards, the response increased significantly until around 700 mm at the resonance frequency. Considering that this peak response is more than thrice of the static condition, it is very likely that the structure will completely be destroyed at this stage. The load factor of civil aircraft itself commonly around 1.5-2.5. Hence, to have a reasonable investigation, the crack propagation investigation is limited to the maximum displacement around twice the static condition, at around 2 Hz.

In the present case, a crack initiation is given in the bottom skin near the root, as the largest stress concentration is within this area. Utilising the XFEM module, it is found that until around 1.7 Hz excitation, as expected, the crack is not propagating due to the low-stress level. However, above 1.7 Hz, the crack starts to propagate. When 2 Hz excitation is applied, the crack propagates the most. Figure 4 shows the crack is still not propagating as the displacement is still around 270 mm. However, at the 1st oscillation cycle when it reaches the amplitude, the crack starts to propagate from the lowest layer of the bottom skin as displayed in Figure 5. At the 2nd cycle, the crack propagated longer and grown to the highest layer of the bottom skin before it stopped. The final condition of the crack is shown in Figure 6.

5. Conclusion

A novel scheme to evaluate crack propagation on the wing structure exerted by oscillating aerodynamic load has been developed. The results presented in this paper has shown the ability of the scheme to evaluate the crack propagation of a wingbox structure. Future work will aim to include a more realistic aeroelastic condition with unsteady aerodynamic loads coupling.

Acknowledgments

The authors wishing to acknowledge the support Indonesia national institutions: LAPAN, IAe and BPPT. We also grateful for the funding provided by Ministry of Higher Education Malaysia (MOHE) and International Islamic University Malaysia (IIUM).

References

[1] Bristow J W and Irving P 2007 *Eng. Fail. Anal.* **14** 459 – 470
[2] Schijve J 1994 *Int. J. Fatigue* **16** 21 – 32

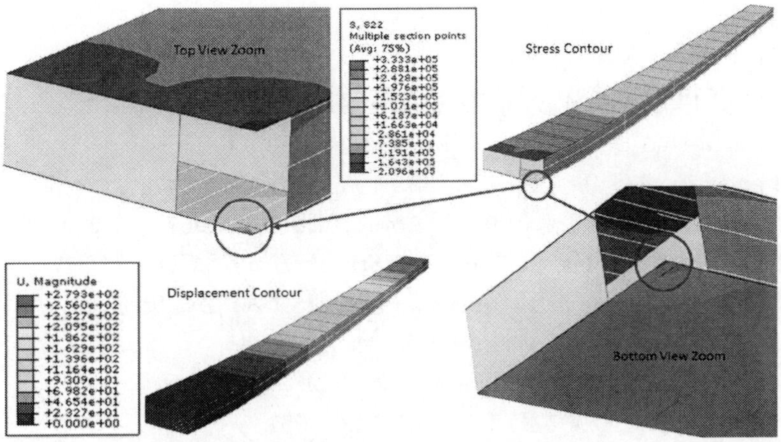

Figure 4: 2 Hz excitation: Stress & displacement contours when the crack is still not growing

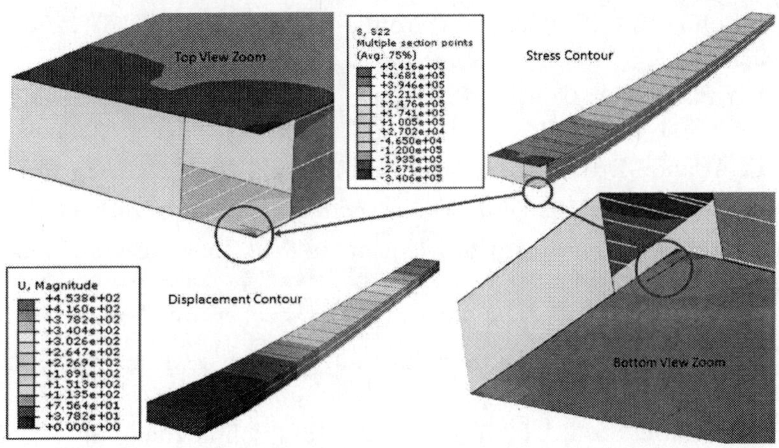

Figure 5: 2 Hz excitation: Stress & displacement contours when the crack is start propagating

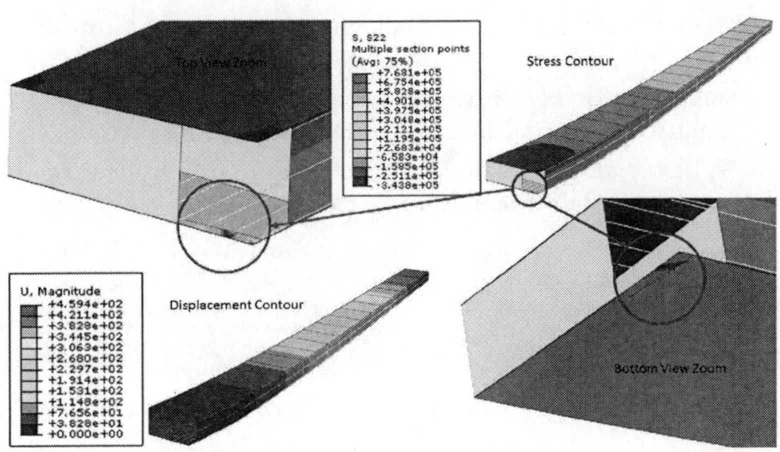

Figure 6: 2 Hz excitation: Stress & displacement contours when the crack is stop propagating

[3] Iyyer N, Sarkar S, Merrill R and Phan N 2007 *Int. J. Fatigue* **29** 1584 – 1607

[4] Schijve J 2009 *Int. J. Fatigue* **31** 998 – 1011

[5] Newman Jr J C 1992 *NASA STI/Recon Technical Report N* **92**

[6] Grbovic A and Rasuo B 2012 *Eng. Fail. Anal.* **26** 50 – 64

[7] Chen W H and Lin H C 1985 *AIAA J.* **23** 795 – 801

[8] Lin K J, Lu P J and Tarn J Q 1991 *J. Aircraft* **28** 899 – 907

[9] Wang K, Inman D J and Farrar C R 2005 *Struct. Health Monit.* **4** 377–392

[10] Abdullah N A, Curiel-Sosa J L and Akbar M 2018 *Compos. Struct.*

[11] Pidaparti R 1997 *Compos. Struct.* **38** 477 – 481

[12] Kapania R K and Castel F 1990 *AIAA J.* **28** 329 – 337

[13] Wang K and Inman D J 2007 *AIAA J.* **45** 542 – 551

[14] Georgiou G, Manan A and Cooper J 2012 *Mech. Syst. Signal Pr.* **32** 32 – 43

[15] Strganac T W and Kim Y I 1996 *J. Aircraft* **33** 68 – 73

[16] Pulungan D, Lubineau G, Yudhanto A, Yaldiz R and Schijve W 2017 *Int. J. Solids Struct.* **117** 177 – 190

[17] Pulungan D, Yudhanto A, Goutham Kumar S, Lubineau G, Yaldiz R and Schijve W 2018 *Polym. Test.* **68** 433–445

[18] Hernandez E, Alfano M, Pulungan D and Lubineau G 2017 *Sci. Rep.* **7** 16344

[19] Belytschko T and Black T 1999 *Int. J. Numer. Meth. Eng.* **45** 601 – 620

[20] Mohammadi S 2008 *Extended Finite Element Method for Fracture Analysis of Structures* (Blackwell Publishing Ltd.)

[21] Curiel-Sosa J L and Karapurath N 2012 *Compos. Sci. Technol.* **72** 788 – 791

[22] Moreno M S, Curiel-Sosa J L, Navarro-Zafra J, Vicente J M and Cela J L 2015 *Compos. Struct.* **119** 264 – 271

[23] Abdullah N A, Curiel-Sosa J L, Taylor Z A, Tafazzolimoghaddam B, Vicente J M and Zhang C 2017 *Compos. Struct.* **173** 78 – 85

[24] Curiel-Sosa J, Tafazzolimoghaddam B and Zhang C 2018 *Compos. Struct.* **189** 641 – 647

[25] Belytschko T, Liu W, Moran B and Elkhodary K 2014 *Nonlinear Finite Elements for Continua and Structures* (Wiley)

[26] Syamsuar S, Sampurno B, Mahasti K M, Pratama M B S, Sasongko T W, Kartika N, Suksmono A, Saputro M I A and Eskayudha D B 2018 *J. Phys. Conf. Ser.* **1005** 012040

[27] Abdurohman K, Bayu Sakti R M and Wismoyo P 2015 *International Seminar on Aerospace Science and Technology, Bali, Indonesia*

IOP Conf. Series: Journal of Physics: Conf. Series **1106** (2018) 012015 doi:10.1088/1742-6596/1106/1/012015

Determination of the rigid rotation plastic hinge point in SENB specimens in different strain hardening steels

W Khor[1,2,4] **, P L Moore**[3]**, H G Pisarski**[3] **and C J Brown**[1]

[1] College of Engineering, Design and Physical Sciences, Brunel University London, UB8 3PH, UK

[2] National Structural Integrity Research Centre, CB21 6AL, UK

[3] TWI, CB21 6AL, UK

Abstract. According to some standards, fracture toughness tests loaded under bending are assumed to deform around a fixed plastic hinge point within the ligament ahead of the notch tip. The rotation factor, r_p, defines the proportion of the ligament ahead of the crack tip to where this hinge point is located. In this paper, the concept of an SENB specimen bending about a fixed rotational point under loading was investigated. Experimental SENB tests were carried out on three different strain hardening steels, and the geometrical point of rotation was determined experimentally throughout the tests using a double clip gauge and the similar triangles principle. The experimental results were then used to develop and validate a series of different strain hardening property numerical models. By extracting the rotational factor from the different strain hardening property models, a relationship between strain hardening and a strain hardening corrected rotational factor, $r_{p\ sh}$ was established. This corrected rotational factor function was used to propose an improved equation for the calculation of CTOD and CTOD R-curves, which gave good estimations of CTOD when compared to values measured experimentally from sections through silicone replicas of the specimen crack-tip. The improved R-curve equation will be proposed for future amendments to the ISO 12135 standard.

1. Introduction

The Crack Tip Opening Displacement (CTOD) is one of the best known elastic-plastic fracture toughness parameters in fracture mechanics. The concept was originally developed at TWI, Cambridge in the 1960's[1], and is commonly used in the oil and gas industry. CTOD is advantageous in cases where the concept of linear elastic fracture mechanics is insufficient to account for ductile deformation, such as in pressure vessels, offshore platforms and pipelines.

There are several different definition for CTOD, but all describe CTOD as a material parameter relative to the physical opening at the original crack tip region. In general, CTOD is described as the fracture toughness in a specimen at the point of maximum load, initiation for stable crack extension or unstable crack extension, as appropriate. Notched specimens, sharpened by fatigue pre-cracking are loaded, and the load-displacement data are used in the calculation of CTOD.

Fracture toughness testing (often described simply as 'CTOD testing') became standardised in the 1970s, and is currently represented in a number of standards BS 7448[2], ISO 12135[3], ASTM E1820[4], and WES1108[5]. However, different assumptions about the determination of CTOD are used, which can lead to different values of CTOD. Whilst BS 7448, WES1108 and single point determination of

[4] Corresponding author, weeliam.khor@gmail.com

CTOD in ISO 12134 assume a rotational point, ASTM 1820 and R-curve determination in ISO 12135 do not; instead they derive CTOD from *J*.

It is important to define the value of CTOD with accuracy, particularly when CTOD is being used to determine tolerable flaw sizes for the assessment of structural integrity. Under-estimates of CTOD could lead to either rejection of the material or unnecessary repairs being required, resulting in significant burdens on the cost of fabrication. On the other hand, over-estimates of CTOD might lead to potentially unsafe structures being assessed as fit-for-service. This study examines the effect of strain hardening on the determination of CTOD in Single Edge Notch Bend (SENB) specimens manufactured from three different steels.

2. The concept of rigid rotational factor for the determination of CTOD

When a SENB specimen is loaded under three-point bending, the crack tip experiences tensile stress, whereas there is a region in the un-cracked ligament which experiences compression. In the calculation of the plastic component of CTOD in BS 7448-1, WES1108, earlier versions of ISO 12135, ASTM E1290-93 and E1820-01, it is assumed that the specimen rotates about a stationary point within the un-cracked ligament ahead of the crack tip. The rotation point is a distance ahead of the crack tip equal to $r_p \times B_0$, where r_p is the rotational factor and B_0 is the remaining ligament ahead of the crack tip. This concept was introduced by Dawes[6], based on a 2-D derived plastic hinge model assuming plane strain condition in the equation[7].

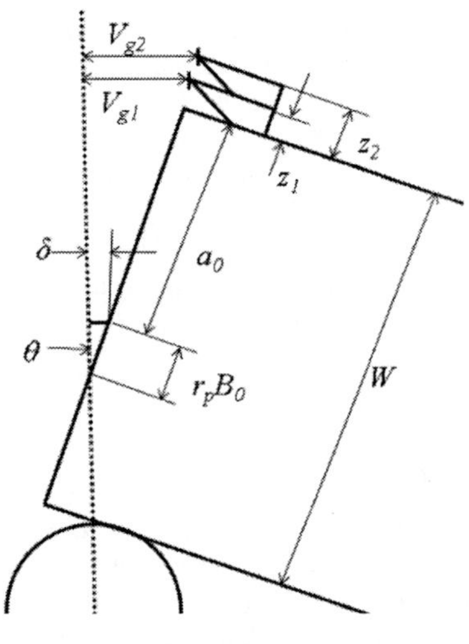

Figure 1. Diagram for the evaluation of the geometrical based CTOD.

Consider the symmetry of a deformed SENB specimen (**Figure 1**), the distance of the rotational point from the crack mouth opening displacement (CMOD), H is defined in terms of rotational factor, $r_pB_0 + a_0$. From **Figure 1**, V_{g1} and δ can be related using the similar triangles assumption, given below:

$$V_{g1} = 2\left[r_p(W - a_0) + a_0 + z_1\right] \sin\theta \quad (1)$$

$$\delta = 2r_p(W - a_0)\sin\theta_p \quad (2)$$

Rearranging Equation (1) into Equation (2) gives[7]

$$\delta = \frac{r_p(W - a_0)V_{g1}}{\left[r_p(W - a_0) + a_0 + z_1\right]}$$

The similar triangles assumption is used for the determination of the plastic CTOD in BS 7448 Parts 1, 4 and ISO 12135 used a constant value of r_p= 0.4 in the calculation of CTOD.

3. Methodology

3.1. *Finite element modelling methods*

A Geometrically and Materially Non-linear Analysis (GMNA) FE model was used to predict CTOD in SENB specimens with the same geometry and material properties as the experimental tests. Fully three-dimensional quarter SENB models were simulated using commercially available software (ABAQUS v6.14) with a blunted crack tip of 0.03mm radius. **Figure 2** shows the outline geometry of the SENB specimen in the model. 8-noded linear brick elements (C3D8R) were used to model the SENB specimens as no difference in results were observed using C3D20R[8].

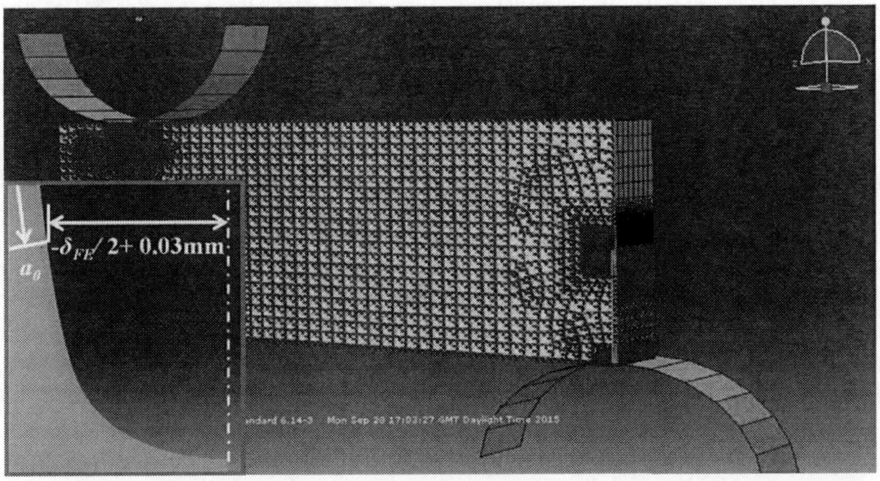

Figure 2. Quarter SENB model for a_0/W= 0.5, showing boundary conditions and measurement technique for CTOD, adjusted for the 0.03mm crack radius used in the models.

A standard convergence test was performed based on varying the element size distributed across the crack tip[8]. CTOD was determined from the numerical model based on both the opening of the original crack tip, which is the original concept of CTOD and equivalent to the experimentally-cast silicone replica measurements, described below (**Figure 2**).

Additional FE analyses were undertaken based on the techniques described above with idealised tensile properties. The idealized tensile properties were generated employing the modified Ramberg-Osgood power law for 0.44≤ σ_{ys}/σ_{uts}≤ 0.98. The equation[9] used is given in Equation (3)

$$\varepsilon = \frac{\sigma}{E} + \alpha\left(\frac{\sigma}{\sigma_{ys}}\right)^{n-1} \tag{3}$$

Where α= 0.002, E= 207GPa and σ_{ys}= 400MPa. Relatively, decreasing n would lead to the increase of strain hardening (decreasing tensile ratio). The true stress-strain curve obtained based on Equation (3) is shown in **Figure 3**.

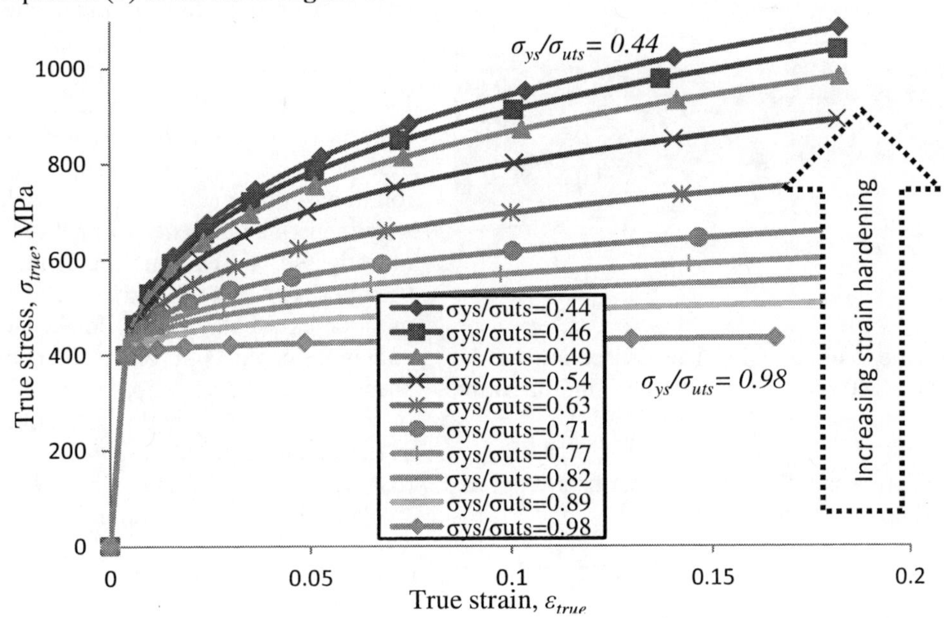

Figure 3. Idealised true stress-strain curve based on the modified Ramberg-Osgood power law.

3.2. Experimental methods

For the experimental tests, three different steels were chosen to cover a range of strain hardening behaviour:

- M01 Low strain hardening: High strength steel SA-543-GrB-Cl1, σ_{ys}= 850MPa and σ_{ys}/σ_{uts}= 0.93
- M02 Medium strain hardening: Structural steel *S355J2*, σ_{ys} = 421MPa and σ_{ys}/σ_{uts}= 0.72
- M03 High strain hardening: Austenitic stainless steel *SS316*, σ_{ys} = 286MPa and σ_{ys}/σ_{uts}= 0.45

SENB specimens of cross-section 20mm x 40mm were machined from the steel plates, notched and fatigue pre-cracked to give a nominal a_0/W ratio of 0.5. A total of eight specimens were tested: two each for M01 and M03, four from M02. A crack casting method using silicone replica compound was used to cast the crack for physical measurement of CTOD. The specimens were loaded to the selected displacement points based on the clip gauge opening, then held in displacement control. The silicone compound was slowly injected into the specimen notch from one side of the specimen and allowed to cure for 5 minutes. After curing time, the specimen was loaded to the next point of interest and held in constant displacement again. The replica was then extracted and the casting procedure was repeated. Typically, around ten silicone replica casts (SRCs) were obtained at different displacement levels for each test.

3.3. Physical measurements of CTOD

The SRCs extracted from the test specimens were sliced in the middle and CTOD was measured using an optical microscope. The location of the initial fatigue crack tip was identified from the initial crack length, and the width of the silicone replica was measured at this point to give the value of CTOD from the replica, δ_{SRC} (**Figure 4**). This method takes CTOD at the mid specimen thickness, which corresponds to the plane strain condition assumed in the equations[10,11]. The CTOD values obtained from the SRCs were considered to represent the actual physical CTOD, and thus were treated as the baseline for comparison to other methods.

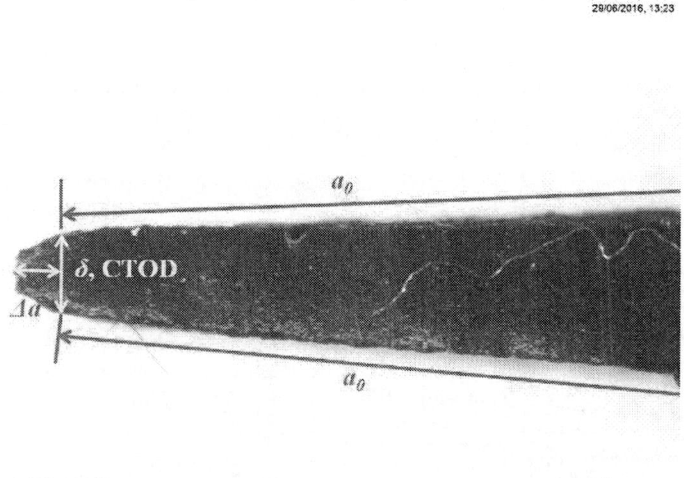

Figure 4. Measurement on the sliced cross section of the silicone crack replica (M03-05, V_{g1}= 2.601mm, middle of the crack).

4. The determination of the rotational factor based on the similar triangles method

The concept of a rigid rotational point has been successfully implemented for the determination of CTOD, where the fixed value of r_p= 0.4 is found to be most accurate[10,12] for material within the range of 0.65< σ_{ys}/σ_{uts}< 0.7. To investigate the rotational factor, r_p for SENB specimens, an opened crack was investigated based on the similar triangles concept (**Figure 1**). This method extrapolates the crack face angles into the unbroken ligament ahead of the crack tip, where the intersection of the angles is defined as the rotational point.

To simplify the derivation, the following terms are defined: $r_pB_0= Y$, $z_1+ a_0= C$, and $a_0+ z_2= D$. To relate the lower and upper clip gauge opening, V_{g1} and V_{g2} respectively to the point of rotation,

$$\sin\theta = \frac{V_{g1}}{C+Y} = \frac{V_{g2}}{D+Y}$$

Leading to

$$\frac{V_{g2}}{V_{g1}} = \frac{D+Y}{C+Y}$$

Expanding D and factoring $C+ Y$ gives

$$\frac{V_{g2}}{V_{g1}} = \frac{C + Y + (z_2 - z_1)}{C + Y} = 1 + \frac{(z_2 - z_1)}{C + Y}$$

Rearranging the equation leads to

$$Y = r_p B_0 = \left(\frac{z_2 - z_1}{V_{g2}/V_{g1} - 1} \right) - (z_1 + a_0)$$

Where the rotation factor, r_p based on V_{g1} and V_{g2} is given as

$$r_p = \left[\left(\frac{z_2 - z_1}{V_{g2}/V_{g1} - 1} \right) - (z_1 + a_0) \right] \times \frac{1}{B_0} \quad (4)$$

Equation (4) allow the rotational factor to be calculated based on two clip gauges positioned at different heights above the crack mouth in standard laboratory tests. Similarly, the rotational factor based on the plastic displacement can be obtained by simply replacing the lower and upper clip gauge displacement, V_{g1} and V_{g2} with the plastic lower and upper clip gauge displacement, V_{p1} and V_{p2}.

To include the effects of strain hardening, $r_{p\,sh}$ is extracted based on the intersection the line extrapolated from the lower clip gauge opening or CMOD and CTOD to the symmetry line (**Figure 5**). Equation (4) was therefore modified for the calculation of $r_{p\,sh}$, described as

$$r_{p\,sh} = \left[\left(\frac{z_1 + a_0}{(CMOD/\delta) - 1} \right) \right] \times \frac{1}{B_0} \quad (5)$$

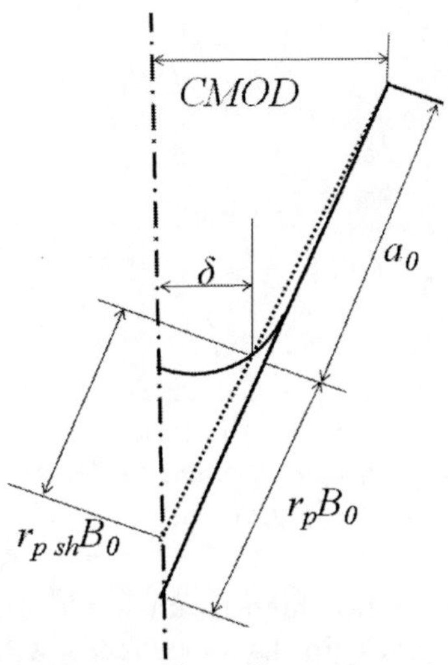

Figure 5. The effect of crack tip blunting due to strain hardening.

5. Results and discussion

Conceptually, the strain hardening based rotational factor, $r_{p\,sh}$ accounts for the effects of strain hardening based on the measured CTOD, which allows the rigid rotational plastic hinge concept to estimate CTOD for any given material strain hardening property. Based on Equation (5), $r_{p\,sh}$ was derived using the CTOD from the FE models with idealised tensile properties (**Figure 6**). Data were extracted for 0.2mm< FE CTOD< 1.0mm to minimise the influence of the elastic CTOD and large deformation on the model results, while ensuring sufficient loads such that the point of rotation has stabilised.

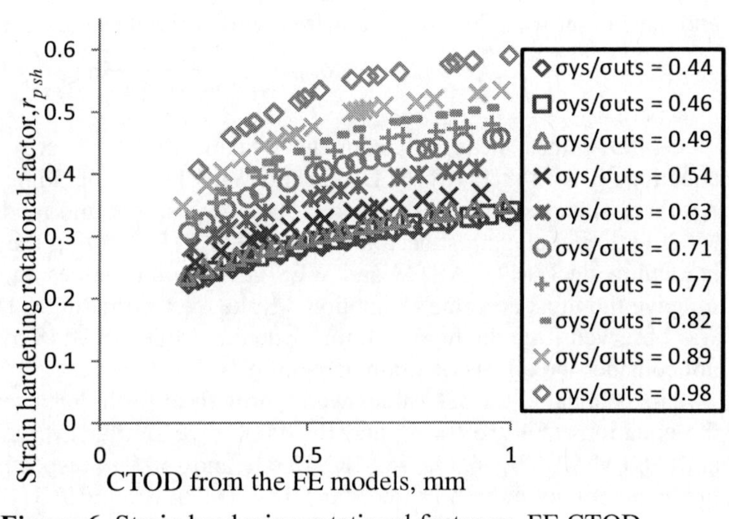

Figure 6. Strain hardening rotational factor vs. FE CTOD.

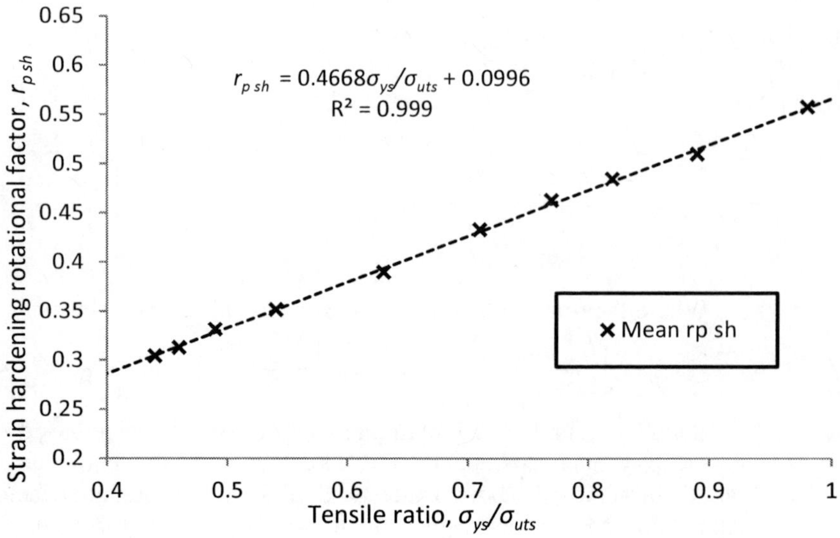

Figure 7. Relationship between the strain hardening rotational factor, $r_{p\ sh}$ and tensile ratio, σ_{ys}/σ_{uts}

Figure 6 shows the dependency of $r_{p\ sh}$ on tensile ratio, as the overall $r_{p\ sh}$ decreases with the increase of strain hardening, where strain hardening is simply expressed as the tensile ratio (σ_{ys}/σ_{uts}). Based on the mean $r_{p\ sh}$ for the respective tensile ratio, a linear relationship was obtained to describe the effect of $r_{p\ sh}$ on strain hardening for the range of materials studied (**Figure 7**), given as:

$$r_{p\ sh} = 0.4668\frac{\sigma_{ys}}{\sigma_{uts}} + 0.0996 \quad (6)$$

The implication of this equation is that for steels with low σ_{ys}/σ_{uts} ratio (high strain hardening steels), SENB specimens plastically rotate about a point closer to the crack tip than specimens with high σ_{ys}/σ_{uts} ratio (low strain hardening steels).

Based on the $r_{p\ sh}$ findings, a CTOD equation considering strain hardening was obtained for SENB specimens with a_0/W of 0.5, using the elastic CTOD from WES 1108 and plastic CTOD based on $r_{p\ sh}$ and the similar triangles concept, thereafter described as δ_{sh}, given as

$$\delta_{sh} = K^2\frac{(1-v^2)}{m_{JWES}\sigma_{ys}E} + \frac{r_{psh}B_oV_p}{r_{psh}B_o+a_0} \quad (7)$$

where m_{JWES}= 4.9- 3.5(σ_{ys}/σ_{uts})

CTOD values measured from the silicone replicas were used to validate the standard CTOD equations (BS 7448-1, ASTM E1820 and WES 1108), including CTOD from Equation (7) (**Figure 8**). All three standards underestimated CTOD for the low and medium strain hardening steels, M01 and M02, with δ_{sh} being most accurate and ASTM E1820 giving the lowest estimation. In the higher strain hardening steel, M03, ASTM and WES gave similar values and lower estimation of CTOD, whereas δ_{sh} gave the most accurate estimation despite overestimating CTOD in several instances. Some scatter was observed from the higher strain hardening M03. BS 7448 overestimated CTOD for M03 as it does not consider the effects of strain hardening[10,11,12].

The calculated CTOD values were normalised to the measured CTOD to indicate the accuracy of the equations (**Figure 9**). δ_{sh} and BS 7448 gave similar average estimation, followed by WES 1108 and ASTM E1820 (-6.4%, -6.8%, -15.6% and -30.9% respectively). δ_{sh} gave the best accuracy and precision compared to all three standards. The WES and ASTM give lower estimations of CTOD as they were validated to determine CTOD from *J* based on the 45 degree intercept concept of CTOD, which gives lower values to CTOD from the original crack tip[10].

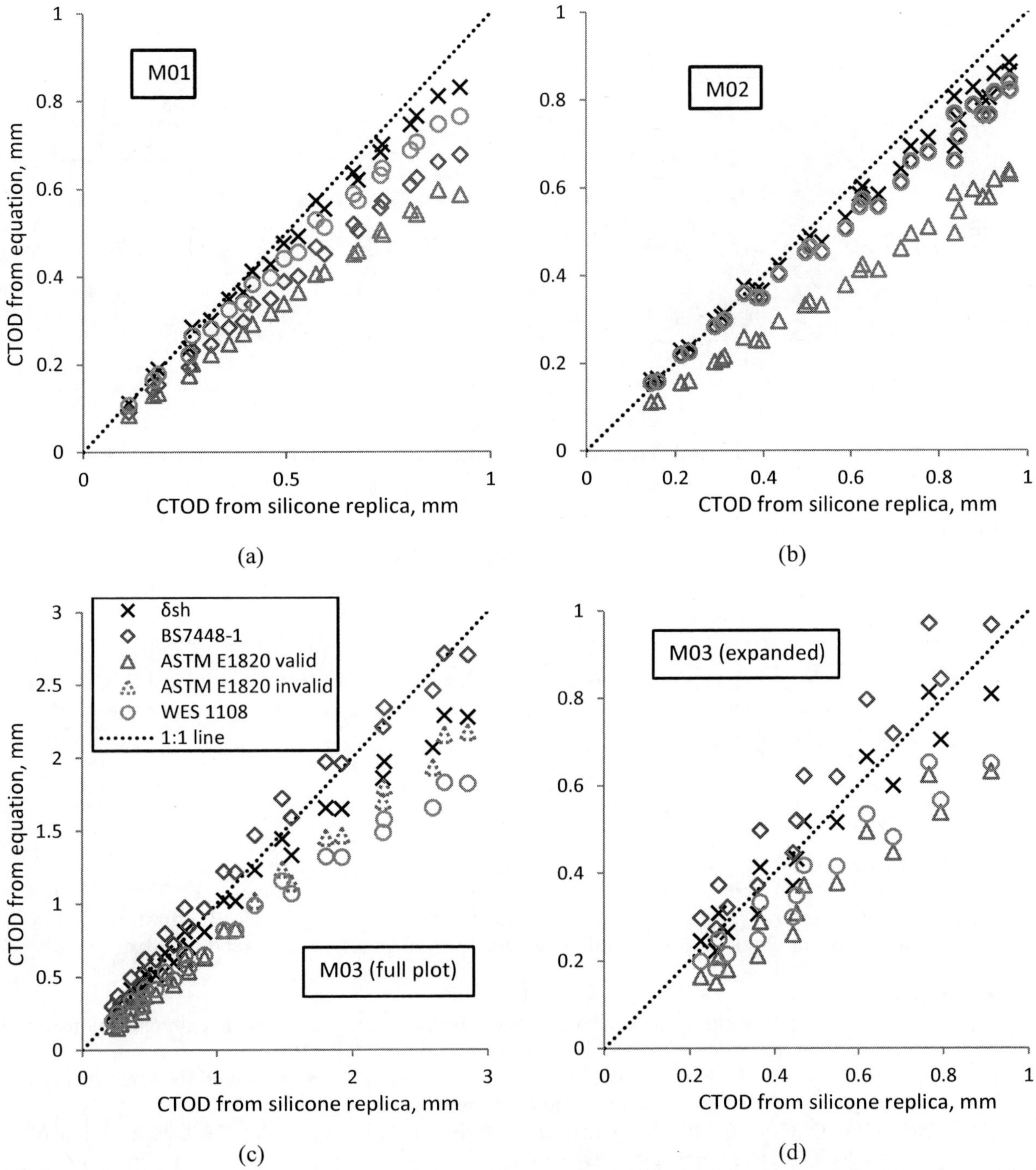

Figure 8. CTOD from silicone replica vs. CTOD from equations, M01, M02 and M03, full plot and expanded view (a, b, c and d respectively)

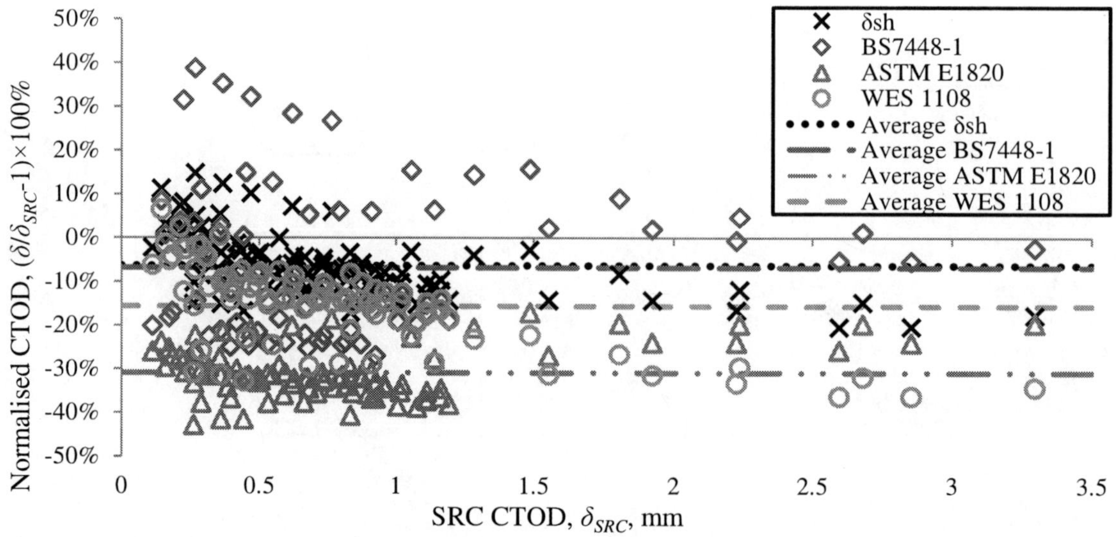

Figure 9. Normalised CTOD for the low, medium and high strain hardening M01, M02 and M03.

6. Conclusion

Based on the experimental measurements of CTOD, it was found that:

- The rigid rotational concept is practical method for the estimation of CTOD
- The strain hardening corrected rotational factor, $r_{p\ sh}$ increases linearly from 0.30 at σ_{ys}/σ_{uts}= 0.44 to 0.56 at σ_{ys}/σ_{uts}= 0.98, given as

$$r_{p\ sh} = 0.4668\frac{\sigma_{ys}}{\sigma_{uts}} + 0.0996$$

- In steel SENB specimens with a nominal crack length of a_0/W= 0.5, the estimated CTOD with the highest accuracy and precision compared to BS 7448-1, ASTM E1820 and WES 1108 is given by the following equation:

$$\delta_{sh} = K^2 \frac{(1-v^2)}{m_{JWES}\sigma_{ys}E} + \frac{r_{psh}B_o V_p}{r_{psh}B_o + a_0}$$

Where m_{JWES}= 4.9- 3.5(σ_{ys}/σ_{uts})

7. Acknowledgements

The research presented here would not have been possible without the funding of TWI's NSIRC programme. The authors are grateful to Dan Bloom, Phillip Cossey, Jack Bradford and Jerry Godden for their contribution to performing the fracture toughness testing reported here.

8. References

[1] Wells AA. Crack opening displacements from elastic-plastic analyses of externally notched tension bars. Engineering Fracture Mechanics. 1969;1(3):399-410.

[2] BSI. BS 7448-1:1991 - Fracture mechanics toughness tests — Part 1: Method for determination of K_{Ic}, critical CTOD and critical J values of metallic materials. BSI; 1991.

[3] ISO. ISO 12135 - 16 Metallic materials - Unified method of test for the determination of quasistatic fracture toughness. ISO; 2016.

[4] ASTM. ASTM E1820-13 - Standard Measurement of Fracture Toughness. ASTM; 2014. p. 1-54.

[5] JWES. WES 1108: 2014 Standard test method for Crack-Tip Opening Displacement (CTOD) fracture toughness measurement. The Japan Welding Engineering Society. 2014.

[6] Dawes MG. Elastic-Plastic Fracture Toughness Based on the COD and J-Contour Integral Concepts. ASTM STP 668. 1979:307-33.

[7] Lin IH, Anderson TL, Derit R, Dawes MG, DeWit R, Dawes MG. Displacements and rotational factors in Single Edge Notched Bend specimens. 1982;20:R3-R7.

IOP Conf. Series: Journal of Physics: Conf. Series **1106** (2018) 012015 doi:10.1088/1742-6596/1106/1/012015

[8] Khor W. Crack tip opening displacement (CTOD) in single edge notched bend (SEN(B)). Brunel University London: Brunel University London; 2018.

[9] Ramberg W, Osgood WR. Description of stress-strain curved by three parameters. National Advisory Committee for Aeronautics 1943 Contract No.: Technical Note No. 902.

MacDonald M, Rhodes J, Taylor GT. Mechanical properties of stainless steel lipped channels. 15th International Specialty Conference on Cold-Formed Steel Structures; October 19-20; St. Louis, Missiuri, U.S.A2000. p. 673-86.

[10] Khor W, Moore PL, Pisarski HG, Brown CJ. Comparison of methods to determine CTOD for SENB specimens in different strain hardening steels. Fatigue and Fracture of Engineering Materials and Structures. 2018;41(3):551-64.

[11] Khor W, Moore PL, Pisarski HG, Haslett M, Brown CJ. Measurement and prediction of CTOD in austenitic stainless steel. Fatigue & Fracture of Engineering Materials & Structures. 2016;39:1433-42.

[12] Moore PL, Khor W. The effect of material strain hardening on the determination of CTOD R-curves using SENB specimens. 12th International Pipeline Conference; September 24-28, 2018; Calgary, Alberta, Canada: ASME; 2018.

Destructive Testing of Open Rotor Propeller Blades for Extreme Operation Conditions

Dacho Dachev[1], Jialin Tang[1], Jamil Kanfoud[2], Cristinel Mares[1], Slim Soua[2]

1. Brunel University, Mechanical and Aerospace Department, UK
2. TWI, Cambridge, UK

0910750@brunel.ac.uk; Jialin.Tang@brunel.ac.uk; jamil.kanfoud@brunel.ac.uk; cristinel.mares@brunel.ac.uk; slim.soua@brunel.ac.uk

Abstract. Due to the complexity of composite material, accurate manufacturing is very complicated and carry over a large number of uncertainties, the process of optimization and validation being of primary importance.

In this paper as part of design technology for rotor blades, the process of mechanical testing against severe working conditions using Acoustic Emission (AE) is discussed. As part of this process, the dynamic behaviour is analysed with validation and model updating of the structure and material properties, followed by structural tests under loads specific to the operational conditions.

The results presented show how the appropriate AE technique has been implemented in order to obtain relevant data and focus on specific areas of interest for the design development, for correlation between modelling and experimental testing, the structure being tested to the ultimate load.

Keywords: Acoustic emission, Composites, Resonance Frequency, Tensile, Flexion

1. Introduction

In aerospace, future aircraft structures require airframes that are lighter, easier to maintain and more durable than those in current operation. All these requirements point out the use of composite materials, especially for the build of the primary structures, aiming for specific tailored properties and significant weight reduction. Currently laminated composite materials are largely used due to their properties, weight, stiffness and strength, being manufactured from unidirectional plies with thickness and specific angle of fibre orientation. Due to the complexity of these materials, accurate manufacturing is very complicated and carry over large number of uncertainties, due to misalignment of ply orientation, incomplete curing of the resin, and excess resin between plies, porosity resulting from machine, human, and manufacturing process inaccuracy. In addition to that, especially in the aerospace industry, the structure is pre-twisted to serve specific design purposes. For critical operational conditions, understanding the material behaviour, the damage initiation and propagation, and possible transition stages must be supported by both modelling and experimental programs combined with proper definition of the loading conditions to provide the highest level of confidence and improve the structural design and material development. Constrains such as mass, rotating speed, material specifications, length, geometry shock and impact require a robust design to increase the stability and fracture resistance this reflecting directly on the mass and volume of the structure. In addition, the environmental conditions with temperatures which can vary from -60oC to +90oC, rapid cooling and heating during the flight envelope, creating conditions for crack occurrence and propagations or critical structure failures. Additional factors that can affect blade performance are humidity, liquids (rain, snow or ice), as well as dust and debris in the air. Previous studies carried out on wind turbine blades have described the fatigue damage modes at the root of the blade Li,Wu and Zang [1] [2] [3] or Hak, Min and Park [4] as shown in Figure 1.

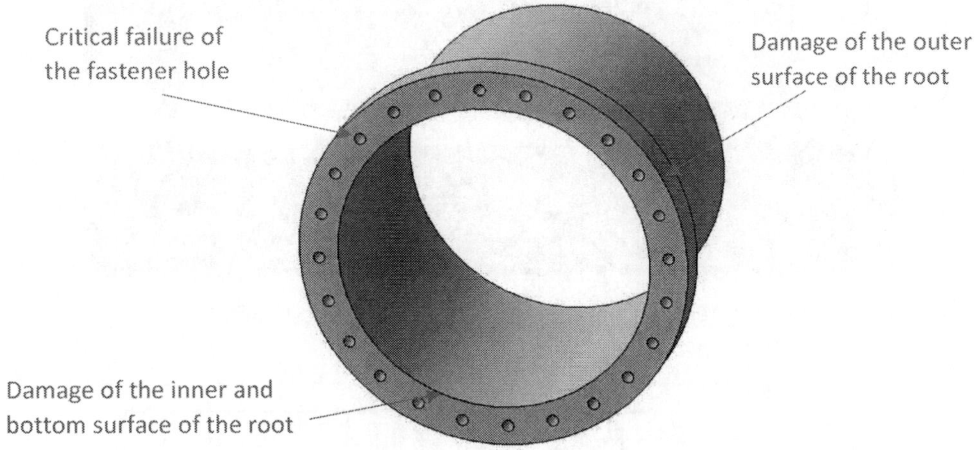

Figure 1: Failures at the end of the blade root

It was shown that the motion of the blade has resulted in different load distribution at different locations, which cannot be predicted accurately by the numerical simulations. As a result of the increased load, partial separation have been observed at the root end connection bolts which may cause the blade to be ripped off the turbine when in motion.

In this paper some phases of the testing program of a rotor blade structure is described highlighting the use of AE to monitor the structural behaviour under loads derived from operational conditions for High and Low Cycle Fatigue (HCF) life estimation. The test dynamic loads are based on the centrifugal, bending and twisting operational conditions with the objective of available stress margin estimation required for HCF avoidance. Some results obtained for two test cases are presented with the aim to present the use of AE measurements and its efficiency compared against the initial FEA analysis and predictions.

2. Experimental investigation and predictions

In the next section the preparation and the results obtained from a tensile test are described. The blade manufacturing characteristics require that the blade is adapted to the tensile machine in a specific manner. For this purpose a hole needs to be cut in the blade, using water jet cutting, to be able to allow the blade to go under tensile loading. Preliminary numerical modelling simulations were carried out to determine an optimal hole size and location with minimal stress concentration but at operational stresses as close as possible to reality. In all simulations equivalent material properties were used requiring the analysis of uncertainties and offsets when simulation results were compared with the experimental data. An AE calibration is carried out on a healthy blade before the introduction of the hole on the blade in order to check the wave velocity and determine a sensor set-up on the top and bottom surface of the blade. A final set of 10 sensors are used with a Vallen pulsing system (Figure 2).

IOP Conf. Series: Journal of Physics: Conf. Series **1106** (2018) 012016 doi:10.1088/1742-6596/1106/1/012016

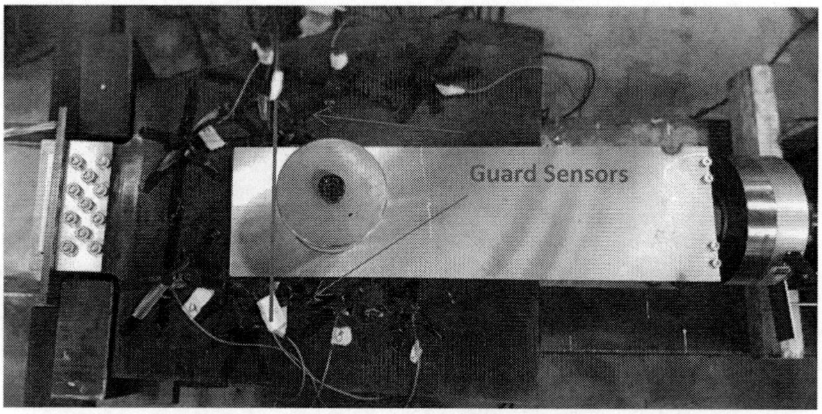

Figure2: Acoustic emission strategy for tensile test monitoring

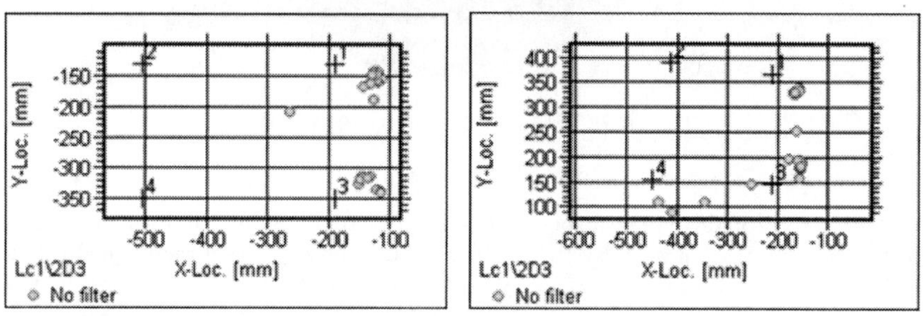

a) Top face

b) Bottom face

Figure 3: Calibration results for both surfaces of the blade

The test measures the velocity response between the mounted sensors by sending an impulse one to another through PLB and measuring the response with signal attenuation analysis and an average velocity calculation. The results of calibration localization of the generated events on both surfaces are accurate, validating the sensor set-up, as shown in Figure 3a, b. Four sensors are monitoring the root of the blade, where majority of the events are expected to occur based on initial analysis. Another set of four sensors are monitoring the mounting hole, the second most stressed area. Finally, two guarding sensors are introduced to be able to "split" the blade on two sections as shown with the red line on Figure 2. This is going to prevent signals to cross over and to be detected from both sets of sensors.

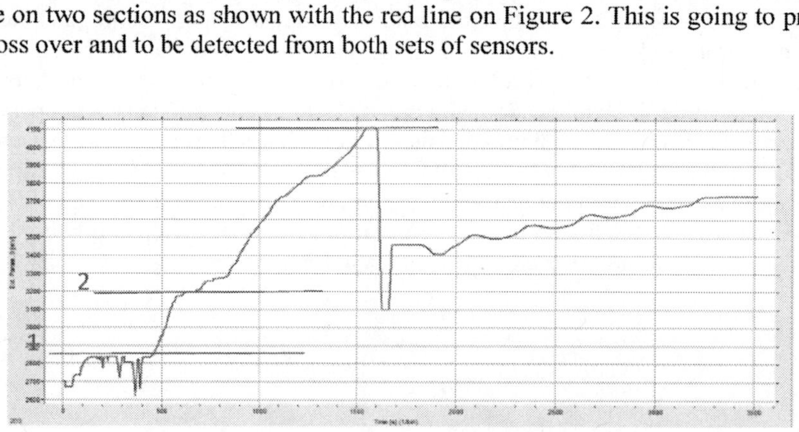

Figure 4: Loading schedule for tensile test

The tensile test is performed on the blade with a step loading of 1kN schedule from an initial force of 18.42kN increased until safety factors of 3.5 and subsequently of 7 are achieved (Figure 4). In the AE signals the majority of the events detected appear at the root of the blade, with less than expected events detected at the top side of the blade, where most of the events cluster around the hole and may be a result of the contact at the pin connection with the blade. Unexpectedly, events appear at the trailing edge, which could possibly be explained with slightly twisted set-up for the test, which resulted in a stress slightly off the central axis of the blade due to the very complex shape of the blade.

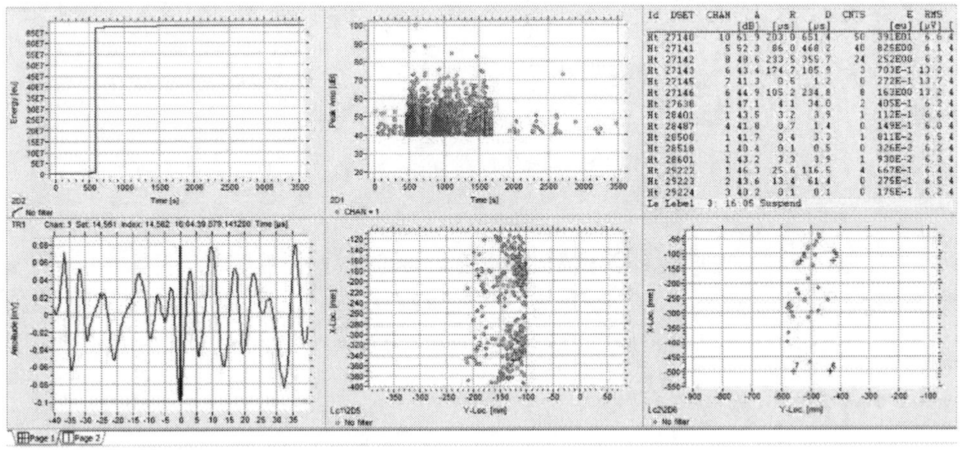

Figure 5: AE test results during the loading program

A second tensile test with a step of 5kN and the time of action of 40 seconds is presented in Figure 6.

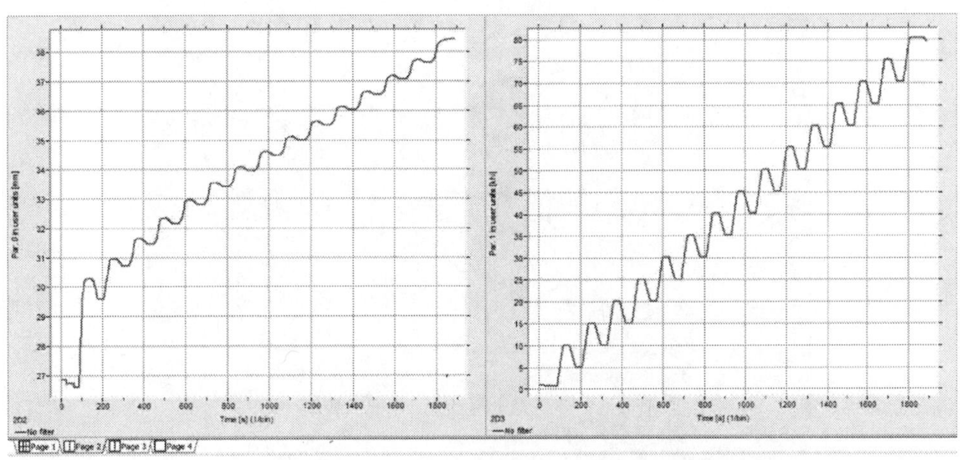

Figure 6: Second test tensile step loading

AE signal analysis results from this test are presented in Figure 7. As predicted, the majority of the events appear at the root of the blade. In contrast, the number of events detected in more controlled environment is much smaller. Noise and events associated to the pin contact are minimized under

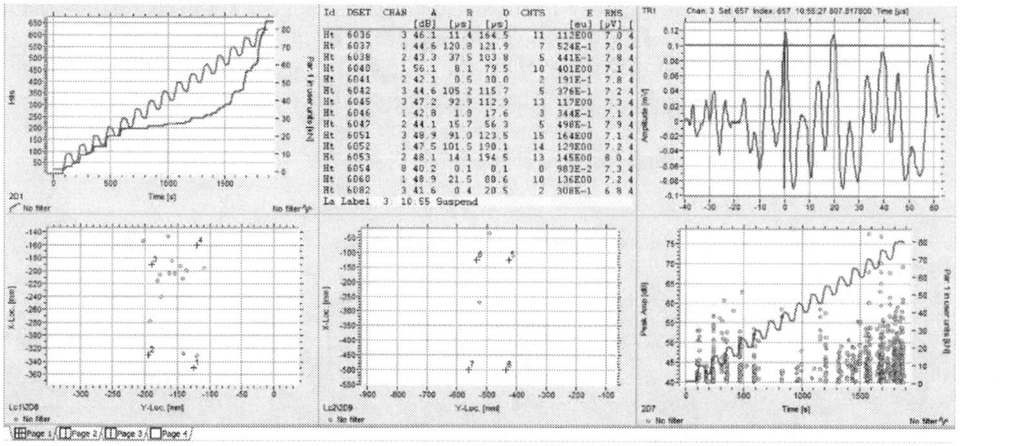

Figure 7: AE test results of the second tensile test

these test conditions. In the second phase a bending test was carried out, so that the blade can be validated with safety factor 3.5 with four sensors positioned to capture and locate any event that occurs in the root area (Figure 8).

Figure 8: Test set-up and sensor configuration for the flexural test

Because the load is going to be applied over relatively small area and the surface of the blade is not perfectly flat, noise can be produced in the area of contact, due to pin slip or damage as result of the concentrated stress. For that reason two guarding sensors, 5 and 6 in Figure 8, have been positioned above the main four, so any noise generated at the stress application point cannot be captured by the main sensors. Finally, two additional sensors 7 and 8 on Figure 8 have been added to monitor the top area of the blade, above the guard sensors, in case that event occur in an area not predicted by the numerical modelling simulation. A load program shown and the measured signals are shown in Figure 9.

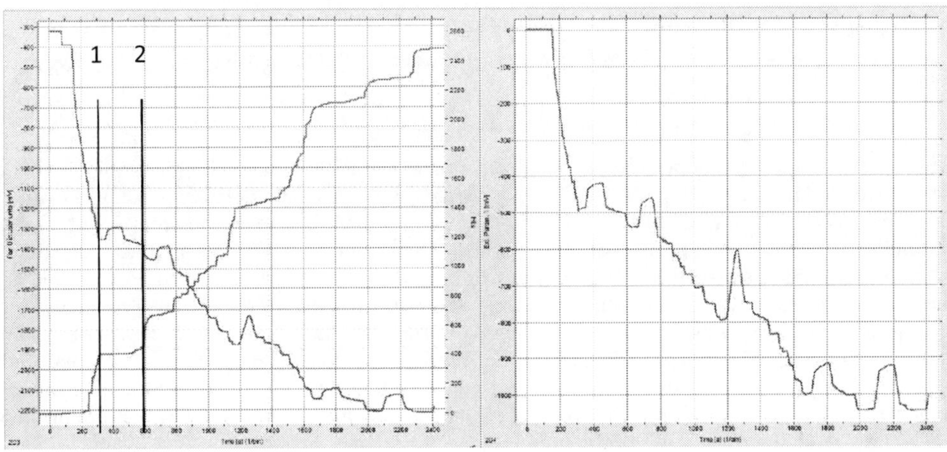

Figure 9: Load and AE cumulative hits variation with identification of the Kaiser Effect

These graphs allow a clear correlation of the AE signals with the load variation and the identification of the Kaiser Effect.

3. Conclusions

A rotor blade design has been validated successfully against a predefined tensile load program derived from the operational conditions of the system. The paper presents the use of the AE signal analysis in validation during testing of the predicted stress states using a 3.5 safety factor.
During the blade experiments, the Kaiser Effect was put in evidence, this being typically observed in testing of metallic elements but not usual for composites.

References

1. F. Li, Y.T. He, H.P. Li, K.M. Ma, Failure analysis for compressor blade of the fourth stage in an aero-engine, J. Mater. Eng. S1 (2006) 382–384.
2. W.M. Wu, Y.X. Gu, W.J. Du, Fracture analysis of the rotor blade of stage I on a compressor, Fail. Anal. Prev. 01 (2006) 61–64.
3. Q.Q. Liu, Fracture analysis on rotor blades of compressor I for a series engines, Fail. Anal. Prev. 02 (2007) (34-6+15).
4. Hak Gu Lee , Min Gyu Kang, Jisang Park, Fatigue failure of a composite wind turbine blade at its root end, Wind Turbine Technology Research Center, Korea Institute of Materials Science, 797 Changwondaero, Changwon, Gyeongnam 7 August 2015 641-831, Republic of Korea
5. McCrory, J.P., Al-Jumaili, S.K., Crivelli, D., Pearson, M.R., Eaton, M.J., Featherston, C.A., Guagliano, M., Holford, K.M. & Pullin, R. 2015, "Damage classification in carbon fibre composites using acoustic emission: A comparison of three techniques", Composites Part B, vol. 68, pp. 424-430.
6. Scott, I.G. 1991, Basic acoustic emission, Gordon and Breach SA, New York.

IOP Conf. Series: Journal of Physics: Conf. Series **1106** (2018) 012017 doi:10.1088/1742-6596/1106/1/012017

An eigenvalue approximation for parameter–dependent undamped gyroscopic systems

Nataliia Gavryliuk and Atul Bhaskar

Faculty of Engineering and the Environment, University of Southampton, SO17 1BJ, UK

E-mail: ng1v14@soton.ac.uk

Abstract. Parameter–dependent eigenvalue problem occurs in a host of engineering contexts. Structural design under dynamic loading is concerned with the evaluation of eigenvalues for a large number of structures, each evaluation corresponding to a combination of design parameters. Exact calculation of the natural frequencies of all the models considered is computationally expensive. Here structural problems that possess gyroscopy, typically encountered in the analysis of rotating elastic structures, are considered. Approximate but inexpensive calculations are sought for this class of problems. In the present work, an algorithm for approximating the natural frequencies of undamped gyroscopic systems is presented. Numerical examples show excellent accuracy, while affording significant computational economy compared to exact calculations. The computational gain is found to be relatively more notable when the size of the considered problem is large.

1. Introduction

Eigenvalue problems in terms of a skew–symmetric matrix arise in the dynamics of different kinds of rotating structures like spinning disks, hard drive, bend saw, helicopter blades, vacuum cleaner, washing machine, etc. They all possess some common mathematical features that are generic to this class of problems. The governing equation of motion involves two real and symmetric matrices associated with the inertia and stiffness of the structure whereas gyroscopy, arising from the linear term within the kinetic energy, leads to a skew–symmetric matrix. In a design scenario, several design alternatives need to be analysed. Solving several eigenvalue problems leads to a large number of evaluation, which is computationally expensive. Here we propose approximation for a parameter–dependent eigenvalue problem, accurately and computationally economically.

The equation of free motion for undamped gyroscopic system is expressed

$$\mathbf{M}\ddot{\mathbf{q}}(t) + \mathbf{G}\dot{\mathbf{q}}(t) + \mathbf{K}\mathbf{q}(t) = \mathbf{0}, \tag{1}$$

where $\mathbf{K}$ and $\mathbf{M}$ are real symmetric stiffness and mass matrices of size $n \times n$ respectively, $\mathbf{G}$ is $n \times n$ skew–symmetric gyroscopic matrix. Seeking a synchronous solution for (1) leads to a λ–matrix problem. Instead, (1) can be cast in the so–called state–space of double the size of the configuration space. Free vibration problem associated with the first order state–space equation is then given by the following generalised eigenvalue problem

$$\mathbf{G}^*\mathbf{u} = -\lambda\mathbf{M}^*\mathbf{u}, \tag{2}$$

where $\mathbf{G}^* = \left[\begin{array}{c|c} \mathbf{G} & \mathbf{K} \\ \hline -\mathbf{K} & \mathbf{0} \end{array}\right]$ is $2n \times 2n$ skew–symmetric matrix, $\mathbf{M}^* = \left[\begin{array}{c|c} \mathbf{M} & \mathbf{0} \\ \hline \mathbf{0} & \mathbf{K} \end{array}\right]$ is a positive–definite symmetric matrix [1], $\mathbf{u}$ is an eigenvector and λ the associated eigenvalue. The geometry of engineering structures depends on a set of design parameters, such as thickness, length, height, material properties, etc. For such a practical design situation (2) can be expressed as a parameter–dependent eigenvalue problem

$$\mathbf{G}^*(p)\mathbf{u}(p) = -\lambda(p)\mathbf{M}^*(p)\mathbf{u}(p), \qquad p_0 \leq p \leq p_f, \tag{3}$$

where p is a design parameter.

Several approaches for approximation of interval eigenvalue problems in terms of symmetric [2–5] and asymmetric matrices [6–8] have been suggested. These of the methods approximately predict the interval range for eigenvalues. By contrast, the proposed method provides an approximation for each value economically and with good accuracy. Previous work done by us [9–11] proved to be promising for symmetric and asymmetric parameter–dependent eigenvalue problems. The algorithms presented there were based on interpolated eigenvector. They showed excellent accuracy and computational efficiency in numerical examples. This idea is taken forward here and an extension of the same to skew–symmetric matrices is presented. The interpolated modes method is now extended to an approximation of a skew–symmetric parameter–dependent eigenvalue problem.

The eigenvalue problem associated with gyroscopic systems has several mathematical features that are different from that for non–gyroscopic problems [12, 13]. Meirovitch [14] introduced a method of solving an eigenvalue problem for gyroscopic system by transforming it to a standard form in term of one real symmetric and one skew–symmetric matrices. The solution of such eigenvalue problems contain complex eigenvalues and eigenvectors [1, 15]. Modal analysis for the response of gyroscopic systems [16–18]. Zheng et al. [19] introduced the method of how to solve a generalised eigenvalue problem of damped gyroscopic systems without finding the left eigenvector. However, the approximation of the parameter–dependent eigenvalue problem for rotating structures does not appear to have been considered before.

2. A mode interpolation algorithm for eigenvalue approximation in terms of a skew–symmetric matrix

Consider the generalised eigenvalue problem of the form (3), where $\mathbf{G}^*$ is a skew–symmetric matrix. The main idea of the proposed method to solve two eigenvalue problems exactly at the ends of a parameter interval. Eigenvalues for the rest of the points inside the interval are then calculated economically but approximately. Before an interpolation takes place, the following eigenvalue problems need be solved exactly

$$\mathbf{G}_0^*\mathbf{u}_0 = -\lambda\mathbf{M}_0^*\mathbf{u}_0, \qquad \mathbf{G}_f^*\mathbf{u}_f = -\lambda\mathbf{M}_f^*\mathbf{u}_f, \tag{4}$$

where $p_0 \leq p \leq p_f$, $\mathbf{G}_0^* = \mathbf{G}^*(p = p_0)$, $\mathbf{M}_0^* = \mathbf{M}^*(p = p_0)$, $\mathbf{G}_f^* = \mathbf{G}^*(p = p_f)$ and $\mathbf{M}_f^* = \mathbf{M}^*(p = p_f)$.

The eigenvalues of an eigenproblem in terms of a skew–symmetric matrix are purely imaginary. Complex numbers do not appear in an order. Therefore, the eigenvalues need to be sorted accordingly to some criteria. Eigenvectors need to be arranged correspondingly.

The approximate eigenvalues are calculated by Rayleigh quotient for a skew–symmetric matrix pencil

$$\widehat{\lambda}(p) = -\frac{\widehat{\mathbf{u}}^H(p)\mathbf{G}^*(p)\widehat{\mathbf{u}}(p)}{\widehat{\mathbf{u}}^H(p)\mathbf{M}^*(p)\widehat{\mathbf{u}}(p)}, \qquad p_0 \leq p \leq p_f, \tag{5}$$

with *exact* matrices at each value of p, i.e. using $\mathbf{G}^*(p)$ and $\mathbf{M}^*(p)$ respectively and a trial vector being taken as the interpolated vector given by

$$\hat{\mathbf{u}}(p) = \frac{(p_f - p)\mathbf{u}_0 + (p - p_0)\mathbf{u}_f}{p_f - p_0}, \qquad p_0 \leq p \leq p_f, \tag{6}$$

which is a linear interpolation between two exact eigenvectors from initial and final eigenvalue problems (4).

3. A numerical example

A computer code for approximating the eigenvalues for a generalised parameter–dependent eigenvalue problem (3), in terms of a skew–symmetric and a symmetric matrices, was developed in the MATLAB environment [20].

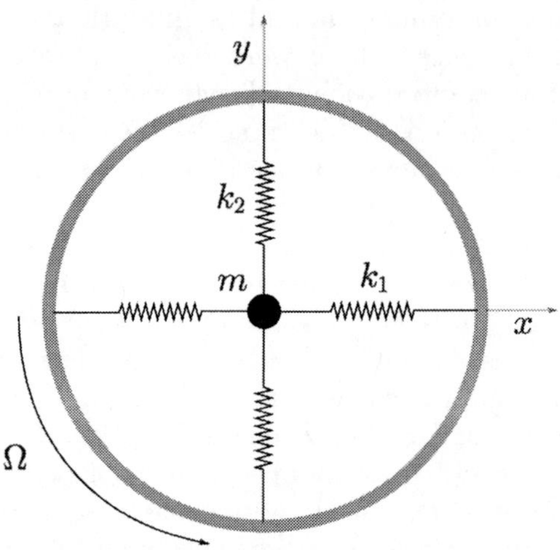

Figure 1. Rotating vibrating structure.

Consider a toy problem that possess the mathematical and physical features of rotating flexible structures as shown in Figure 1. The structure spins with different angular velocity Ω, k_1 and k_2 are the values of total stiffness of the springs along x and y directions respectively. The angular velocity Ω is considered as a parameter p in a parameter–dependent eigenvalue problem (3) and varies from 0.1 to 0.3, i.e. $\Omega_0 = 0.1$ and $\Omega_f = 0.3$. Here, matrices $\mathbf{G}^*(\Omega)$ and $\mathbf{M}^*(\Omega)$ of order $2n \times 2n$ are given by

$$\mathbf{G}^*(\Omega) = \left[\begin{array}{c:c} \mathbf{G}(\Omega) & \mathbf{K}(\Omega) \\ \hdashline -\mathbf{K}(\Omega) & \mathbf{0} \end{array}\right], \qquad \mathbf{M}^*(\Omega) = \left[\begin{array}{c:c} \mathbf{M} & \mathbf{0} \\ \hdashline \mathbf{0} & \mathbf{K}(\Omega) \end{array}\right], \tag{7}$$

where $n \times n$ gyroscopic, mass and stiffness matrices are given by

$$\begin{aligned} \mathbf{G}(\Omega) &= \Omega \begin{bmatrix} 0 & -2m \\ 2m & 0 \end{bmatrix}, \qquad \mathbf{M} = \begin{bmatrix} m & 0 \\ 0 & m \end{bmatrix}, \\ \mathbf{K}(\Omega) &= \begin{bmatrix} k_1 & 0 \\ 0 & k_2 \end{bmatrix} - \Omega^2 \begin{bmatrix} -m & 0 \\ 0 & -m \end{bmatrix} = \mathbf{K} - \Omega^2 \mathbf{M}. \end{aligned} \tag{8}$$

A number of subdivisions of parametric range equals to 10 taken as the number of designs to be assessed. For this numerical example, the values of the model parameters are taken as

IOP Conf. Series: Journal of Physics: Conf. Series **1106** (2018) 012017 doi:10.1088/1742-6596/1106/1/012017

$k_1 = 1$, $k_2 = 2$ and $m = 1$. All the values are non–dimensional and hence only their numeric values are quoted here.

It is of our interest to compare results obtained by proposed method with the results obtained by a Rayleigh quotient approximation based on fixed reference modes. While Rayleigh quotient approximation is well known, here we provide trial vectors *that are parameter–dependent*, and not fixed, which turn out to be rich in the direction of the actual parameter–dependent eigenvector. Moreover, this calculation of the proposed trial vectors is extremely economical. The performance of the proposed approximation with respect to a fixed mode Rayleigh quotient

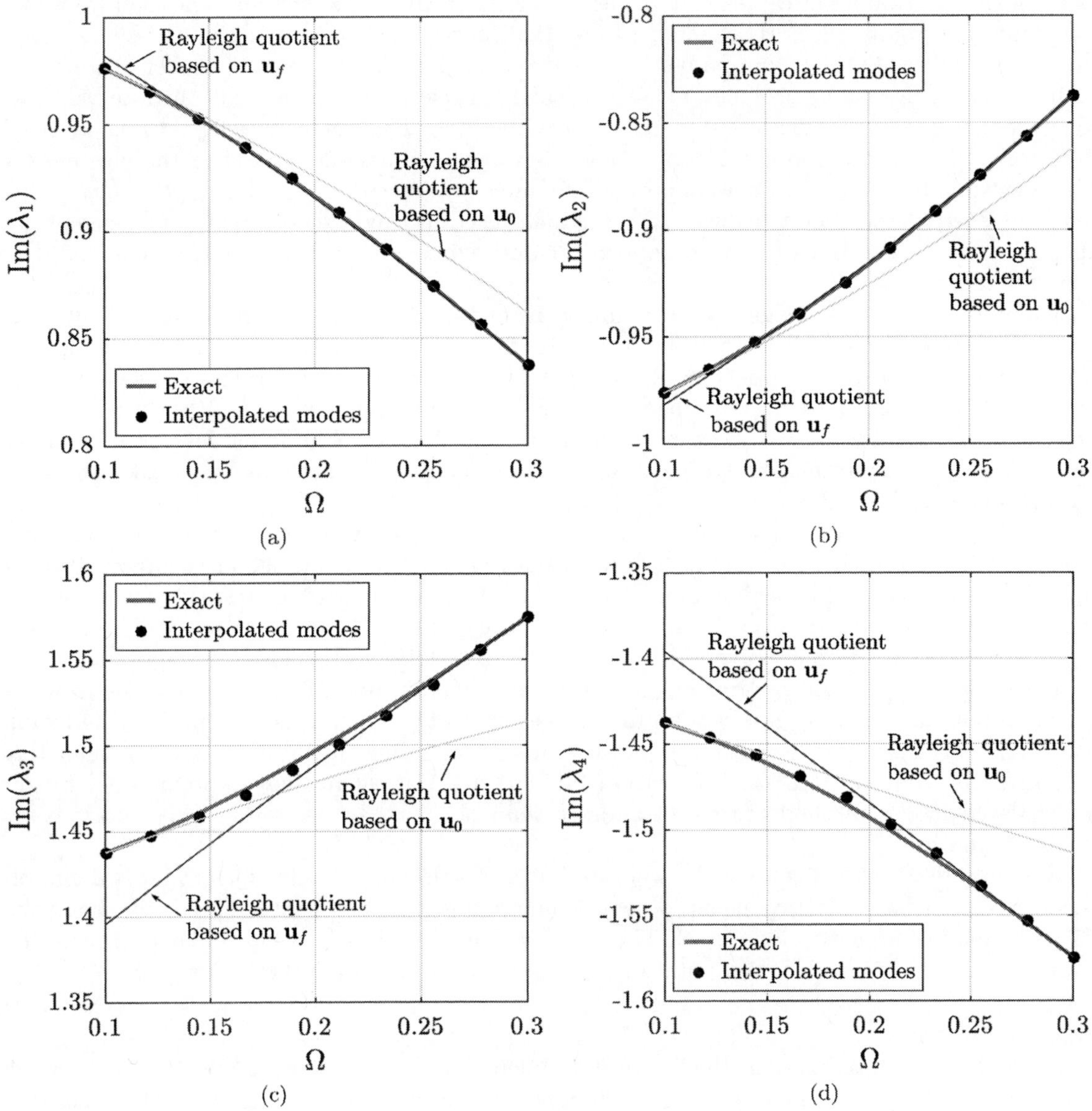

Figure 2. Imaginary parts of the first four eigenvalues as functions of parameter Ω of a parameter–dependent eigenvalue problem computed exactly (thick red line), by the interpolated modes approach (black dots) and reference fixed mode based Rayleigh quotient (thin lines as labelled).

IOP Conf. Series: Journal of Physics: Conf. Series **1106** (2018) 012017 doi:10.1088/1742-6596/1106/1/012017

approximation is compared next. As parameter p of the system changes, the eigenvalues along all the parametric range, approximated by Rayleigh quotient based on reference trial vectors at the two ends of parametric range, are given by

$$\widetilde{\lambda}(\Omega) \approx -\frac{\mathbf{u}_0^H \mathbf{G}^*(\Omega)\mathbf{u}_0}{\mathbf{u}_0^H \mathbf{M}^*(\Omega)\mathbf{u}_0}, \qquad \widetilde{\widetilde{\lambda}}(\Omega) \approx -\frac{\mathbf{u}_f^H \mathbf{G}^*(\Omega)\mathbf{u}_f}{\mathbf{u}_f^H \mathbf{M}^*(\Omega)\mathbf{u}_f}, \tag{9}$$

where $\mathbf{u}_0$ and $\mathbf{u}_f$ are the exact eigenvectors from eigenproblems (4) respectively, $\Omega_0 \leq \Omega \leq \Omega_f$.

The approximations can be made for a certain number of modes that are required. In this example, all eigenvalues are approximated as the size of the problem is small. The eigenvalues are calculated in several ways: (i) exactly, (ii) by the interpolated modes method presented here, (iii) by Rayleigh quotient approximation based on fixed reference modes. Exact calculations require solving eigenvalue problems exactly at each parametric value along all the interval. Two eigenproblems, i.e. those expressed in (4), were solved first. The eigenvalues from these problems were sorted in ascending order by the absolutes values of imaginary parts of the eigenvalues. The corresponding eigenvectors were normalised and sorted according to their eigenvalues. After this, the interpolated vector (6) was calculated and used in Rayleigh quotient (5) for eigenvalue approximation. Equations (9) were applied for approximation by Rayleigh quotient based on fixed references.

The imaginary parts of eigenvalues as functions of parameter are shown in Figure 2. Figures for real parts are omitted as they are equal to zero, i.e. eigenvalues are purely imaginary. The obtained eigenvalues appear as 2 conjugate pairs, where imaginary parts are equal but have opposite signs. The first conjugate pair is shown in Figures 2(a) – 2(b), the second is presented in Figures 2(c) – 2(d). Results obtained using the interpolated modes method marked by black dots are in excellent agreement with those calculated exactly as marked by thick red lines (Figures 2(a) – 2(d)).

The Rayleigh quotient approximation based on exact eigenvectors $\mathbf{u}_0$ and $\mathbf{u}_f$ from fixed modes, which is presented in Figure 2 by thin lines as labelled, shows good accuracy closer to the taken references in the beginning and the end of the parameter range respectively. However, the accuracy deteriorates away from those reference points. The use of interpolated modes improves the accuracy impressively with respect to fixed trial vector based calculations.

The maximum percentage error of the interpolated modes method in this numerical example based on obtained results, shown in Figure 2, reaches 0.3 %. The errors in approximation while using reference fixed mode Rayleigh quotient are 3.9 % which is 13 times greater than that obtained from interpolated modes method. Therefore, the eigenvalues approximated by the interpolated modes method are more accurate than obtained by references fixed mode based Rayleigh quotient.

The proposed method is not only accurate but also computationally efficient. It is definitely cheaper to calculate only two eigenproblems from ends of parameter interval exactly than solve all problems inside parameter range. The approach presented here is particular of interest for large–scale problems, as it is computationally expensive to calculate them.

4. Conclusions

A method for approximating the natural frequencies of undamped gyroscopic systems is presented. While considering several design alternatives of a structure, one needs to solve a parameter–dependent eigenvalue problem. Solving several large–scale problems is computationally expensive. The proposed method, using only two exact eigenvectors from the ends of parametric interval and Rayleigh quotient, approximates the eigenvalues throughout parameter range economically.

The interpolated modes method was compared with Rayleigh quotient approximation, based on a fixed reference mode, to demonstrate the efficiency of the presented method. The approximate methods were applied to a rotating vibrating structure. The results obtained demonstrate excellent accuracy in eigenvalue approximation by the interpolated modes method, which was more accurate compared to the Rayleigh quotient approximation using fixed modes.

The computational gain will be very significant if the proposed method is applied to the large–scale problems, which is in future implementation.

References

[1] Meirovitch L 1980 *Computational Methods in Structural Dynamics* (Alphen aan den Rijn: Sijthoff & Noordhoff International publishers, The Netherlands)

[2] Deif A 1991 The interval eigenvalue problem *Journal of Applied Mathematics and Mechanics* vol 71 **1** pp 61 – 64

[3] Deif A and Rohn J 1992 On the range of eigenvalues of an interval matrix *Computing* vol 47 **3–4** pp 373 –377

[4] Seif N P, Hashem S and Deif A S 1992 Bounding the eigenvectors for symmetric interval matrices *Journal of Applied Mathematics and Mechanics* vol 72 pp 233 –236

[5] Behnke H 1992 Bounds for eigenvalues of parameter–dependent matrices *Computing* vol 49 pp 159 –167

[6] Goldhirsch I, Orszag S A and Maulik B K 1987 An efficient method for computing leading eigenvalues and eigenvectors of large asymmetric matrices *Journal of Scientific Computing* vol 2 **1** pp 33 – 58

[7] Saad Y 1984 Chebyshev Acceleration Techniques for Solving Nonsymmetric Eigenvalue Problems *Mathematics of Computation* vol 42 **166** pp 567 –588

[8] Bai Z 1995 Progress in the Numerical Solution of the Nonsymmetric Eigenvalue Problem *Numerical Linear Algebra with Applications* vol 2 **3** pp 219 –234

[9] Bhaskar A 2000 Gross modifications in structural dynamics via interpolated modes *Proceedings of the 7th International Conference on Recent Advances in Structural Dynamics* (Southampton) pp 61–73

[10] Gavryliuk N and Bhaskar A 2017 A mode shape based approximation for parameter-dependent eigenvalue problem arising in structural vibration *Proceedings of the 24th International Congress on Sound and Vibration* (London)

[11] Gavryliuk N, Ghandchi–Tehrani M and Bhaskar A 2018 An eigenvector interpolation based approximation for parameter–dependent asymmetric eigenvalue problems *Proceedings of 1th International Conference on Advances in Aerospace Structures, Systems & Technology* (London)

[12] Lancaster P 2013 Stability of linear gyroscopic systems: A review *Linear Algebra and its Applications* **439** pp 686 –706

[13] Paardekooper M H C 1970 An eigenvalue algorithm for skew–symmetric matrices *Numerische Mathematik* vol 17 **3** pp 189 –202

[14] Meirovitch L 1974 A new method of solution of the eigenvalue problem for gyroscopic systems *American Institute of Aeronautics and Astronautics Journal* vol 12 **10** pp 1337 – 1342

[15] Huseyin K 1978 *Vibrations and Stability of Multiple Parameter Systems* (Alphen aan den Rijn: Sijthoff & Noordhoff International publishers, The Netherlands)

[16] Meirovitch L 1975 A modal analysis for the response of linear gyroscopic systems *Journal of Applied Mechanics* vol 42 **2** pp 446 –450

[17] Meirovitch L 1976 A stationarity principle for eigenvalue problem for rotating structures *AIAA Journal* vol 14 **10** pp 1387–1394

[18] Meirovitch L and Ryland II G 1979 Response of slightly damped gyroscopic systems *Journal of Sound and Vibration* vol 67 **1** pp 1 –19

[19] Zheng Z, Ren G and Williams F W 1997 The eigenvalue problem for damped gyroscopic systems *International Journal of Mechanical Sciences* vol 39 **6** pp 741 – 750

[20] The MathWorks Inc. *MATLAB and Statistics Toolbox Release R2016a* (Natick, Massachusetts, United States)

IOP Conf. Series: Journal of Physics: Conf. Series **1106** (2018) 012018 doi:10.1088/1742-6596/1106/1/012018

Critical buckling predictions for plates and stiffened panels from natural frequency measurements

David Kennedy, Kwok Ip Lo

School of Engineering, Cardiff University, Queen's Buildings, The Parade, Cardiff CF24 3AA, United Kingdom

KennedyD@cardiff.ac.uk

Abstract. The paper establishes a linear relationship between the square of natural frequency and applied compressive loading for a simply supported plate. This relationship enables the critical buckling load to be predicted by extrapolating from the natural frequencies obtained when the plate is lightly loaded. If the fundamental natural frequency and the critical buckling load have different mode shapes, buckling predictions can be made by extrapolating from higher natural frequencies. Shear loading causes a continual change in the mode shape as the loading is increased, and the critical buckling load can be estimated approximately by curve fitting and extrapolation. Similar results are obtained for simply supported stiffened panels.

1. Introduction

Plates and stiffened panels are used in many engineering applications, including aerospace, road and rail vehicles, ships and bridges. Initial imperfections, cracks, delaminations and other types of damage reduce their strength and stiffness properties, including their critical buckling loads. However buckling tests for built-up structures such as aircraft wings are impractical, while predictions from analytical models tend to be imprecise due to difficulties in detecting the location and extent of any damage.

Doyle [1] discussed the transverse vibration of axially compressed bars and frame structures, establishing a linear relationship between the applied load and the square of natural frequency such that the latter reduces to zero at the critical buckling load. Singer et al. [2] developed the Vibration Correlation Technique (VCT) which predicts the critical buckling load by extrapolating the natural frequencies obtained from non-destructive testing as the structure is progressively loaded at pre-buckling levels. The technique has been successfully applied to plates and shells [3], but it has been shown that the linear relationship breaks down in the presence of geometric imperfections [4, 5].

This paper explores the validity of the linear relationship for isotropic simply supported plates and stiffened panels with varying aspect ratios and subjected to different kinds of in-plane loading, including shear. Section 2 outlines the underlying vibration and buckling theory for a simply supported plate and establishes relationships between the natural frequencies and critical buckling loads, taking account of any possible changes in mode shapes as the applied load is increased. The predictions are validated by numerical results in section 3 and are shown also to apply to stiffened panels. The conclusions in section 4 include a discussion of how the critical buckling load can be estimated in cases where the linear relationship breaks down.

2. Vibration of loaded plates

The governing differential equation for an isotropic rectangular plate of length a, width b, mass per unit area μ, flexural rigidity D, loaded in longitudinal compression N_x, transverse compression N_y and in-plane shear N_{xy} per unit width and vibrating with frequency ω is [6]

$$D\left(\frac{\partial^4 w}{\partial x^4}+2\frac{\partial^4 w}{\partial x^2 \partial y^2}+\frac{\partial^4 w}{\partial y^4}\right)+\left(N_x\frac{\partial^2 w}{\partial x^2}+2N_{xy}\frac{\partial^2 w}{\partial x \partial y}+N_y\frac{\partial^2 w}{\partial y^2}+\mu\frac{\partial^2 w}{\partial t^2}\right)=0 \quad (1)$$

2.1. *Simply supported plate loaded in compression*

If all four edges of the plate are simply supported and $N_{xy}=0$, the solution to equation (1) takes the form

$$w(x,y,t)=w_{mn}\sin\frac{m\pi x}{a}\sin\frac{n\pi y}{b}\sin\omega t \quad (2)$$

with the natural frequencies ω_{mn} given by

$$\mu\omega_{mn}^2=D\pi^4\left(\frac{m^2}{a^2}+\frac{n^2}{b^2}\right)^2-\pi^2\left(N_x\frac{m^2}{a^2}+N_y\frac{n^2}{b^2}\right) \quad (3)$$

Setting $N_x=N_y=0$ in equation (3) gives the natural frequencies $\bar{\omega}_{mn}$ of an unloaded plate, of which the lowest is $\bar{\omega}_{11}$, i.e. the fundamental vibration mode has one half-wave in both the longitudinal and transverse directions. Setting $\omega_{mn}=0$ and $N_y=0$ [$N_x=0$] gives buckling loads $\bar{N}_{xmn}$ [$\bar{N}_{ymn}$] in longitudinal [transverse] compression, of which the lowest has $n=1$ [$m=1$] and a value of m [n] which depends on the aspect ratio of the plate, so that the critical buckling mode may differ from the fundamental vibration mode.

Equation (3) can be written as

$$\omega_{mn}^2=\bar{\omega}_{mn}^2\left(1-\frac{N_x}{\bar{N}_{xmn}}-\frac{N_y}{\bar{N}_{ymn}}\right) \quad (4)$$

indicating a linear relationship between ω_{mn}^2 and N_x [N_y] such that the natural frequency is reduced to zero when the loading reaches the buckling load $\bar{N}_{xmn}$ [$\bar{N}_{ymn}$]. Thus if the critical buckling load has $m=n=1$ it can be extrapolated from the fundamental natural frequencies ω_{11} obtained when the plate is loaded at lower levels.

If the critical buckling mode has $m\neq 1$ or $n\neq 1$, it differs from the fundamental vibration mode. The critical buckling load can now be extrapolated from changes in the higher natural frequency ω_{mn}. As the load is gradually increased from zero there will be one or more discrete changes in the fundamental vibration mode which are accompanied by shifts between the straight line relationships associated with different values of m and/or n.

2.2. *Other loading and edge conditions*

If the plate is loaded in shear ($N_{xy}\neq 0$) then equation (2) must be replaced by a series solution in order to approximate the skewed deflection pattern. The pure vibration mode is sinusoidal, but as the load is gradually increased the skewing becomes more pronounced, indicating a continual change in mode shape towards the critical buckling mode. Thus the relationship between the square of natural frequency and the applied loading is no longer linear. In section 3 an attempt will be made to estimate the critical buckling load approximately from the natural frequencies by curve fitting and extrapolation.

An alternative to equation (2) is also required in order to satisfy alternative edge conditions, including the often uncertain elastic restraints associated with imperfect connections to adjacent parts of a built-up structure. Here there is also a non-linear relationship between the square of natural frequency and the applied loading. Similar considerations are needed in the presence of thickness variations, geometric imperfections or localized damage, which will be explored in detail in future publications.

3. Numerical results

Vibration and buckling analysis was performed on plates and stiffened panels using the exact strip software VICONOPT [7], which employs a transcendental stiffness matrix derived from analytical solutions of the governing differential equations of the component plates. Undamped natural frequencies and critical buckling loads and are found using the Wittrick-Williams algorithm [8, 9]. In the simplest (VIPASA) form of the analysis [10], the vibration or buckling mode is assumed to vary sinusoidally in the longitudinal direction, with a half-wavelength which divides exactly into the panel length, so that exact solutions are obtained for simply supported isotropic and orthotropic plates and panels without shear loading. For panels that are anisotropic or loaded in shear, the VICON analysis [11] employs a series solution to couple such modes, using Lagrangian multipliers to approximate the end conditions.

3.1. Simply supported plate loaded in compression

Figure 1 shows the square of the lowest natural frequency for simply supported plates with different aspect ratios a/b loaded in longitudinal compression. The plots are normalised with respect to the fundamental natural frequency for an unloaded plate $\bar{\omega}_{11}^2$ and the critical buckling load $\bar{N}_{xmn}$. For plates with aspect ratio 1 and 1.2, the critical buckling mode has $m=n=1$ and a linear relationship is observed. For plates with higher aspect ratios, the critical buckling mode has $m>1$ and $n=1$ and the relationship is piecewise linear, confirming the analysis of [6]. For example, when $a/b=3$, the critical buckling mode has $m=3$ and $n=1$. There is a mode shift from $m=1$ to $m=2$ when $N_x=0.64\bar{N}_{x31}$ and a further shift to $m=3$ when $N_x=0.86\bar{N}_{x31}$, see figure 2 which also shows contour plots of the mode shapes.

It is noted that if VCT is applied at low load levels, extrapolating the plot of $(\omega_{11}/\bar{\omega}_{11})^2$ will fail to predict the critical buckling load $\bar{N}_{x31}$, but it can be predicted accurately by extrapolating $(\omega_{31}/\bar{\omega}_{31})^2$.

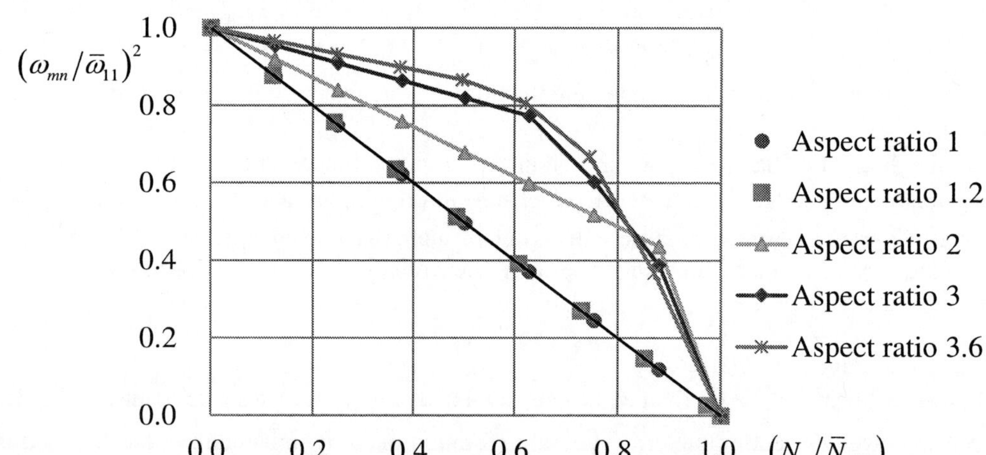

Figure 1. Non-dimensional plot of lowest natural frequency squared versus longitudinal compression for plates of different aspect ratios.

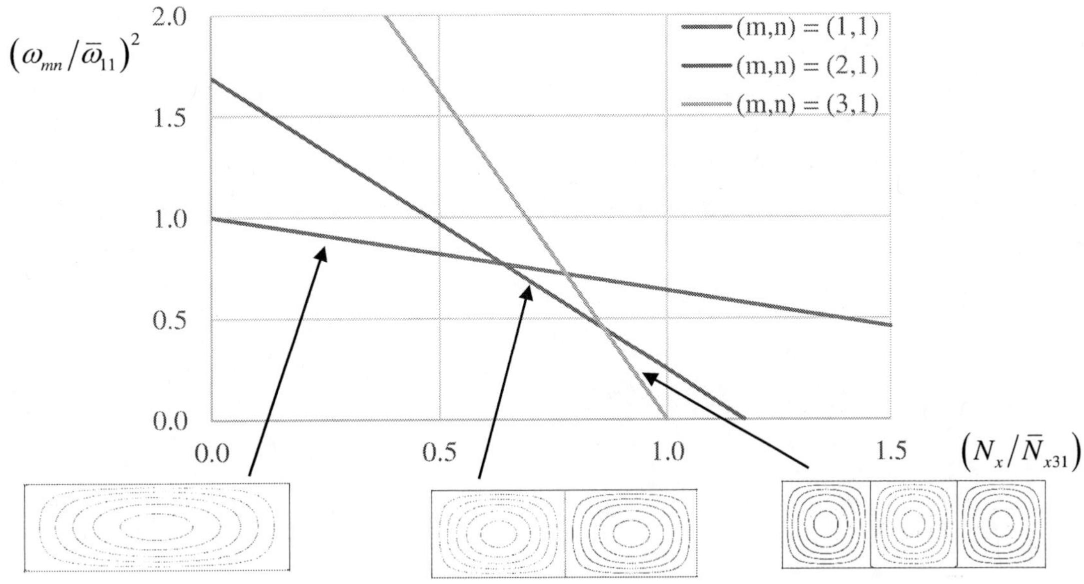

Figure 2. Non-dimensional plot of natural frequency squared versus longitudinal compression for plate of aspect ratio 3, showing changes in mode shapes.

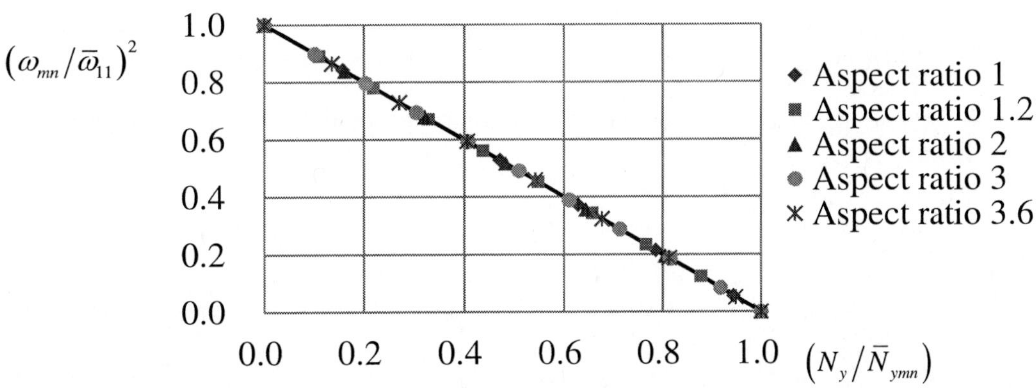

Figure 3. Non-dimensional plot of lowest natural frequency squared versus transverse compression for plates of different aspect ratios.

The fundamental vibration mode and the critical buckling mode for loading in transverse compression both have $m = n = 1$ for plates with aspect ratio $a/b \geq 1$, so it is easy to predict the critical buckling load from changes in the lowest natural frequency (see figure 3). Mode shifts would however occur for plates with aspect ratios sufficiently less than 1.

3.2. Simply supported plate loaded in shear

When a simply supported plate is loaded in shear, the vibration and buckling modes are skewed (see figure 4) and the sinusoidal solution of equation (2) must be replaced by a series solution. The (m,n) classification is replaced by a longitudinal response parameter [7]. Figure 5 illustrates the development of skewing as the shear load is increased, showing that the fundamental vibration mode ($N_{xy} = 0$) and the critical buckling mode ($N_{xy} = \bar{N}_{xy}$) differ substantially. The relationship between the square of the lowest natural frequency and the applied load is highly nonlinear (see figure 6), so the critical buckling load $\bar{N}_{xy}$ cannot be found by linear extrapolation.

The VCT was simulated by considering the first r points ($r = 2,3,4,5$) in figure 6 for the plate with aspect ratio 1. A polynomial curve of order $r-1$ was fitted through the r points and extrapolated to find the load $\bar{N}_{xy}^{est}$ at which the lowest natural frequency becomes zero. This is an estimate of the true critical buckling load $\bar{N}_{xy}$. Table 1 shows that acceptable accuracy is achieved by considering the natural frequencies at 5 points for which the shear loading lies below 25% of the critical buckling load.

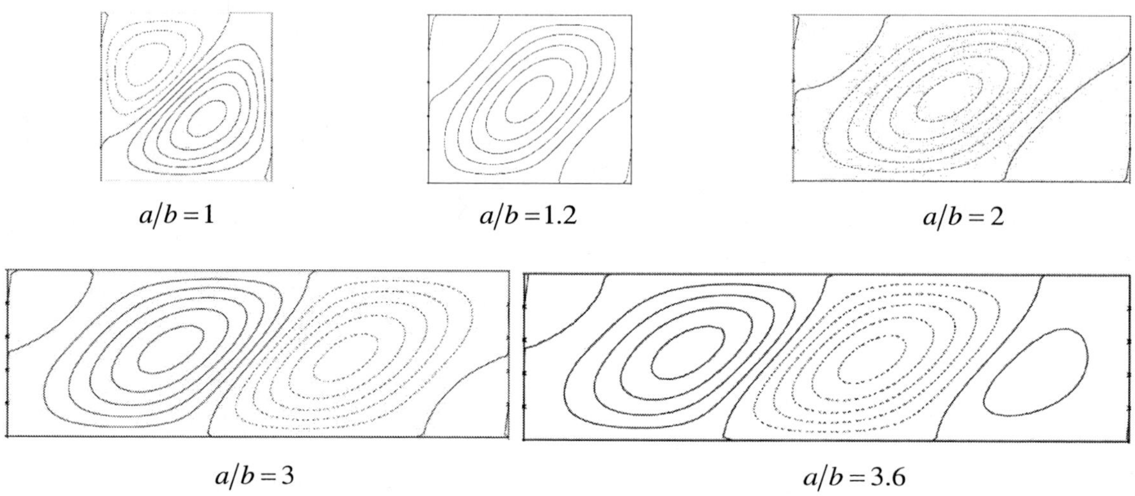

Figure 4. Critical buckling modes for plates of different aspect ratios a/b loaded in shear.

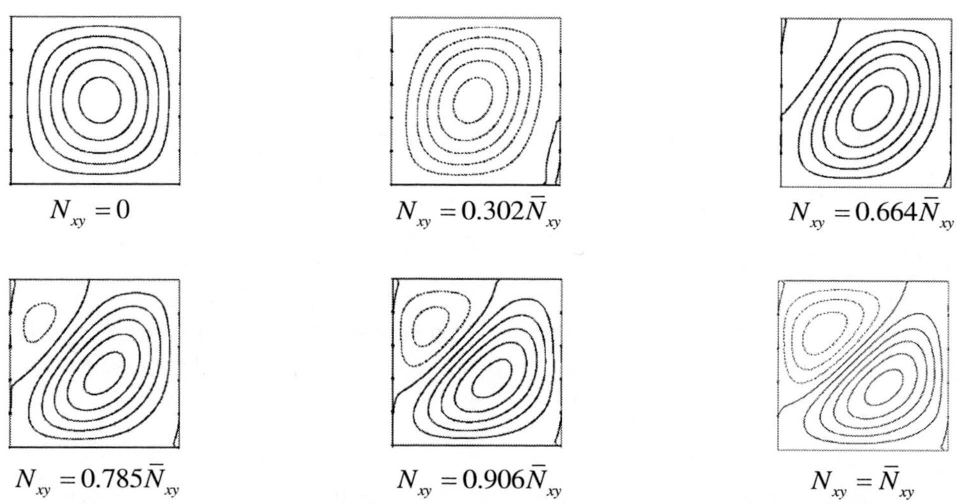

Figure 5. Vibration ($0 \le N_{xy} < \bar{N}_{xy}$) and buckling ($N_{xy} = \bar{N}_{xy}$) modes of a simply supported square plate loaded in shear N_{xy}.

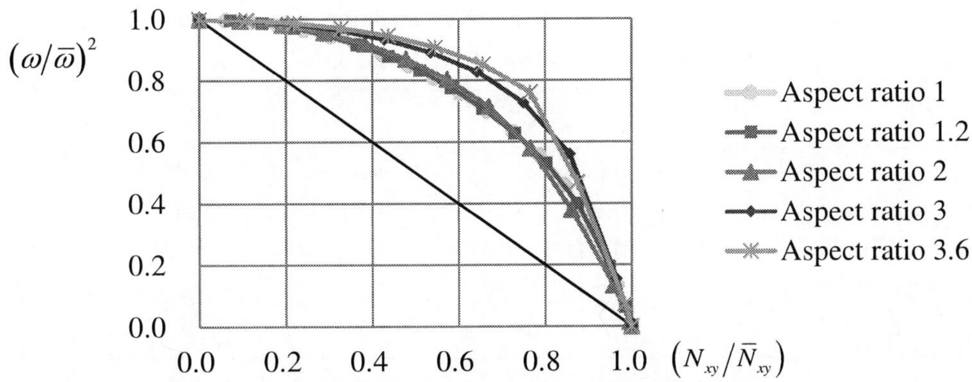

Figure 6. Non-dimensional plot of lowest natural frequency squared versus shear load for plates of different aspect ratios.

Table 1. Non-dimensional estimates $\bar{N}_{xy}^{est}/\bar{N}_{xy}$ of critical shear buckling load of a simply supported square plate, obtained by extrapolating from the first r points in figure 6.

Point (r)	$N_{xy}/\bar{N}_{xy}$ at point r	$N_{xy}^{est}/\bar{N}_{xy}$
2	0.060	19.773
3	0.121	1.092
4	0.181	1.063
5	0.242	1.006

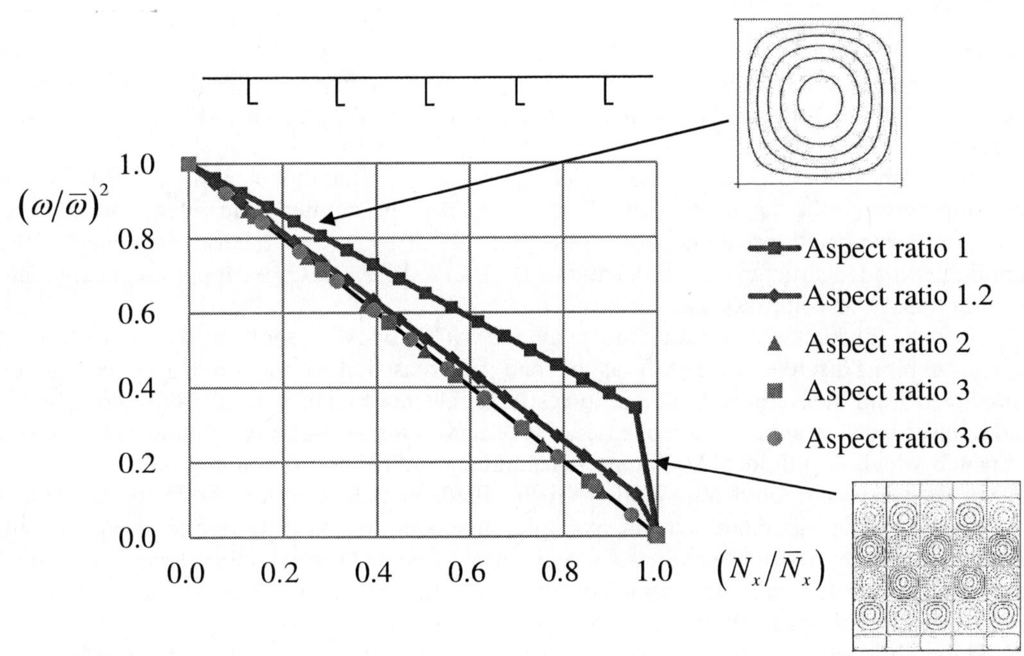

Figure 7. Non-dimensional plot of lowest natural frequency squared versus longitudinal compression for stiffened panels of different aspect ratios.

Table 2. Non-dimensional estimates $\bar{N}_{xy}^{est}/\bar{N}_{xy}$ of critical shear buckling load of a simply supported square stiffened panel, obtained by extrapolating fitted polynomials of order $r-1$.

Point (r)	$N_{xy}/\bar{N}_{xy}$ at point r	$N_{xy}^{est}/\bar{N}_{xy}$
2	0.066	35.714
3	0.131	1.524
4	0.197	1.357
5	0.263	1.109
6	0.328	1.037

3.3. Stiffened panel loaded in compression or shear

Figure 7 shows the square of the lowest natural frequency for the stiffened panel whose cross section is shown in figure 7. The stiffeners have half the thickness of the skin, web height $0.06b$, flange width $0.02b$ and spacing $0.2b$. Panels with different aspect ratios a/b are loaded in longitudinal compression. The plots are normalised with respect to the square of the fundamental natural frequency for an unloaded panel $\bar{\omega}^2$ and the critical buckling load $\bar{N}_x$. The figure shows that, when $a/b=1$ or 1.2, the lowest natural frequency has a global mode while critical buckling occurs in a local mode. There is a single mode shift as the critical buckling load is approached. For higher aspect ratios the fundamental natural frequency and critical buckling load have identical global modes and a linear relationship is observed.

When the panel is loaded in shear there is a gradual change of mode shape as the loading is increased, and the natural frequencies vary in a similar way to those of a single plate. Table 2 illustrates that the critical buckling load can be estimated with reasonable accuracy by extrapolating from the natural frequencies obtained as the load is increased to 33% of its critical value.

4. Conclusions

This study has confirmed the Vibration Correlation Technique for predicting the critical buckling load of a simply supported plate or stiffened panel loaded in compression or shear. Three kinds of behaviour have been identified.

Firstly, when the fundamental vibration mode and the critical buckling mode are identical, there is a linear relationship between the square of natural frequency and the applied load. Hence the critical buckling load can be accurately predicted from measurements of the lowest natural frequency of the unloaded and lightly loaded structure. Such behaviour is observed for a simply supported square plate in longitudinal or transverse compression.

Secondly, when the fundamental vibration mode and the critical buckling mode differ, but the lowest natural frequency exhibits discrete mode shifts as the load is increased, the critical buckling load can be accurately predicted from measurements of the lowest and selected higher natural frequencies of the unloaded and lightly loaded structure. Such behaviour is observed for a long simply supported plate and for stiffened panels which exhibit local buckling modes in longitudinal compression.

Thirdly, when there is a smooth transition from the fundamental vibration mode to the critical buckling mode as the loading is increased, for example on account of skewing due to shear loading, there is a non-linear relationship between the square of natural frequency and applied load. The critical buckling load can be predicted with reasonable accuracy by extrapolating a polynomial curve fitted to the squares of natural frequencies obtained when the structure is lightly loaded.

Although all the illustrative results presented here have been for perfect structures, the methods are also expected to prove useful for non-destructive monitoring and assessment of damaged structures whose natural frequencies and critical buckling loads have been degraded.

IOP Conf. Series: Journal of Physics: Conf. Series **1106** (2018) 012018 doi:10.1088/1742-6596/1106/1/012018

References

[1] Doyle JF 1973 Approximate modifications of stiffness and carry-over factors for axial loads and vibrations *Struct. Eng.* **51** 183–7
[2] Singer J, Arbocz J and Weller T 2002 *Buckling Experiments: Experimental Methods in Buckling of Thin-Walled Structures* vol 2 (New York: Wiley) ch 15.2
[3] Singer J and Abramovich H 1979 Vibration correlation techniques for definition of practical boundary conditions in stiffened shells *AIAA J.* **17** 762–9
[4] Rosen A and Singer J 1974 Effect of axisymmetric imperfections on the vibrations of cylindrical shells under axial compression *AIAA J.* **12** 995–7
[5] Abramovich H and Kalnins K 2017 The vibration correlation technique revisited *Proc.11th Int. Symp. Vib. Contin. Syst. (Llanberis)* http://www.isvcs.org/ISVCS11-Proceedings.pdf pp 1–3
[6] Leissa AW 1969 *Vibration of Plates* Technical Report SP-160 (Washington: NASA) ch 10
[7] Kennedy D, Williams FW and Anderson MS 1994 Buckling and vibration analysis of laminated panels using VICONOPT *ASCE J. Aerosp. Eng.* **7** 245–62
[8] Wittrick WH and Williams FW 1971 A general algorithm for computing natural frequencies of elastic structures *Q. J. Mech. Appl. Math.* **24** 263–84
[9] Wittrick WH and Williams FW 1973 An algorithm for computing critical buckling loads of elastic structures *J. Struct. Mech.* **1** 497–518
[10] Wittrick WH and Williams FW 1974 Buckling and vibration of anisotropic or isotropic plate assemblies under combined loadings *Int. J. Mech. Sci.* 209–39
[11] Anderson MS, Williams FW and Wright CJ 1983. Buckling and vibration of any prismatic assembly of shear and compression loaded anisotropic plates with an arbitrary supporting structure *Int. J. Mech. Sci.* **25** 585–96

IOP Conf. Series: Journal of Physics: Conf. Series **1106** (2018) 012019 doi:10.1088/1742-6596/1106/1/012019

On the use of measured rotational responses in structural dynamic analysis

Tiago A N Silva*, **Raquel A B Almeida and António P V Urgueira**

NOVA UNIDEMI, Universidade NOVA de Lisboa, Campus de Caparica, 2829-516 Caparica, Portugal.

E-mail: `tan.silva@fct.unl.pt`

Abstract. Testing is a very important task in structural dynamics and on its applications, namely on the characterization of the dynamic behaviour of structures or on the use of the structural responses to indirectly determine dynamic properties or parameters. The increasing availability of sensors, capable of measuring rotational responses, leads to the improvement of existing methods and to the development of new ones in several fields of application. In this work, an evaluation of the applicability and performance of the use of rotational responses is carried out considering applications ranging from the determination of rigid body properties to damage identification. Data acquisition using MEMS and piezoelectric sensors is covered and a set of results is discussed.

1. Introduction

Over the years, the structural dynamics research community has studied the dynamic behaviour of structures from different perspectives. In this field of engineering, the use of testing procedures along with modelling and simulation techniques has became standard, regarding the experimental and theoretical description of the structure under study. The dynamic characteristics of a structure are often derived from a set of experimentally acquired frequency response functions (FRF), the spatial model usually contains a more refined set of information, in terms of the available degrees of freedom (DOF) and response types. Thus, spatial models, often derived numerically using the finite element (FE) method, allow for the computation of physically inaccessible FRFs, although they require model validation, as there is always lack of knowledge, at some extent, concerning modelling assumptions and parameters. Whenever test and theoretical responses must be matched, the issue of model incompleteness arises, due to the fact that experimental data is usually collected at a few translational DOFs due to applied forces, using hammer or shaker exciters, over a limited frequency range. The limitations of testing, regarding inaccessible responses due to different aspects, continues to present a major challenge for the scientific community. This is due to the fact that the knowledge of a complete response model could enhance several procedures in different areas of application. Hence, one must cope with either the FE model reduction to the set of experimental points or the expansion of the experimental data over the FE model DOFs [1, 2]. Regarding the subject of model incompleteness, over the years, several model reduction techniques have been proposed and successfully applied [3–5], although experimental data expansion strategies are significantly fewer and the demand for efficient techniques is still an open issue [6].

IOP Conf. Series: Journal of Physics: Conf. Series **1106** (2018) 012019 doi:10.1088/1742-6596/1106/1/012019

Pursuing a complete response model, one must be able to measure and to excite the entire set of DOFs in the structure to acquire the complete FRF matrix. This implies the acquisition of translational and rotational responses due to excitations by forces and pure moments, although the measurement and excitation of rotational DOFs is still not standard. This limitation in testing leads to an incomplete response model. Using conventional testing techniques, one can only access at most 25% of the complete response model, assuming that one can measure all the translational DOFs and excite them with applied forces, which is still an unpractical assumption. Details on measuring and exciting rotational DOFs can be found in [7]. In the work [6], one can find a technique for expanding translational measured FRFs over the entire set of DOFs. This technique is based upon a modified Kidder's method and the principle of reciprocity, and it avoids the need for modal identification, as it directly uses the measured translational FRFs to estimate the entire response model. An experimental application of it is given in [8].

The demand for a complete response model has been very active in the structural coupling community, where the significance of assessing FRF at rotational DOFs was addressed in many studies, e.g. [9, 10]. Additionally, the knowledge of rotational responses proved to be essential on the identification of the dynamic characteristic of joints, the identification of unknown forces from measured responses or vibration isolation, among others. The use of piezoelectric accelerometers, with wide-spread usage in applications of structural dynamics, as well as the less used Microelectromechanical systems (MEMS) sensors have been considered by several researchers to acquire rotational responses [11, 12].

Aiming at stressing the capabilities of using rotational responses, this paper presents part of the work developed by a research group at NOVA UNIDEMI. The use of piezoelectric as well as MEMS sensors is covered, discussing some advantages and limitations of both, but highlighting their combined use on applications to determine the rigid body properties of a structure, to identify damage or to couple substructures. Some results of these are given and the reader is directed to the published works for further details.

2. Acquiring rotational responses

The authors of [12] presented a comparison between rotational FRFs indirectly measured using the T-block technique [13] and rotational FRFs directly measured using piezoelectric rotational accelerometers. From the presented results, the application of direct rotational piezoelectric accelerometers is promising, although this type of piezoelectric sensors is limited to measure the rotation in relation to a single axis and its commercial cost is high, comparing to the cost of MEMS sensors or even to the one of conventional translational single-axis piezoelectric accelerometers.

The group at NOVA UNIDEMI has invested in a home-made acquisition system that allows for the simultaneous and synchronous acquisition of different kinds of sensors, e.g., conventional translational piezoelectric accelerometers, strain gauges or tri-axial MEMS accelerometers and MEMS gyroscopes, all triggered by a conventional instrumented impact hammer. In this framework, our group has developed the capability to acquire rotational FRFs directly measured using MEMS gyroscopes [14]. The MEMS sensors are generally lighter and have a very lower power consumption when compared to piezoelectric ones. However, the MEMS gyroscopes are limited to a lower bandwidth, only up to 140Hz. This is a significant limitation for several applications, as modal studies, but it is not paramount regarding the interest in the rigid body properties, which is one of the applications covered in this paper (section 3.1). Note that this limited bandwidth is referred to the MEMS gyroscopes only, as the tri-axial MEMS accelerometers present a larger bandwidth, up to around 2kHz. The referred limitation can be mitigated by the expansion of translational FRFs, of course this has a price to pay, as the experimental data expansion is base upon a numerical model. Some details are given in section 3.3.

To summarize, the acquisition system of NOVA UNIDEMI is able to acquire FRFs at different points and directions on a structure, using piezoelectric and/or MEMS sensors, and an instrumented impact hammer, which triggers the experimental data acquisition. The acquisition system was built using a National Instruments (NI) DAQ 9172 chassis that collects the data from all sensors and a LabView implementation that is responsible for the signal management and processing, allowing the experimental data to be exported and then treated by other software for further developments.

In terms of sensors, the group at NOVA UNIDEMI has available tri-axial translational sensors, piezoelectric accelerometers (B&K 4506 B) and MEMS ones (ADXL326), and rotational sensors, MEMS gyroscope (IDG500). Note that the MEMS sensors are considerably cheaper, when compared to their piezoelectric equivalent, providing accurate enough measures in a certain measuring domain.

3. Cases of Application

In this section, the use of rotational responses regarding different cases of application is explained and results are discussed.

3.1. Determination of rigid body properties

The knowledge of the rigid body properties of a structure is very important in different fields, namely in multibody dynamics or flight control. Inertia Restraint methods (IRM) are a set of methods developed to determine the rigid body properties of a structure [15]. The IRM are based on the principle that the dynamic response of structures in free-free conditions are characterized, in the low frequency region, by a constant term designated as inertial restraint or mass line. This approach is suitable for structures where rigid and flexible modes are well separated. Note that the mass line is the modal constant at zero Hz of a typical accelerance (FRF with the response in acceleration units). As discussed in [16], the accuracy of the IRM heavily depends on the selection of the measuring locations, as well as, the conditions of the excitation. This work review a modified version of the IRM given in [17] to cope with rotational FRFs.

Consider a vector with the measured translations and rotations w.r.t. all the three axis at a measuring point i defined as,

$$\ddot{\mathbf{x}}_{m_i} = \left\{ \ddot{u}\ \ddot{v}\ \ddot{w}\ \ddot{\alpha}\ \ddot{\beta}\ \ddot{\gamma} \right\}_{m_i}^{T} \tag{1}$$

If one considers more that one sensor, one can define the vector with all the responses at k measuring points as,

$$\ddot{\mathbf{x}}_{m} = \left\{ \ddot{\mathbf{x}}_{m_1}^{T}\ \ddot{\mathbf{x}}_{m_2}^{T}\ \cdots\ \ddot{\mathbf{x}}_{m_k}^{T} \right\}^{T} \tag{2}$$

To transform the vector of experimental responses to the origin 0 of a global referential in the structure, one may use the following transformation,

$$\ddot{\mathbf{x}}_{m} = \mathbf{A}\mathbf{G}_{m0}\ddot{\mathbf{x}}_{0} \tag{3}$$

where $\mathbf{A}$ is a boolean diagonal matrix that turns on and off channels of the sensors, allowing to use sensors of different kinds; $\mathbf{G}_{m0}$ is a transformation matrix from each local referential to the global one; and $\ddot{\mathbf{x}}_0$ is the vector of experimental responses at the origin of the global referential. Note that $\mathbf{A}\mathbf{G}_{m0}$ is often a rectangular matrix and the pseudo-inverse must be considered to obtain $\ddot{\mathbf{x}}_0$.

Hence, for a given force vector $\mathbf{f}_0$, defined in relation to the origin of the referred global referential, one has the equilibrium equation,

$$\mathbf{f}_{0} = \mathbf{M}_{0}\ddot{\mathbf{x}}_{0} \tag{4}$$

where $\mathbf{M}_0$ is a generalized mass matrix,defined at the origin of the global referential, containing all the inertia terms, namely: mass, coordinates of the centre of mass and inertia tensor.

Details on the modification of the IRM given in [17] can be found in [18]. In that work, a experimental application, using i) tri-axial piezoelectric accelerometers and ii) tri-axial MEMS accelerometers and MEMS gyroscopes, shows that the accuracy of the estimated rigid body properties is quite similar for both implementations, showing no evidences to prefer the use of conventional, more expensive sensors. Moreover, one of the major advantages of using rotational responses with the IRM is related to the independence of the measuring locations, refuting the well-known restrictions given in [16]. Additionally, it was observed that one only needs to use six responses, three translations and three rotations, that can be obtained using just one MEMS sensor, in the limit. Instead of using at least three tri-axial piezoelectric accelerometers, with a distribution in the structure that must agree to the referred set of restrictions.

3.2. Damage identification

Within the framework of damage identification, the group at NOVA UNIDEMI considered the use of well-establish frequency domain damage indicators with rotational FRFs. The transmissibility damage indicator (TDI) and the weighted damage indicator (WDI) proposed in [19, 20] are based on the response vector assurance criterion (RVAC) [21]. The TDI and WDI are successfully used to detect damage and they are usually computed from a set of translational FRFs. In our research, these damage indicators may now be computed also for a set of measured rotational FRFs, following the standard procedure.

From the observation of fig. 1, it is evident that the use of rotational responses to compute the referred damage indicators enhance the sensibility of them to detect damage. This is also true, when the difference between rotational FRFs for the healthy and for the damaged structure is computed, aiming at localizing damages.

This results are obtained with simulated test data, but they are promising even when noisy data is considered. The experimental application of these adapted indicators is an ongoing work.

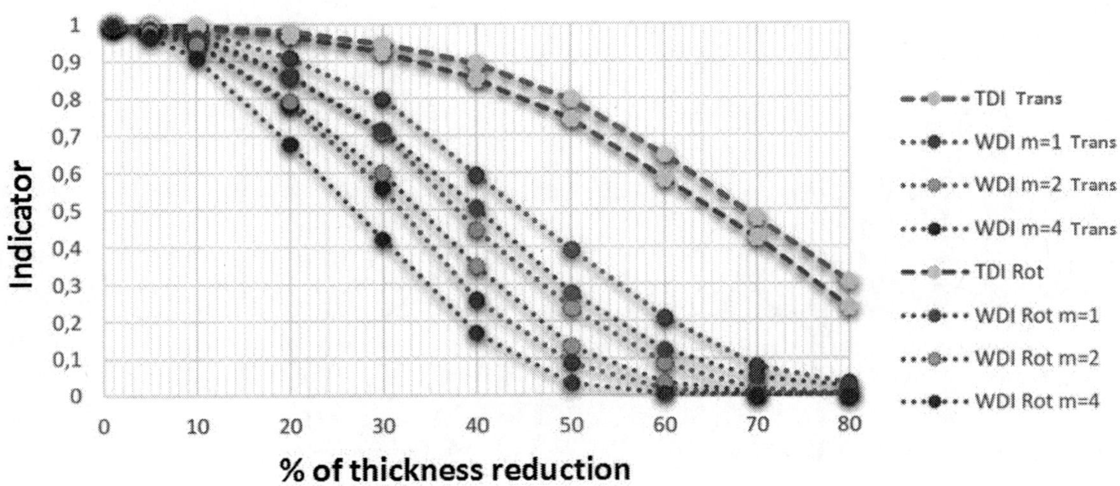

Figure 1. TDI and WDI for different exponents m.

IOP Conf. Series: Journal of Physics: Conf. Series **1106** (2018) 012019 doi:10.1088/1742-6596/1106/1/012019

3.3. Estimation of rotational FRFs

As referred, the MEMS gyroscopes have a limited bandwidth. In several applications, this limitation makes this kind of sensors useless. On the other hand, the measured rotational FRFs are usually only due to exciting forces. Thus, one cannot access the complete response model. In this sense, a method to expand experimental FRFs was proposed and successfully used to estimate rotational FRFs, due to exciting forces and moments [6]. This technique is based upon a modified Kidder's method and the principle of reciprocity, and it avoids the need for modal identification, as it directly uses the measured FRFs $\mathbf{H}_{pj}(\omega)$. Hence, if one acquires a set of FRFs, it can be shown that a FRF vector expanded for the unmeasured degrees of freedom (DOF) is given by,

$$\mathbf{H}_{sj}(\omega) = -\left(\mathbf{K}_{ss} - \omega^2\mathbf{M}_{ss}\right)^{-1}\left(\mathbf{K}_{sp} - \omega^2\mathbf{M}_{sp}\right)\mathbf{H}_{pj}(\omega) \tag{5}$$

where $\mathbf{K}$ and $\mathbf{M}$ are the stiffness and mass matrices, p denotes the measured DOFs and s the unmeasured ones, and j sets a given force location. Note that damping can also be considered in this process.

This expansion method is robust to experimental noise and it can be applied with just a reduced set of measured translational FRFs, as shown in [8]. It is quite relevant to mention that the correctness of the numerical model used to expand the data is not a paramount issue, although it must have a certain degree of similarity, regarding the structural behaviour. The estimated or expanded FRF shown in fig. 2 was obtained for a steel beam, using only two translational piezoelectric accelerometers (two measuring locations) and ten impact points (rover hammer test). Details on this case of application are given in [8]. Figure 2 shows the comparison between a direct rotational FRF (H_{θ_a,τ_a}), that is the relation between a rotational response θ_a due to a pure moment τ_a, obtained via: i) experimentally using a T-block (dotted line); ii) a damped finite element (FE) model, used in the expansion process (dashed line); and iii) the expansion process, using only experimantal translational FRFs as reference (solid line).

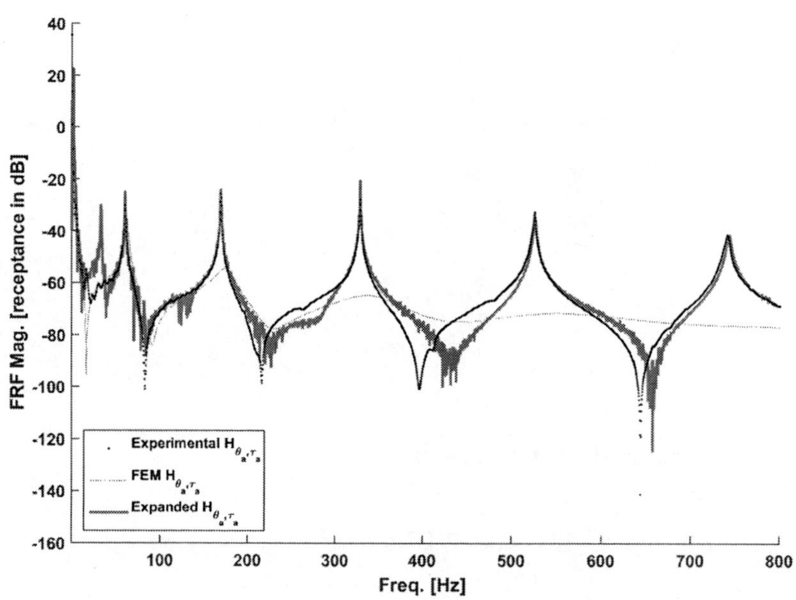

Figure 2. Direct rotational FRF (H_{θ_a,τ_a}).

This expansion technique was then implemented in conjunction with the frequency based structuring (FBS) method [22] in the context of structural coupling, with quite good results (work to appear in proc. of VETOMAC XIV, 2018). Note that this work is similar to the one of [10], but one avoids modal identification and the use of expensive rotational piezoelectric accelerometers. Figure 3 shows the comparison between a transferred rotational FRF (measured rotation and excitation with pure moment at different locations). The agreement between both curves is quite good in a broad frequency range. In this application, one assumes that the rotational FRFs are inaccessible, and by this the comparison is only made based on the reference complete FE model responses.

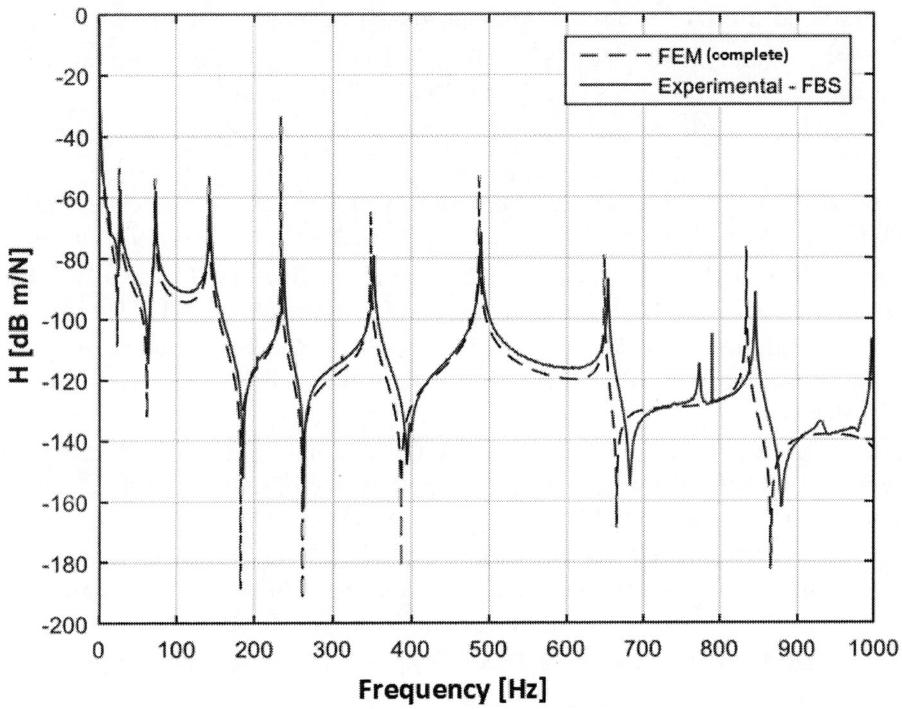

Figure 3. Rotational FRFs due to a exciting moment: numerical FRF of the complete structure (dashed line); estimated FRF by the FBS method using experimental data (solid line).

4. Conclusion

This paper summarizes part of the the work developed by a research group at NOVA UNIDEMI, regarding different applications of measured rotational responses in structural dynamics. The use of less conventional sensors and a non-commercial data acquisition system is briefly presented and some results are discussed, stressing some advantages and limitations of the use of both translational and rotational sensors. It is argued that the use of MEMS gyroscopes enhance the identification of rigid body properties of a structure, as it makes this process independent from any specific location of measuring points. In this case, one must only to assure that all the rigid body modes are excited. Additionally, concerning damage detection, the use of rotational responses with well-established damage indicators increase the sensitivity of those indicators to damage. On the other hand, knowing one of the major limitations of MEMS gyroscopes, the limited maximum frequency, one discussed the capability to estimate rotational FRFs from a set of measured translational ones. The estimated FRFs were experimentally assessed and successfully used in the context of structural coupling.

Acknowledgments

The authors acknowledge the collaboration of M.Sc. Pedro Riscado and M.Sc. Nuno Venâncio, and the financial support of the Portuguese Foundation for Science and Technology under the project UNIDEMI Pest-OE/EME/UI0667/2014.

References

[1] Friswell M I and Mottershead J E 1995 *Finite Element Model Updating in Structural Dynamics* Solid Mechanics and its Applications (Dordrecht: Kluwer Academic Publishers)

[2] Maia N M M and Silva J M M 1997 *Theoretical and Experimental Modal Analysis* Mechanical Engineering Series (Research Studies Press Limited)

[3] Qu Z Q 2004 *Model Order Reduction Techniques* (London: Springer London) ISBN 978-1-84996-924-6

[4] Koutsovasilis P 2009 *Model Order Reduction in Structural Mechanics: Coupling the Rigid and Elastic Multi Body Dynamics* Ph.D. thesis Technische Universität Dresden

[5] Besselink B, Tabak U, Lutowska A, van de Wouw N, Nijmeijer H, Rixen D J, Hochstenbach M E and Schilders W 2013 *Journal of Sound and Vibration* **332** 4403–4422

[6] Silva T A N and Maia N M M 2015 *Conference Proceedings of the Society for Experimental Mechanics Series* vol 6 ed Allemang R J (Springer International Publishing) pp 35–48 ISBN 9783319069883

[7] Maia N M M, Urgueira A P V, Almeida R A B and Silva T A N 2019 *Handbook of Experimental Structural Dynamics* ed Allemang R J and Avitabile P (Springer-Verlag New York) ISBN 978-1-4614-4546-3

[8] Silva T, Maia N, Urgueira A and Riscado P 2016 *Proceedings of ISMA 2016 - International Conference on Noise and Vibration Engineering and USD2016 - International Conference on Uncertainty in Structural Dynamics* ISBN 9789073802940

[9] Batista F and Maia N M M 2012 *ICSV19 - 19th International Congress on Sound and Vibration* (Vilnius, Lithuania)

[10] Drozg A, Čepon G and Boltežar M 2018 *Mechanical Systems and Signal Processing* **98** 570–579

[11] Urgueira A, Venâncio N, Riscado P, Almeida R and Silva T 2018 *MATEC Web of Conferences* vol 148 ISSN 2261236X

[12] Drozg A, Rogelj J, Čepon G and Boltežar M 2018 *Measurement* **127** 292–298

[13] Ewins D J and Sainsbury M G 1972 *The Shock and Vibration Bulletin* **42** 105–122

[14] Leondes C T 2006 *MEMS/NEMS Handbook: Techniques and applications* (Boston, MA: Springer US) ISBN 978-0-387-24520-1

[15] Schedlinski C and Link M 2001 *Mechanical Systems and Signal Processing* **15** 189–211

[16] Lee H, Lee Y B and Park Y S 1999 *Mechanical Systems and Signal Processing* **13** 571–592

[17] Urgueira A P V 1995 *International Modal Analysis Conference (IMAC XIII)* (Nashville, Tennessee USA) pp 1479–1483

[18] Venâncio N 2017 *Identification fo rigid body properties using MEMS sensors (in portuguese)* M.Sc. thesis, DEMI, FCT NOVA, Lisboa

[19] Urgueira A P V, Almeida R A B and Maia N M M 2011 *Mechanical Systems and Signal Processing* **25** 940–951

[20] Almeida R A B, Urgueira A P V and Maia N M M 2011 *ICEDyn 2011 International Conference on Structural Engineering Dynamics* (Tavira, Portugal)

[21] Heylen W, Lammens S and Sas P 2007 *Modal Analysis: Theory and Testing* (Heverlee, Belgium: Katholieke Universiteit Leuven - Department Werktuigkunde) ISBN 90-73802-61-X

[22] de Klerk D, Rixen D J and de Jong J 2006 *IMAC-XXIV: Conference & Exposition on Structural Dynamics* (SEM) ISBN 0912053941 ISSN 21915644

IOP Conf. Series: Journal of Physics: Conf. Series **1106** (2018) 012020 doi:10.1088/1742-6596/1106/1/012020

Tribological and vibrational effects of laser surface texturing on steel-steel sliding contact

S Wang[1,*], A Chen[1], L Li[1], and X Zhang[1]

[1] Department of Mechanical Engineering, University of Shanghai for Science and Technology, Shanghai, 200093, China
[*]Corresponding author: wsw@usst.edu,cn

Abstract. Laser surface texturing (LST) not only can significantly improve the tribological performance of various tribological contacts, but also can significantly reduce the frictional noise and vibrations generated from sliding contacts. In order to investigate the tribological effect of laser surface texturing, sixteen stainless steel specimens with different surface textures, and different textured area ratios are designed and processed on the surfaces of specimens by means of laser surface texturing technique. Four different operation speeds are carried out using a standard friction and wear testing machine (UMT-TriboLab) under rich oil lubrication condition. The experimental results show that the diameter of surface circular dimple had the most significant effect on the tribological properties of steel-steel sliding contacts, followed by texture depth, texture area density, and sliding speed. The effect of laser surface texturing on the frictional noise and vibration generated from steel-steel in dry sliding contacts is also investigated. Laser surface texture has significant effect on the reduction of noise generated from steel-steel dry sliding contacts in both time and frequency domains. The conclusion is that laser surface texturing has significant effects on the tribological, vibrational, and other physical performance of functional surfaces. To get the best tribological and vibrational performance, the functional surface texture should be optimized considering the coupled effects of surface texture, lubrication, and contact condition.

1. Introduction

Studies have shown that the rougher surfaces with certain micro-structures have better friction, wear, lubrication, and vibration and noise reduction performance. Etsion et al. [1-4] have carried out various studies on the tribological effects of LST in automotive applications. Martin and Andres [5] used the laser marking technique to manufacture circular pits and intermittent grooves in the end face of the ring block, and studied the effect of two micro surface shapes on frictional behavior under spherical point contact interface. Schreck and Zum Gahr [6] reported that compared with the polished steel and ceramic surfaces the coefficient of friction (COF) of laser micro-textured surfaces can be reduced by 30%. Wang et al [7] investigated the effects of shapes and textured area ratio of LST on friction and wear of steel ball reciprocating sliding on flat surface of cast iron block. The effects of residual stress due to LST on the wear resistance of cast iron was also investigated in this study.

Sudeen et al [8, 9] experimentally investigated the friction and vibration behaviors of lubricated concentrated point contacts with ground, lapped and textured lapped flat surfaces under reciprocating motions. In Sudeen's study, the coefficient of friction reduced by 30% in some of the cases and vibrations associated with lubricated point contacts generated between textured surfaces/balls reduced significantly at resonance frequency in comparison to polished surfaces/balls.

In the presenting study, in order to optimize the surface textures the tribological effects of circular

IOP Conf. Series: Journal of Physics: Conf. Series **1106** (2018) 012020 doi:10.1088/1742-6596/1106/1/012020

dimple diameter, depth, and area density are studied. Also, the noise generated from steel-steel dry sliding contacts with and without laser surface texture is compared in order to investigate the effect of laser surface texture on noise reduction.

2. Experimental Study

2.1 Test specimen preparation

The tribo-pair is composed of a circular stainless steel plate and a stainless steel ball. The steel ball (ϕ5mm) is the upper stationary specimen, and the circular plate (ϕ70mm) is the lower rotation specimen. Sixteen lower specimen with different surface textures were manufactured by means of a laser machine. The dimensional parameters of surface textures are listed in Table 1. According to the principle of DOE (Design of Experiment), sixteen specimens were prepared and their topographies are shown in Fig. 1.

Table 1. Dimensional parameter of surface texture and sliding speed.

	A Texture Diameter (μm)	B Texture depth (μm)	C Texture area density (%)	D Sliding speed (m/s)
1	100	5	10	0.25
2	120	10	20	0.50
3	140	15	30	0.75
4	160	20	40	1.0

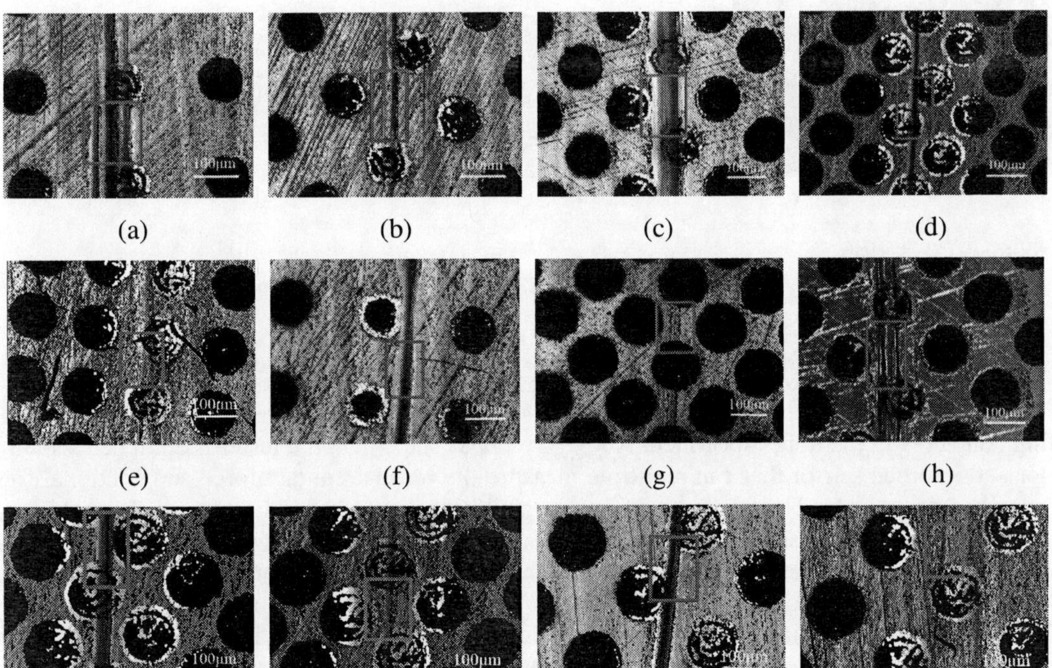

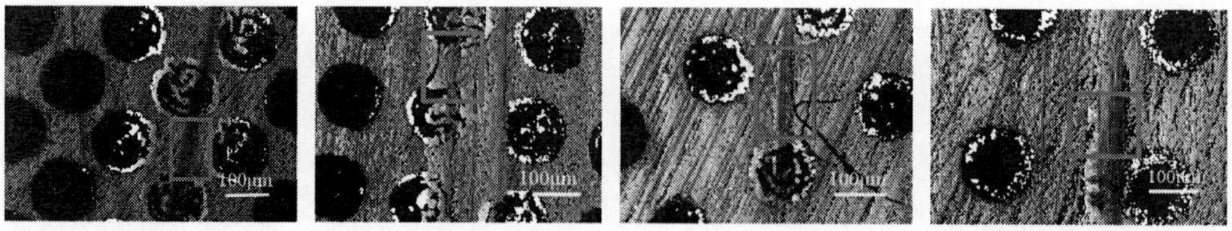

(m) (n) (o) (p)

Fig 1. Topography of laser surface texture (1-16 specimen).

2.2 Test rig

Figure 2 and 3 shows the UMT-TriboLab friction and wear testing machine and the contact of friction pair, respectively.

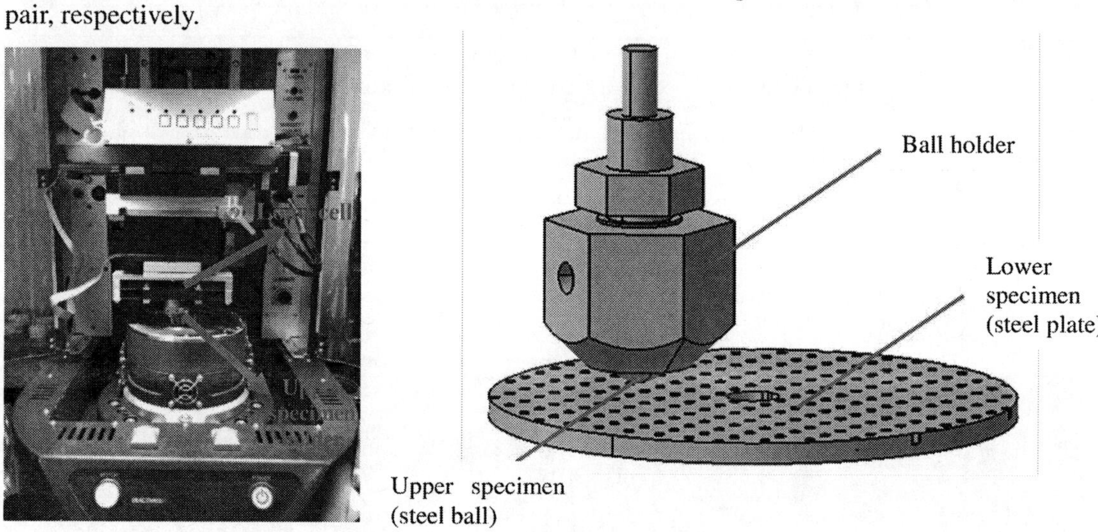

(a) (b)

Figure 2. The UMT-TriboLab (a) testing machine and (b) sliding friction pair.

2.3 Experimental methodology

For the UMT-TriboLab, the upper specimen is stationary, but the lower specimen is rotating and sliding contact with the upper specimen. Above the upper specimen, the tester is equipped with a force sensor and a torque sensor that can real-time measure the normal contact force and frictional force of the contact, respectively. The friction coefficient can be obtained via the data acquisition system. After tests, specimens were ultrasonic cleaned and the wear losses (depths and volumes) of both upper and lower specimens were measured using a white light interference surface profilometer (Bruker Contour GT-K 3D Optical Microscope).

Experiments were carried out in both lubrication and dry conditions. In lubrication condition, the lower specimen was fully covered by automotive lubrication oil. The tests were carried out at room temperature, and each experiment last for the same contact distance 100 meter under the same normal load of 10 N in lubrication condition, while each experiment last for the same contact distance of 20 meter under 2 N normal load in dry condition. Each experiment was repeated for three times at the same operation condition with lubrication, but repeated for two times under dry condition. After all tests, the tested specimens were ultrasonic cleaned, dried and weighed.

During the tribological tests, the noise generated by sliding contacts was also measured using a sound pressure, in order to investigate the effect of laser surface texture on the noise generation from steel-steel sliding contacts under both lubrication and dry conditions.

The orthogonal design of the experiments (ODOE) was carried out, and four factors (texture diameter, depth, textured area density, and relative sliding speed) and four levels in each factor were used. Table 2 presents the detailed arrangement of the ODOE.

Table 2. The orthogonal design of experiments $L_{16}(4^5)$

	A Texture Diameter (μm)	B Texture depth (μm)	C Texture area density (%)	D Sliding speed (m/s)	E Empty
1	100(1)	5(1)	10(1)	0.25(1)	-(1)
2	100(1)	10(2)	20(2)	0.50(2)	-(2)
3	100(1)	15(3)	30(3)	0.75(3)	-(3)
4	100(1)	20(4)	40(4)	1.0(4)	-(4)
5	120(2)	5(1)	20(2)	0.75(3)	-(4)
6	120(2)	10(2)	10(1)	1.0(4)	-(3)
7	120(2)	15(3)	40(4)	0.25(1)	-(2)
8	120(2)	20(4)	30(3)	0.50(2)	-(1)
9	140(3)	5(1)	30(3)	1.0(4)	-(2)
10	140(3)	10(2)	40(4)	0.75(4)	-(1)
11	140(3)	15(3)	10(1)	0.50(2)	-(4)
12	140(3)	20(4)	20(2)	0.25(1)	-(3)
13	160(4)	5(1)	40(4)	0.50(2)	-(3)
14	160(4)	10(2)	30(3)	0.25(1)	-(4)
15	160(4)	15(3)	20(2)	1.0(4)	-(1)
16	160(4)	20(4)	10(1)	0.75(3)	-(2)

2.4 Experimental results and analysis

2.4.1 Friction and wear from lubricated tests. Figure 3 demonstrates the measured coefficients of friction (COFs) of all tests including one test without surface textures and sixteen tests with various surface textures. It can be seen that the surface textures significantly reduced the COFs in all tests although they were not so significant in the first 20 s of the tests, because the first 20 s was the transit state of sliding contacts. In comparison, the average COF of the test 16 was the lowest one and followed by the test 15. For the wear loss of lower specimens, the wear loss of specimen 15 was the smallest one and followed by 12.

IOP Conf. Series: Journal of Physics: Conf. Series **1106** (2018) 012020 doi:10.1088/1742-6596/1106/1/012020

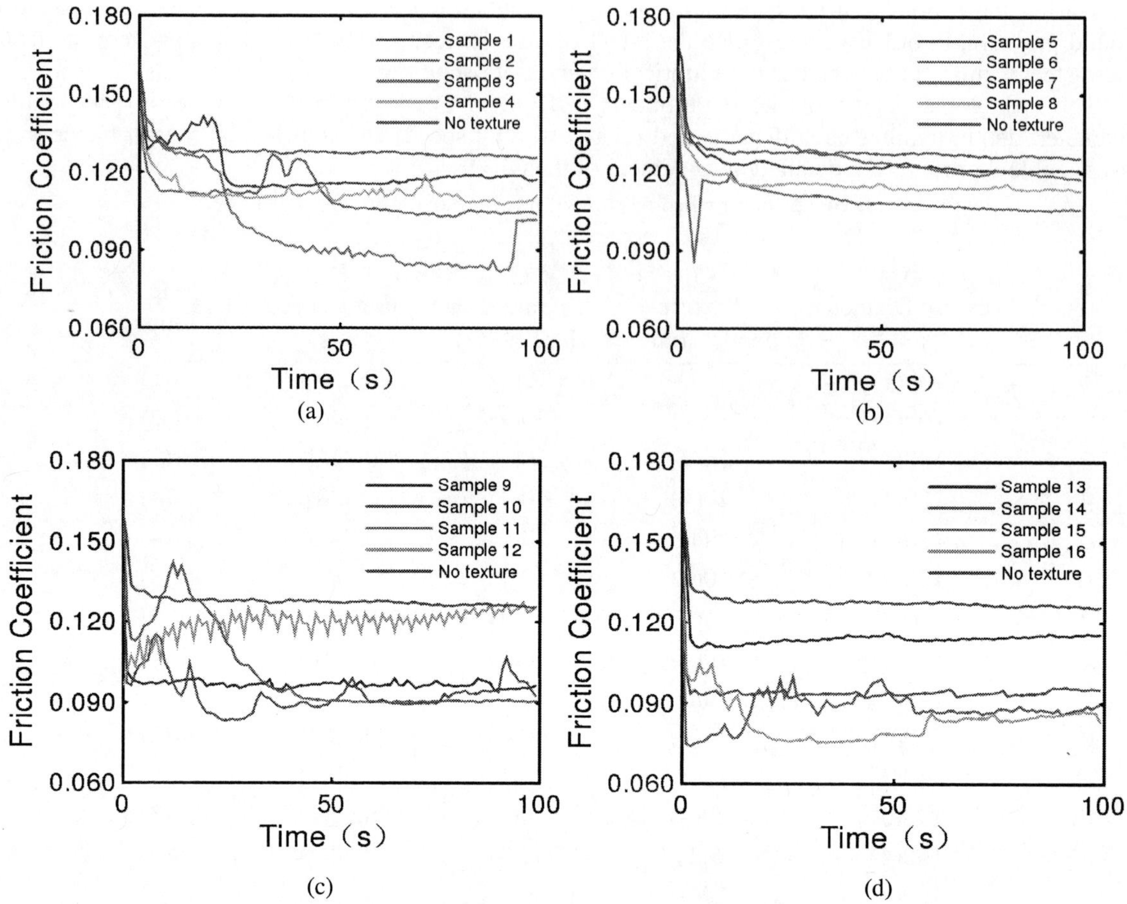

Figure 3. Measured COFs from surfaces without texture and with textures of samples 1-16.

A white light interference surface topography instrument was used to measure the wear losses (depths) of the tested specimens. Figure 4 shows the wear losses (depths) of different surface textured specimens from the UMT-TriboLab tests. A detailed analysis of the effects of surface texture parameters on the tribological properties (friction and wear) will be presented in section 2.5.

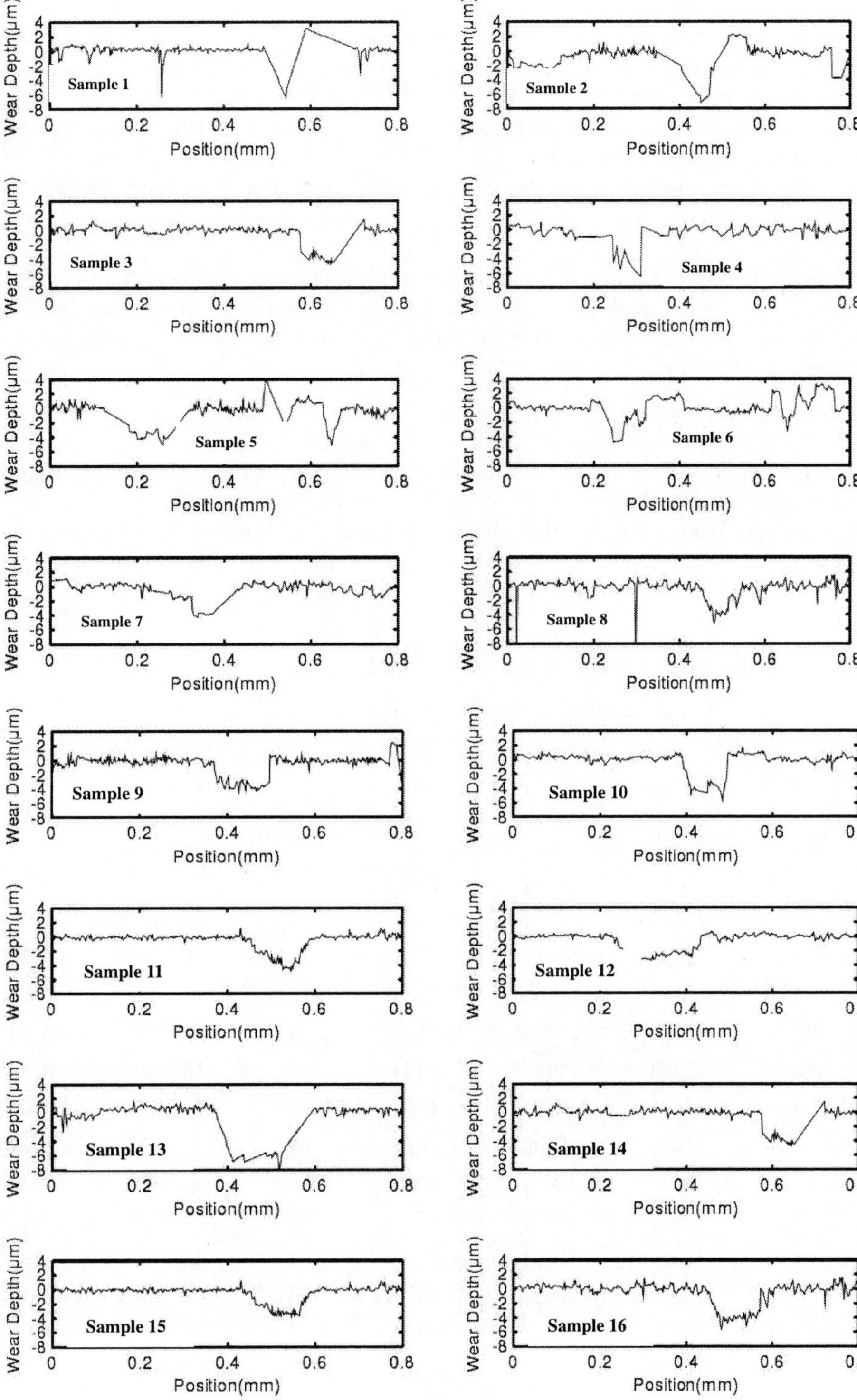

Figure 4. Measured wear loss (depth) of samples 1-16.

2.4.2 Analysis of tribological effect of laser surface texture. After obtaining the friction coefficients of all sixteen tests, the average steady friction coefficient was taken for the analysis, as shown in Table 3. Since the experimental results have two parameters, the average friction coefficient and the wear depth, in order to take into account the effects of the two parameters, a comprehensive scoring method was used to determine the total score of each test. The total score was used for the analysis of the tribological effect of laser surface textures.

A 5-point system was used to score the friction coefficient and wear depth for all tests. For the coefficient of friction (COF), 5-point, 4-point, 3-point, 2-point, and 1-point represents the COF of 0.075 to 0.085, 0.085 to 0.095, 0.095 to 0.105, 0.105 to 0.115, and 0.115 to 0.125, respectively. For the wear depth, 5-point, 4-point, 3-point, 2-point, and 1-point represents the wear depth of 3.0 to 4.0, 4.0 to 5.0, 5.0 to 6.0, 6.0 to 7.0, and 7.0 to 8.0 micrometer, respectively. Based on the above scoring principle and the experimental results shown in Table 2, the average score of each factor in each level may be obtained from Eq. (1):

$$\bar{K}_{ij} = K_{ij} / s \tag{1}$$

where s is the number of occurrences of the i-level on column j. Since the number of occurrences at each level of this study was 4, then s = 4.

The optimal texture parameter may be determined based on the range R_j, which can be calculated according to Eq. (2):

$$R_j = \underset{i}{Max}\{\bar{K}_{ij}\} - \underset{i}{Min}\{\bar{K}_{ji}\} \tag{2}$$

The calculated results of the range R_j are shown in Table 4.

Table 3. Friction and wear test results.

	A Texture diameter (μm)	B Texture depth (μm)	C Areal density (%)	D Speed (m/s)	E Vacant column	Test Results Average friction coefficient	Test Results Wear depth (μm)
1	100(1)	5(1)	10(1)	0.25(1)	-(1)	0.1172	6.03
2	100(1)	10(2)	20(2)	0.5(2)	-(2)	0.1039	7.06
3	100(1)	15(3)	30(3)	0.75(3)	-(3)	0.1016	4.39
4	100(1)	20(4)	40(4)	1.0(4)	-(4)	0.1113	5.86
5	120(2)	5(1)	20(2)	0.75(3)	-(4)	0.1208	4.74
6	120(2)	10(2)	10(1)	1.0(4)	-(3)	0.1048	4.62
7	120(2)	15(3)	40(4)	0.25(1)	-(2)	0.1077	4.06
8	120(2)	20(4)	30(3)	0.5(2)	-(1)	0.1131	5.17
9	140(3)	5(1)	30(3)	1.0(4)	-(2)	0.0963	4.29
10	140(3)	10(2)	40(4)	0.75(3)	-(1)	0.0925	4.43
11	140(3)	15(3)	10(1)	0.5(2)	-(4)	0.1025	4.32
12	140(3)	20(4)	20(2)	0.25(1)	-(3)	0.0905	3.35
13	160(4)	5(1)	40(4)	0.5(2)	-(3)	0.1155	6.68
14	160(4)	10(2)	30(3)	0.25(1)	-(4)	0.0951	4.39
15	160(4)	15(3)	20(2)	1.0(4)	-(1)	0.0896	3.27
16	160(4)	20(4)	10(1)	0.75(3)	-(2)	0.0831	4.38
17	0	0	0	0	0	0.1211	5.87

IOP Conf. Series: Journal of Physics: Conf. Series **1106** (2018) 012020 doi:10.1088/1742-6596/1106/1/012020

Table 4. Range analysis.

	A Texture diameter	B Texture depth	C Areal density	D Speed
K_{1j}	19	18	26	25
K_{2j}	23	26	27	19
K_{3j}	31	29	26	29
K_{4j}	28	28	22	28
$\overline{K}_{1j}$	4.75	4.5	6.5	6.25
$\overline{K}_{2j}$	5.75	6.5	6.75	4.75
$\overline{K}_{3j}$	7.75	7.25	6.5	7.25
$\overline{K}_{4j}$	7	7	5.5	7
R_j	3	2.75	1.25	2.5
Optimal Level	A_3	B_3	C_2	D_3

It can be seen from Table 4 that the best combination of factors and levels (texture parameters) is $A_3B_3C_2D_3$, i.e. texture diameter is 140 μm, texture depth is 15 μm, texture areal density is 20%, and sliding speed is 0.75 m/s. Table 4 also shows that the influence of each factor on the tribological property of steel-steel sliding contacts is different. The diameter of circular dimple has the most significant effect on the tribological performance of steel-steel sliding contact, followed by texture depth, sliding speed, and texture areal density under the operation conditions of this study.

2.4.3 Noise generated from dry sliding contacts. During tribological tests the noise generated from steel-steel sliding contacts under both lubrication and dry conditions was measured. The methodology of the noise measurements are described in Section 2.3. Since the noise generated in lubrication condition was less significant than that in dry condition, here only the noise generated in dry sliding condition is presented, as shown in Figs. 5 and 6.

It can be seen from Fig. 5 that the noise (RMS of sound pressure) generated from steel ball sliding contacted with textured stainless steel circular plates (sample 1-16) was much smaller than that generated from steel ball sliding contacted with surface polished stainless steel plate (sample 0) without surface texture.

It has been observed from Fig. 6 that there were two significant resonant peaks of noise generated from steel ball dry sliding contacted with polished steel plate without surface texture, 250 Hz with 0.85 Pa and 1450 Hz with 0.75 Pa. However, there was only one significant resonant peak of noise generated from most steel balls dry sliding contacted with steel plates with surface textures.

Table 5 demonstrates the RMSs of noise generated from dry sliding contacts in both time and frequency domains, and the maximum reduction of RMS of noise in time and frequency domain was 84% and 89%, respectively.

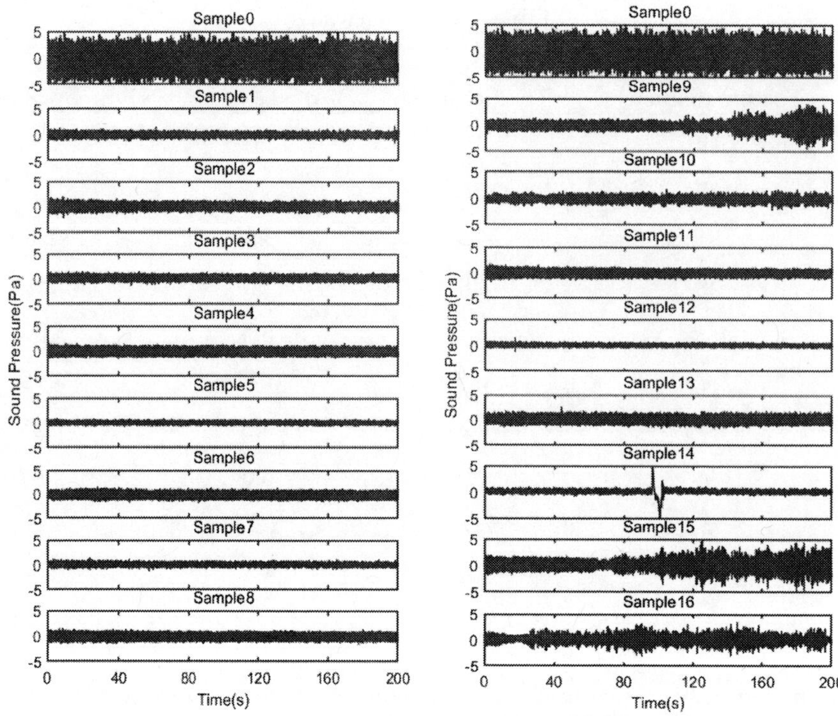

(a) Frictional noise of sample 1-8 (b) Frictional noise of sample 9-16

Figure 5. Frictional noise in time domain (a) from test 1-8 and (b) from test 9-16.

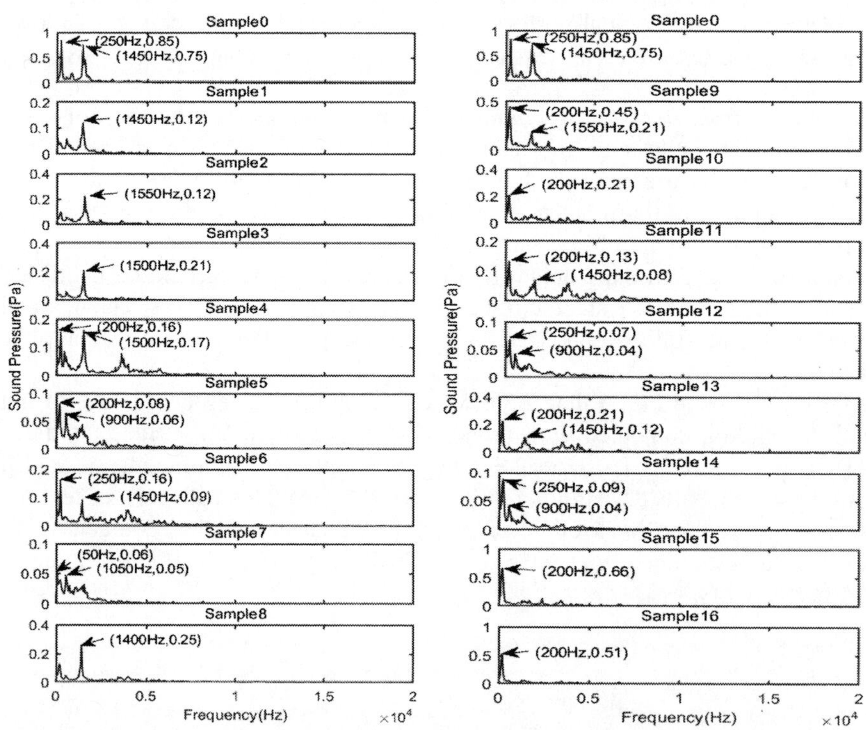

(a) Frictional noise of sample 1-8 (b) Frictional noise of sample 9-16

Figure 6. Frictional noise in frequency domain (a) from test 1-8 and (b) from test 9-16.

Table 5. RMSs of noise generated from dry sliding contacts in both time and frequency domains

Test No.	RMS (Time domain)	RMS (Frequency domain)	Noise reduction (%) Time domain	Noise reduction (%) Frequency domain	Resonant frequency/ Amplitude (Pa)
0	1.1693	0.0685	0	0	250Hz/0.85 1450Hz/0.75
1	0.2423	0.0111	79.28	83.8	1450Hz/0.12
2	0.3710	0.0184	68.27	73.14	1550Hz/0.22
3	0.3247	0.0149	72.23	78.25	1500Hz/0.21
4	0.3164	0.0180	72.94	73.72	200Hz/0.16 1500Hz/0.17
5	0.1830	0.0095	84.35	86.13	200Hz/0.08
6	0.2870	0.0149	75.46	78.25	250Hz/0.16 1450Hz/0.09
7	0.2154	0.0075	81.58	89.05	50Hz/0.06
8	0.3411	0.0191	70.83	72.12	1400Hz/0.25
9	0.4727	0.0347	59.57	49.34	200Hz/0.45
10	0.3114	0.0180	73.37	73.72	200Hz/0.21
11	0.2682	0.0147	77.06	78.54	200Hz/0.13
12	0.2210	0.0075	81.1	89.05	250Hz/0.07
13	0.3787	0.0197	67.61	71.24	200Hz/0.21
14	0.5120	0.0086	56.21	87.45	250Hz/0.09
15	0.6248	0.0491	46.57	28.32	200Hz/0.66
16	0.5238	0.0337	55.2	50.8	200Hz/0.51

3. Conclusion

Experimental results from steel-steel sliding contacts show that LST had significant effect on the friction reduction and wear resistance under lubrication conditions. In this study, the diameter of surface texture had the most significant effect on the tribological properties of steel-steel sliding contacts, followed by texture depth, texture area density, and sliding speed.

Laser surface texture has significant effect on the reduction of noise generated from steel-steel dry sliding contacts in both time and frequency domains.

To get the best tribological and vibrational performance, the functional surface texture should be optimized considering the coupled effects of surface texture, lubrication, and contact condition.

4. References

[1] Etsion I and Burstein L 1996, *J. Tribology Transactions.* 39(3), 677.
[2] Ronen A, Etsion I and Kligerman Y 2001, *J. Tribology Transactions*, 44(3), 359.
[3] Kligerman Y, Etsion I and Shinkarenko A 2005, *Transactions of the ASME*, 127(3), 632.
[4] Ryk G and Etsion I 2006, *Wear*, 261(7/8), 792.
[5] Etsion I and Sher E 2009, *Tribology International*, 42(4), 542.
[6] Martin D and Andres L 2008, *Advanced Engineering Materials*, 10(6), 554.
[7] Schreck S and Gahr KHZ 2005, *Applied Surface Science*, 247, 616.
[8] Wang S, Yan F and Chen A 2018, *Industrial Lubrication and Tribology*, 70 (1), 126.
[9] Sudeep U, Pandey RK and Tandon N 2013, *Tribology International*, 62 (3), 198.
[10] Sudeep U, Tandon N and Pandey R K 2015, *J. of Tribology*, TRIB-15-1058.

Acknowledgments
The authors gratefully acknowledge the financial supports from the National Natural Science Foundation of China (No. 51275126). Professors J. Li and X. Zeng and Ms H. Yang from the Green Chemical Engineering Technology Research Center, Shanghai Advanced Research Institute, China Academy of Sciences, are highly appreciated for their helps in completion of all the experiments in this study.

IOP Conf. Series: Journal of Physics: Conf. Series **1106** (2018) 012021 doi:10.1088/1742-6596/1106/1/012021

Modelling and Characterisation of a Servo Self-Piercing Riveting (SPR) System

D Tang[1,4], M Evans[2], P Briskham[2], L Susmel[3] and N Sims[1]

[1] Department of Mechanical Engineering, The University of Sheffield, 36-38 Victoria Street, Sheffield, S3 7QB, UK

[2] Henrob Ltd., Second Avenue, Zone 2, Deeside Industrial Park, Flintshire, CH5 2NX, UK

[3] Department of Civil and Structural Engineering, The University of Sheffield, Mappin Street, Sheffield, S1 3JD, UK

[4] Author to whom any correspondence should be addressed

E-mail: dhtang1@sheffield.ac.uk

Abstract. SPR is a cold mechanical joining process in which multiple sheets of material are riveted together without the need for a predrilled hole. It works by pushing a typically semi-tubular rivet into a target stack of material, during which the plastic deformation of the material and rivet are such that a mechanical lock is formed within the material stack. The process is used extensively in the automotive industry in car body construction, and is a competing technology to more established joining techniques such as resistance spot welding. As part of the ongoing development of the technique, there is a strong need to understand and simulate the dynamics of the process. In this work, a lumped parameter model of the SPR system with a non-parametric model of the joint is presented. Simulated results are compared with experimental data for a given joint configuration. Furthermore, the model is used to highlight the significance of the compliances within the system. It is shown that during rivet insertion, the stiffness of the C-frame structure is an influential factor in determining the dynamic response of the system. The results provide the basis for a more comprehensive sensitivity analysis into the factors which affect the quality of the resulting joint.

1. Introduction

Self-piercing riveting (SPR) is a cold mechanical joining process in which multiple sheets of material are riveted together without the need for a predrilled hole. It works by punching a typically semi-tubular rivet into a target stack of material, during which the plastic deformation of the material and rivet are such that a mechanical lock is formed within the material stack. The rivet does not typically fully pierce the bottom sheet of material, and in many joint designs does not protrude above the top surface. The typical riveting sequence can be visualised in figure 1(a) and a diagram of a SPR setter assembly is shown in figure 1(b).

In the servo SPR process, the upsetting force required to form the joint is provided by the C-frame structure. The punch is actuated by a servo motor via a roller screw mechanism. The initial velocity of the punch is controlled to determine the amount of kinetic energy input into the joining process.

SPR is used extensively in the automotive industry in car body construction as it offers many advantages over traditional joining techniques such as resistance spot welding [1]. Car manufacturers are increasingly using novel materials which are stronger, lighter, but also more difficult to join. These new demands often represent a rise in the loads, stresses, and operating speeds of the fastening equipment. There is a growing necessity for model-based analysis and design, which can drive further development of the SPR technique and offer insight into the sensitivities of the process, while minimising the amount of experimental effort required.

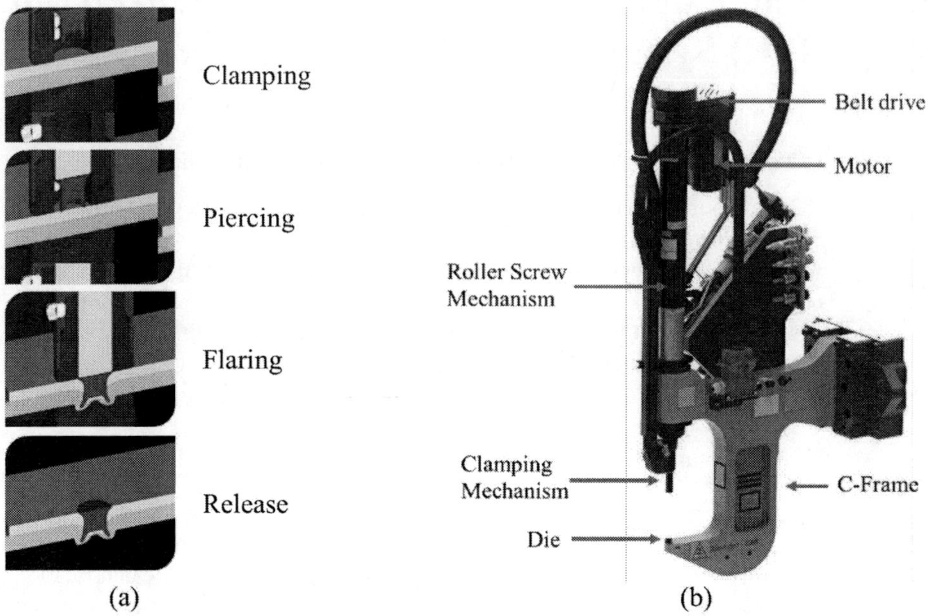

Figure 1. (a) Stages of the rivet insertion process. (b) Illustration of a servo SPR setter assembly.

In the existing literature, the modelling of the SPR process has been predominantly addressed via finite element (FE) methods [2-5]. The number of variables in a typical SPR process makes it a complex subject to model. In [6] Haque compiled a detailed but by no means exclusive list of factors which are known to affect the quality of the joint. However, studies which take a model-based approach to investigating the sensitivity of the process, such as [7] and [8], have focused solely on the interactions at the rivet-material level and do not account for the response of the overall SPR system.

There remains a gap in the existing body of work relating the performance of the full riveting system and the resulting joint. While the effects of the rivet and die properties have traditionally been the subject of investigation, the other components within the system, such as the motor, the mechanical transmission, the C-frame, etc. might have a key influence on the riveting process yet little work has been published in this regard. The knowledge base on SPR would benefit substantially from a comprehensive sensitivity analysis involving the full riveting system.

Works dealing with other joining and forming systems provide useful insight into the identification of models for related processes. The study of a servo-screw press in [9] alluded to the use of force-stroke curves to simulate the dynamic loading of a deep drawing process, and to assess its influence on the drive. In [10], Behrens et al. investigated the interactions between a forming machine and a workpiece, by simulating the respective models separately in their own environments via an "offline-coupling" method. A cold forging press was the subject of investigation in [11], in which the stiffnesses of the press and the tooling were implemented in the FE process simulation as linear spring elements. Parallels can be drawn between the aforementioned processes and the SPR process. The compliances in the SPR system, the model of the workpiece or the joint, and its interaction with the model of the riveting equipment are all critical factors in the accuracy of the simulation.

To better understand the SPR system, it is noted that it is the influence of the machine on the workpiece which is ultimately of interest. A FE approach to the modelling of the SPR system may be impractical. In contrast, a lumped parameter model is considered more suitable for the purposes of gaining insight into the main dynamics and sensitivities of the process. The mechanical, electrical and digital domains can be easily expressed in a single simulation environment, and furthermore, the low computational cost facilitates iterative analyses. Regarding the model of the joint, it remains to be established if anything short of a full FE model would be capable of representing the joint to the desired level of detail. This paper examines the interaction between an analytical model of the riveting equipment and a non-parametric model of the SPR joint. The model of the SPR process is used to predict the response of the system, and furthermore, to show how the compliances within the system affect its response at the riveting interface.

The structure of this paper is as follows: the next section explains the experimental setup. Then, the modelling of the SPR process is presented. This is followed by results and discussion around the model-based analysis. Ultimately, the conclusions and future work are highlighted.

2. Test Setup

The test rig consists chiefly of a servo SPR setter assembly, as illustrated in figure 1(b). Data are collected on various data loggers and synchronised in post-processing. Measurements are taken on a cycle-by-cycle basis, with each cycle being a full sequence of advance, rivet insertion, and retraction of the punch. The main measurements are as follows: angular velocity and displacement of the motor, relative velocity and displacement between the punch and die, the process force and the motor current.

Due to lack of direct visibility of the rivet during the insertion process, the insertion of the rivet into the stack of material is inferred from the motion of the punch relative to the die, which is observed using a high speed camera. It is assumed that the interface between the punch and the rivet head, as well as the bottom sheet of material and the die remain in contact throughout the loading part of the riveting process. The process force is measured using a load cell located below the die.

3. Modelling

The SPR process was considered as three major subsystems: the Control Unit, the Setter Assembly and the Joint. An illustration of the model is presented in figure 2.

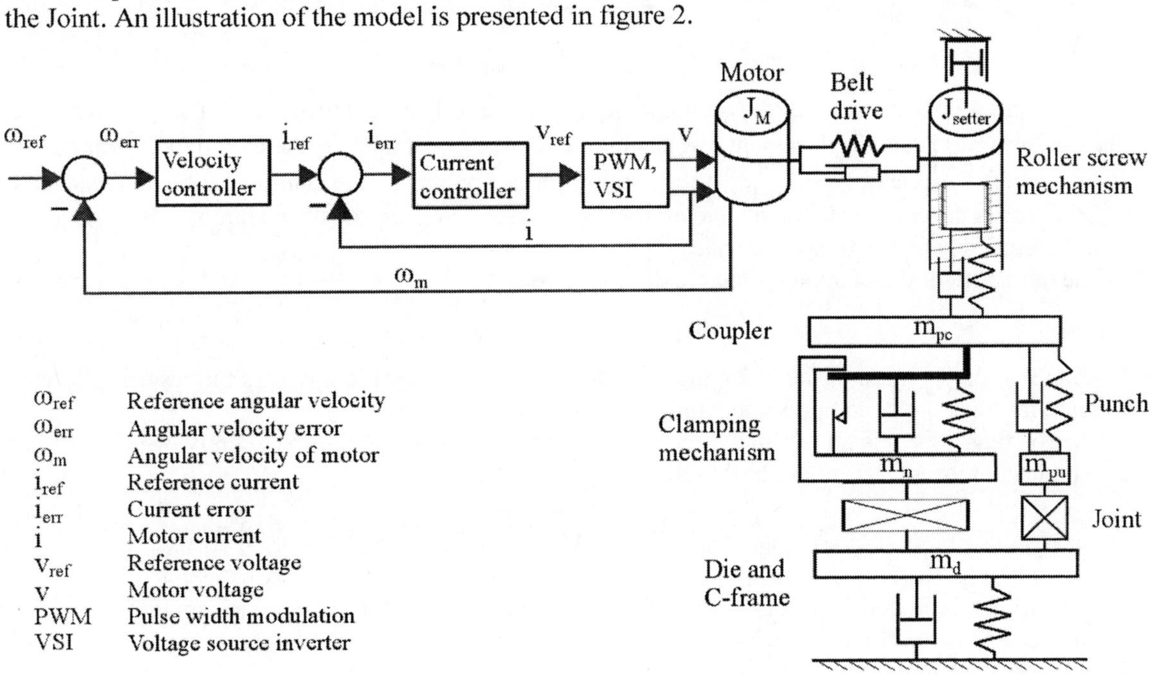

Figure 2. Sketch of the model structure of the system. The control structure is shown on the left with arrows indicating the signal flow, and the setter assembly is shown on the right.

3.1. Setter assembly

The riveting machine was simplified to a six degree-of-freedom system. The model was developed in MATLAB/Simulink via a physics-based approach, in which equations of motion were derived from first principles, and parameter values were either obtained from datasheets or identified experimentally.

The governing equation for the motor is expressed as:

$$J_M\ddot{\theta}_m = T_e - R_1{}^2K_b\left(\theta_m - \frac{R_2}{R_1}\theta_L\right) - R_1^2C_b\left(\dot{\theta}_m - \frac{R_2}{R_1}\dot{\theta}_L\right) \tag{1}$$

Where J_M is the inertia of the motor and pulley, θ_m is angular displacement of the motor, T_e is the electromagnetic torque, R_1 and R_2 are the radii of the driving and driven pulleys respectively, K_b is the belt stiffness, θ_L is the angular displacement of the driven pulley, and C_b is the belt damping constant.

Equation (2) describes the dynamics of the roller screw mechanism.

$$\begin{aligned} J_{setter}\ddot{\theta}_L = R_1R_2K_b\left(\theta_m - \frac{R_2}{R_1}\theta_L\right) + R_1R_2C_b\left(\dot{\theta}_m - \frac{R_2}{R_1}\dot{\theta}_L\right) \\ - \left(\frac{F_LP_h}{2\pi} + \text{sign}(\dot{\theta}_L)\frac{F_LD_0}{2}\mu_{prac} + B_r\dot{\theta}_L\right) \end{aligned} \tag{2}$$

Where J_{setter} is the inertia of the rivet setter, P_h is the lead of the roller screw, D_0 is the nominal diameter of the nut, μ_{prac} is the coefficient of friction, and B_r is the viscous friction coefficient. F_L is the force transmitted axially through the shaft of the roller screw during the riveting operation, defined as:

$$F_L = K_r(x_r - x_{pc}) + C_r(\dot{x}_r - \dot{x}_{pc}) \tag{3}$$

Where K_r and C_r are the effective stiffness and damping constants of the roller screw, x_r is the axial displacement of the roller assembly, and x_{pc} is the displacement of the coupler. The coupler connects the output shaft of the roller screw mechanism to the punch and is also a critical component in the generation of the clamping force. The governing equation for the coupler is formulated as:

$$m_{pc}\ddot{x}_{pc} = F_L - K_{plpu}(x_{pc} - x_{pu}) - C_{plpu}(\dot{x}_{pc} - \dot{x}_{pu}) - F_{clamp} \tag{4}$$

Where m_{pc} is the mass of the coupler, K_{plpu} is the effective stiffness of the plunger-punch subassembly, x_{pu} is the displacement of the end of the punch, and C_{plpu} is the effective damping in the plunger-punch subassembly. F_{clamp} is the clamping force applied by the clamping tube on the stack of material, and is dependent on the relative motion between the coupler and the clamping tube. The details are not included here due to limited space.

The dynamics of the clamping tube are described by:

$$m_n\ddot{x}_n = F_{clamp} - r_{mat} \tag{5}$$

Where m_n is the mass of the clamping tube and $\ddot{x}_n$ is its acceleration. r_{mat} is the restoring force of the material stack under compression, which is defined in section 3.2.

The riveting force r is transmitted directly through the plunger-punch subassembly, which is treated as a single component. Its equation of motion is expressed as:

$$m_{plpu}\ddot{x}_{pu} = K_{plpu}(x_{pc} - x_{pu}) + C_{plpu}(\dot{x}_{pc} - \dot{x}_{pu}) - r \tag{6}$$

Where m_{plpu} is the mass of the plunger-punch subassembly. r is defined in section 3.2.

The C-frame is modelled as mass-spring-damper system, the parameters of which are identified using a high-speed camera. The C-frame and setter are treated as two separate subsystems which interact only via the SPR joint. Equation (7) describes the dynamics of the C-frame.

IOP Conf. Series: Journal of Physics: Conf. Series **1106** (2018) 012021 doi:10.1088/1742-6596/1106/1/012021

$$m_{\mathrm{d}}\ddot{x}_{\mathrm{d}} = r + r_{\mathrm{mat}} - K_{\mathrm{c}}x_{\mathrm{d}} - C_{\mathrm{c}}x_{\mathrm{d}} \tag{7}$$

Where x_{d} is the displacement or deflection of the C-frame.

3.2. SPR joint

The modelling of the SPR joint was framed as a nonlinear system identification problem, to which the restoring force surface (RFS) method was well suited. The technique as described in [12] and its variant in [13], were used as a reference in the development of the model. The individual responses of the system to multiple riveting cycles were collected into a single dataset representing its response to a series of impulse excitations.

Regarding its interactions within the full system model, the joint was considered as a nonlinear element which generated a restoring force according to the extent and rate of its compression, i.e. the displacement and velocity across the nonlinear element, or the punch-die relative displacement.

Variations of the model of the joint were evaluated:

1. The RFS as a three dimensional surface, shown in figure 3(a).
2. A two dimensional polynomial curve defined in force vs. displacement, shown in figure 3(b).

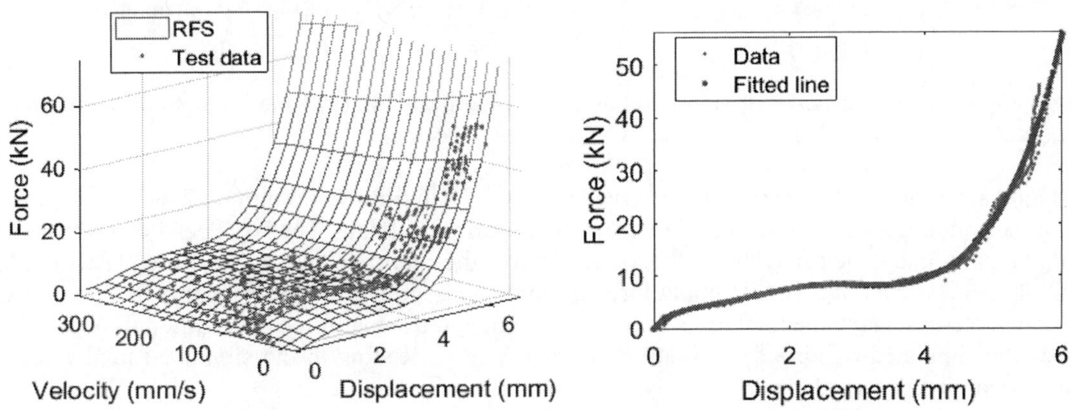

(a) (b)

Figure 3. (a) Restoring force surface. (b) Force-displacement characteristic.

With reference to figure 3(b), the velocity dependence of the given joint configuration was assumed to be negligible, based on the repeatability of the force-displacement profiles across the range of setting velocities 100 – 300 mm/s. Therefore, for characterising the given joint configuration, the simpler model defined in force-displacement was considered adequate. The riveting force r was defined as a seventh-order polynomial function f:

$$r = \mathrm{f}(x_{\mathrm{pu}} - x_{\mathrm{d}}) \tag{8}$$

The restoring force of the material under clamping was defined as:

$$r_{\mathrm{mat}} = K_{\mathrm{mat}}(x_{\mathrm{n}} - x_{\mathrm{d}}) + C_{\mathrm{mat}}(\dot{x}_{\mathrm{n}} - \dot{x}_{\mathrm{d}}) \tag{9}$$

Where K_{mat} is the effective stiffness of the material, x_{n} is the displacement of the clamp tube, and C_{mat} is the effective damping constant of the material.

4. Results and Discussion

4.1. Simulated riveting process

The performance of the full model was assessed via the comparison of experimental and simulated data. Figure 4(a) shows the punch-die relative displacement and force traces for the loading part of a riveting

IOP Conf. Series: Journal of Physics: Conf. Series **1106** (2018) 012021 doi:10.1088/1742-6596/1106/1/012021

process. Figure 4(b) shows the relative velocity and force from the same run plotted against the relative displacement. The relative displacement represents the extent of rivet insertion into the material, and the relative velocity represents the rate of rivet insertion.

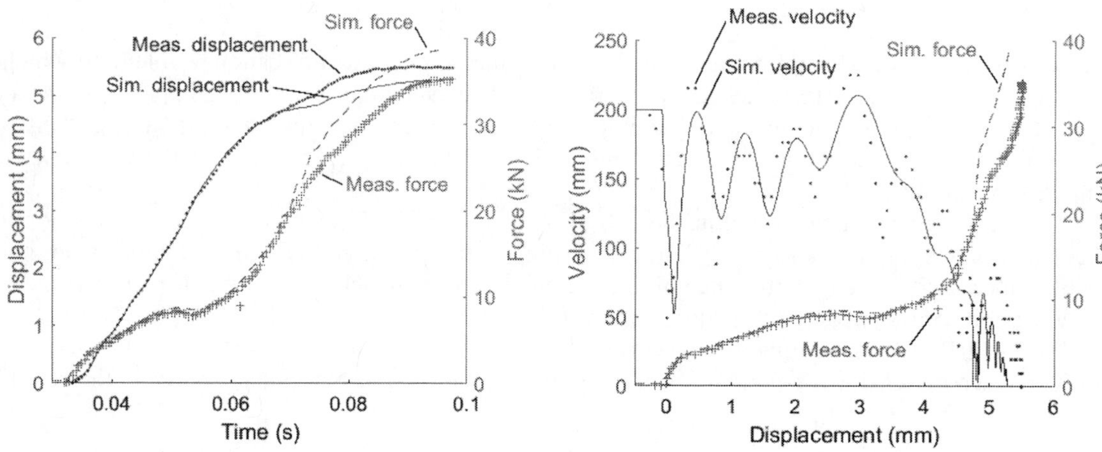

Figure 4. Comparison between simulated and experimental data, (a) Relative displacement and force vs. time, (b) Relative velocity and force vs. relative displacement.

Good agreement in the results are observed up to 0.07 s in time or 4.5 mm in displacement, at which point the simulated traces start to deviate from the experimental data. The issue can be attributed to two main factors: firstly, the behaviour of the material under the clamping load was not adequately modelled, leading to a faster build-up of clamping force through the riveting process. Secondly, the rivet insertion was characterised using the total process force rather than the actual force transmitted through the rivet, hence the simulated riveting force was larger than desired, leading to the simulated total force being higher than that observed experimentally.

In order to improve the performance of the model, the direct measurements of either the riveting force or the clamping force are necessary. Knowing either component of the total process force will allow the other to be calculated. Better characterisation of the material behaviour under clamping would further improve the accuracy of the model.

4.2. Model-based analysis of the effects of the C-frame stiffness

The effect of the static and dynamic properties of the C-frame on the joint was of particular interest. The effective stiffness of the C-frame model was varied to examine how the changes influenced the punch-die relative motion. Three different values were chosen to represent a low, intermediate and high stiffness in the range of stiffnesses of real C-frame types. All other variables in the simulation were kept the same, including the process parameters and the joint configuration. The simulated results are shown in figure 5.

From figure 5(a) it can be seen that at the given setting velocity, the stiffness of the C-frame has a notable effect on the process force. The maximum force increases with C-frame stiffness, which is not surprising when considering the conversion of the kinetic energy of the setter into stored strain energy of the C-frame. In figure 5(b) the traces represent the extent of the simulated rivet insertion. The overall displacement increases for C-frames of larger stiffnesses, as a consequence of the larger process force generated. In figure 5(c), up to 0.055 sec, the oscillatory response of the punch relative to the die is due to the initial riveting load. Taking the case of the intermediate stiffness, K_c: 136646000 N/m, the frequency of the initial oscillations is 200 Hz, which matches the natural frequency of the C-frame. The oscillation of the C-frame implies oscillation of the die, which is expected to directly affect the forming of the joint and the joint quality. The results strongly suggest that the dynamics of the C-frame dominate the response of the system at the riveting interface, and therefore significantly influence the rivet insertion process.

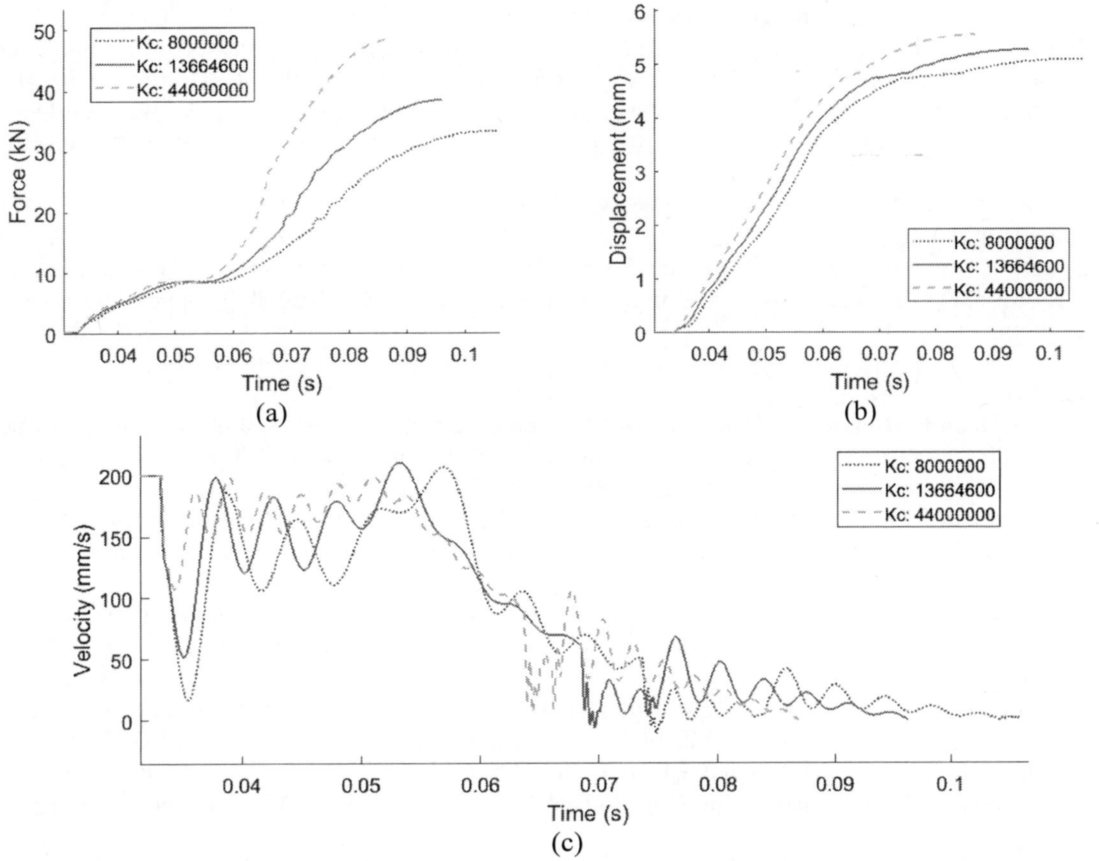

Figure 5. Effects of changing the effective stiffness K_c of the C-frame. (a) Force vs. time, (b) Relative displacement vs. time, (c) Relative velocity vs. time.

In light of the results, a question is raised over the one of the underlying assumptions in the modelling of the joint: that the force-displacement characteristic curve for the joint was independent of the compliance of the system on which it was identified. Having argued that the dynamics of the C-frame significantly influence the riveting process, it can also be argued that the characterisation of a joint configuration on a particular system may yield a model which is specific to that system. Further joint identification on different C-frame types is required to verify the validity of the assumption.

The current study may have implications for the body of work relating to the FE-based modelling of the SPR process, in which the dynamic response of the riveting machine and C-frame are broadly unaccounted for. These approaches may be representative of the SPR process on traditional hydraulic systems, where the mode of rivet insertion can be considered as quasi-static loading. However, in the case of inertia-based servo systems the inertial effects are believed to play an important role, and therefore should be included in modelling of the process.

5. Conclusion

A lumped parameter model of the SPR process has been created in MATLAB/Simulink.

The simulated response of the system has been compared against experimental results for a given joint configuration. Partial agreement indicates that dynamics in the riveting system have been suitably captured in the model during the initial phase of rivet insertion. Assuming that the force-displacement characteristic of the joint is not affected by the compliance of the system on which it is identified, the simulated results suggest that the dynamic response of the machine at the riveting interface is significantly influenced by the dynamics of the C-frame.

Further work should be focused on the identification of the different components of the total process force, and also characterisation of different types of C-frames in order to validate the results of the model-based analysis. Another area of interest would be how the dynamics of the C-frame affect the quality of the joint. The findings of this work form the basis for a more comprehensive sensitivity analysis of the servo SPR system.

Acknowledgements
This work is supported by the UK Engineering and Physical Sciences Research Council (grant number EP/M508135/1). The author would also like to thank Henrob Ltd. for providing support and access to test equipment.

References

[1] Briskham P, Blundell N, Han L, Hewitt R, Young K and Boomer D 2006 Comparison of Self-Pierce Riveting, Resistance Spot Welding and Spot Friction Joining for Aluminium Automotive Sheet (SAE International)
[2] Bouchard P-O, Laurent T and Tollier L 2008 Numerical modeling of self-pierce riveting—from riveting process modeling down to structural analysis *J. Mater. Process. Technol.* **202** 290-300
[3] Casalino G, Rotondo A and Ludovico A 2008 On the numerical modelling of the multiphysics self piercing riveting process based on the finite element technique *Adv. Eng. Softw.* **39** 787-95
[4] Porcaro R, Hanssen AG, Langseth M and Aalberg A 2006 Self-piercing riveting process: An experimental and numerical investigation *J. Mater. Process. Technol.* **171** 10-20
[5] Huang L, Moraes JFC, Sediako DG, Jordon JB, Guo H and Su X 2016 Finite-Element and Residual Stress Analysis of Self-Pierce Riveting in Dissimilar Metal Sheets *J. Manuf. Sci. Eng.* **139** 021007
[6] Haque R 2018 Quality of self-piercing riveting (SPR) joints from cross-sectional perspective: A review *Arch. Civ. Mech. Eng.* **18** 83-93
[7] Mucha J 2011 A study of quality parameters and behaviour of self-piercing riveted aluminium sheets with different joining conditions *Strojniški vestnik-J. Mech. Eng.* **57** 323-33
[8] Jäckel M, Falk T and Landgrebe D 2016 Concept for Further Development of Self-pierce Riveting by Using Cyber Physical Systems *Procedia CIRP* **44** 293-7
[9] Schenke C-C, Wiemer H and Großmann K 2012 Analysis of servo-mechanic drive concepts for forming presses *Prod. Engineer.* **6** 467-74
[10] Behrens B-A, Bouguecha A, Krimm R, Matthias T and Czora M 2013 Consideration of the Machine Influence on Multistage Sheet Metal Forming Processes *Process Machine Interactions*, ed Denkena B, Hollmann F (Berlin: Springer-Verlag Berlin Heidelberg) pp 403-17
[11] Kroiß T, Engel U and Merklein M 2013 Comprehensive approach for process modeling and optimization in cold forging considering interactions between process, tool and press *J. Mater. Process. Technol.* **213** 1118-27
[12] Masri S and Caughey T 1979 A nonparametric identification technique for nonlinear dynamic problems *J. Appl. Mech.* **46** 433-47
[13] Crawley EF and Aubert AC 1986 Identification of nonlinear structural elements by force-state mapping *AIAA J.* **24** 155-62

IOP Conf. Series: Journal of Physics: Conf. Series **1106** (2018) 012022 doi:10.1088/1742-6596/1106/1/012022

Classification-based Damage Localization in Composite Plate using Strain Field Data

R Janeliukstis*, S Rucevskis and A Chate

Institute of Materials and Structures, Riga Technical University, Riga, Latvia

E - mail: Rims.Janeliukstis_1@rtu.lv

Abstract. Problem of damage localization in a cantilevered composite plate using strain data is tackled. The plate is partitioned into a number of zones and a mass equal to 9.43 % of plate's mass is put in each of these zones subsequently. The composite plate is harmonically excited via piezoelectric actuator with a series of driving frequencies equal to natural frequencies of the plate extracted in modal analysis test. In each of these events, the mechanical strains are recorded with strain gauges. Mean values of strain time series are calculated and used as a feature for building a classification model in which zones of the plate serve as classes. Linear discriminant classifier initially yielding the best classification accuracy (90.7 %) is selected. The classification model is validated by selecting 5 query points on the plate and localizing the unknown query points in terms of classes (zones) of the plate. Overall, the unknown query points were classified successfully with only slight misclassification for the point at the boundary between 2 zones.

1. Introduction

The vast field of machine learning has applications in chemistry, medicine, biology, economics and sports among other. Relatively recently, engineers and scientists have discovered the virtues of using the methods and ideas of data science to tackle the challenges in civil, mechanical and aerospace engineering. One of the most prominent problems lies in the identification of damage – a subdomain of structural health monitoring. The underlying idea behind SHM is the installation of particular types of sensors on the structure of interest or embed them within the structure, if possible [1]. The signal of structural response modified by some disturbance may signalize the change of integrity of the structure [2]. It is of essential importance to fuse the sensor data and extract the damage sensitive features to successfully apply the machine learning algorithms in order to, for example, classify if the particular element of the structure is damaged or not and what is the severity of the damage and so on.

The advantages of composite materials over the more conventional ones (concrete, wood, metals, ceramics, etc.) are well known, hence it is crucial to ensure that composite structures maintain their integrity throughout their service life. The vast exploitation of composites in aerospace and their vulnerability to impact damage [3-6] has led researchers to develop damage identification methods based on machine learning approaches. In [7] authors simulated defects of various severities in different locations of a composite plate and used a mixture of support vector machines and decision trees to obtain the probabilities of the position of damage. In [8] researchers studied impacted composite plates equipped with piezoelectric sensor and actuator. The Wavelet Packet Transform was used to extract damage sensitive features and linear discriminant classifier was employed to classify damaged and healthy patterns of the plates. Researchers in [9] applied various *k*-nearest neighbour classifiers to classify different types of defects in aluminium and composite specimens, equipped with

IOP Conf. Series: Journal of Physics: Conf. Series **1106** (2018) 012022 doi:10.1088/1742-6596/1106/1/012022

piezoelectric patches for actuation and sensing. These authors also used machine learning to distinguish between influences of different temperatures in [10]. In [11] authors developed a numerical model for localization of added mass on a cantilevered composite plate equipped with strain sensors. The plate was partitioned into 18 zones and k-nearest neighbours, as well as decision trees classifiers were used to localize simulated query points to belong to one of these 18 zones.

In the present study, the location of added mass on a cantilevered composite plate is found by employing the classification approach of recorded strain signals. The plate is partitioned into 18 zones and in each of these zones a mass of 9.43 % of plate's mass is placed and strains in the form of time series are recorded from 2 strain gauges (one mounted in parallel and the other in perpendicular direction to fibre orientation). The mean values of strains are extracted as a sensitive feature to mass position and used as an input for linear discriminant classifier. After training the classification model with 90.7 % accuracy, the model is validated on 5 query points of mass position on the plate. Results suggest a promising tool for localization of local changes in mass of the structure.

2. Damage localization based on linear discriminant classification

2.1 Linear discriminant classifier

Discriminant classification models assume that different classes y generate data x based on different Gaussian distributions. The space of x values divides into regions where a classification y is a particular value. For linear discriminant analysis the regions are separated by straight lines. The equation for such a boundary between two classes x_i and x_{i+1} is

$$K + L(i) \cdot x_i + L(i+1) \cdot x_{i+1} = 0 \quad (1)$$

where K is the constant term obtained from a linear discriminant classifier model, L is the slope term for the two classes, also obtained from a classifier model.

Classification based on linear discriminant involves the following steps [12]:

- Classifier training – the fitting function estimates the parameters of a Gaussian distribution for each class:
  - the sample mean

$$\hat{\mu} = \frac{\sum_{n=1}^{N} M_{nk} x_n}{\sum_{n=1}^{N} M_{nk}} \quad (2)$$

where M_{nk} is $N \times K$ class membership matrix with the property

$$M_{nk} = \begin{cases} 1 \text{ if observation } n \text{ is from class } k \\ 0 \text{ otherwise} \end{cases} \quad (3)$$

  - the sample covariance is calculated by first subtracting the sample mean of each class from the observations of that class, and taking the empirical covariance matrix of the result

$$\hat{\Sigma} = \frac{\sum_{n=1}^{N} \sum_{k=1}^{K} M_{nk} (x_n - \hat{\mu}_k)(x_n - \hat{\mu}_k)^T}{N - K} \quad (4)$$

(the model has the same covariance matrix for each class, only the means vary).

- Prediction of the new data – the trained classifier finds the class with the smallest misclassification cost [13]

$$\hat{y} = \underset{y=1,\dots,K}{\operatorname{argmin}} \sum_{k=1}^{K} \hat{P}(k|x) C(y|k) \quad (5)$$

where $\hat{y}$ is the predicted classification, K is the number of classes, $\hat{P}(k|x)$ is the posterior probability of class k for observation x. Posterior probability is a product of prior probability (either with uniform or empirical distribution) and a likelihood function (which is usually of multivariate normal density). $C(y|k)$ is the cost of classifying an observation as y when its true class is k.

2.2 Damage localization based on linear discriminant classification

In this study, a non-destructive approach is applied to localize the local changes of mass on the composite plate. The plate is by no means damaged – a pseudo defect is introduced using a small weight of 20 g, comprising 9.43 % of plate's mass. The essential steps of damage localization are depicted in figure 1.

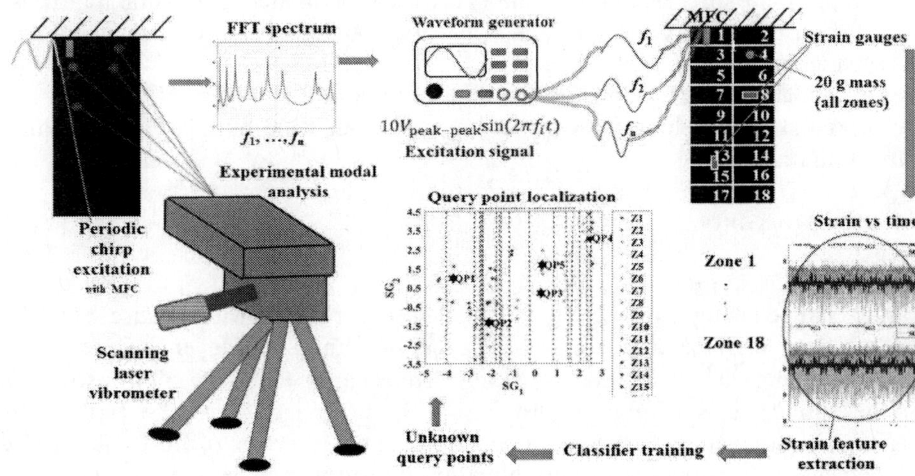

Figure 1. Damage localization scheme based on strain feature classification.

1. The plate is clamped at one end yielding a cantilever configuration. A piezoelectric actuator from MFC (macro-fiber composites) is mounted close to the clamping. This system is placed on rigid floor.
2. MFC element is used to excite the structural vibration in the plate and an experimental modal analysis is carried out with an aid of scanning laser vibrometer. Vibration velocity is measured and Fast Fourier Transform of the velocity is calculated to obtain the vibration spectrum. Natural frequencies (peak picking method) and mode shapes of the plate are extracted using POLYTEC software.
3. The plate is instrumented with 2 strain gauges – one in the longitudinal direction with respect to the fiber orientation and one in transversal. Wave generator, signal amplifier and dynamic strain measurement system Spider 8 are connected to the strain gauges and MFC element.
4. The plate is partitioned into 18 zones and a small weight of 20 g is consequently placed in each of these zones. During this act, the plate is harmonically excited through MFC element with an amplitude of 10 volts peak-to-peak and a driving frequency equal to that identified from modal analysis (step 2). Time series of mechanical strain from both strain gauges is recorded. A total of 3 such measurements are made by placing the mass at each of 18 zones.
5. The strain data is collected and a feature sensitive to the position of the added mass is extracted from this data. By process of trial and error, it is found that mean value over the entire time series

$$\mu_1^{SG1,Z1} = \frac{1}{n}\sum_i^n x_{i,1}^{SG1,Z1},\ \mu_2^{SG1,Z1} = \frac{1}{n}\sum_i^n x_{i,2}^{SG1,Z1},\ \mu_3^{SG1,Z1} = \frac{1}{n}\sum_i^n x_{i,3}^{SG1,Z1} \quad (6)$$

gives the best results. In equation (6) indices 1, 2 and 3 are numbers of measurement sessions, $Z1$ indicates zone 1, n is the number of samples in the timeseries. These mean values are collected from all 18 zones and both strain gauges forming a $(18 \cdot 3) \times 2$ matrix with 2 columns of predictor values and 54 rows of class (zone) labels

$$\begin{pmatrix} \{1\} & \{\mu^{SG1,Z1}\} & \{\mu^{SG2,Z1}\} \\ \dots & \dots & \dots \\ \{18\} & \{\mu^{SG1,Z18}\} & \{\mu^{SG2,Z18}\} \end{pmatrix} \quad (7)$$

IOP Conf. Series: Journal of Physics: Conf. Series **1106** (2018) 012022 doi:10.1088/1742-6596/1106/1/012022

where curvy brackets indicate vectors with 3 components due to 3 measurements per each zone. For example, $\{1\} = \{1 \quad 1 \quad 1\}^T$ and $\{\mu^{SG1,Z1}\} = \{\mu_1^{SG1,Z1} \quad \mu_2^{SG1,Z1} \quad \mu_3^{SG1,Z1}\}^T$.

6. Various classifiers from MATLAB Classifier Toolbox are tested on the matrix presented in step 5. It is found that the linear discriminant classifier gives the best class separation with an accuracy of 90.7 %. Hence, this classifier is used to train a classification model. Confusion matrix is computed to check for the misclassified class labels and to analyse strengths and weaknesses of the model for each class separately.
7. The model is validated using 5 points of mass application in the process of classifier training. The points are picked such to belong not only to one particular zone but also to lie on the boundary between two or more zones.

3. Experimental procedure

3.1 Composite plate with sensors

The photo of the clamped composite plate with installed sensors is shown in figure 2 (a). The material is prepreg carbon fiber laminate consisting of 32 layers of lamina with layer thickness of 89 μm and density of $\rho = 1560$ kg/m^3. The layer stacking sequence is $(90/0)_{8S}$. Two strain gauges are soldered on the surface as shown in figure 2 (b). Strain gauge #1 is the farthest from MFC, perpendicular to fiber orientation, while strain gauge #2, in the direction of fibres, is the closest of the two to MFC element. Thin copper tracks are glued on the surface of the plate in order to ensure the proper electrical conductivity for the system.

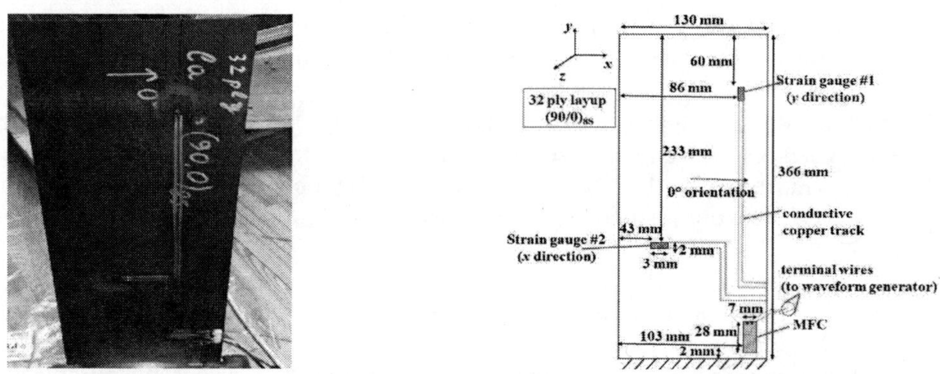

(a) (b)

Figure 2. Cantilevered composite plate: (a) photo; (b) positions of sensors with active dimensions.

3.2 Experimental modal analysis

The procedure is carried out by employing a POLYTEC PSV-400-B scanning laser vibrometer system consisting of a PSV-I-400 LR optical scanning head, OFV-5000 controller, PSV-E-400 junction box, a Bruel & Kjaer type 2732 amplifier, and a computer system with a data acquisition board and PSV software. The first step of experimental modal analysis consists of setting an outer edges and a scanning grid for the measured object. This scanning grid represents the resolution of extracted mode shapes. The size of the scanning grid is taken as 5×15 points. The plate is excited with a piezoelectric MFC (macro-fiber composite) element (model M2807-P1 smart material) glued on the surface of the plate. The excitation signal is periodic chirp with a bandwidth of 5-800 Hz and resolution of 0.25 Hz. POLYTEC system permits measuring either displacement due to vibration, vibration velocity or acceleration in the plane perpendicular to the surface of the measured object at each of the points of the scanning grid. The resonance frequencies and corresponding mode shapes are obtained by taking the magnitude of the fast Fourier transform of the vibration velocity signal.

IOP Conf. Series: Journal of Physics: Conf. Series **1106** (2018) 012022 doi:10.1088/1742-6596/1106/1/012022

3.3 Dynamic strain measurement

The dynamic strain measurement system shown in figure 3 (a) consists of a waveform generator (Agilent 3322A 20 MHz Function/Arbitrary Waveform Generator) which is connected to the MFC actuator for harmonic excitation of the plate, signal amplifier (LE 150/025 Piezomechanik GmbH signal amplifier (230 V AC, serial number: 10902/936), strain measurement acquisition box (Spider 8 600 Hz/ DC HBM with USB adapter USBHBM2903) which is connected to both strain gauges using two channels and the waveform generator. Spider 8 system is connected to personal computer through a USB port. The recorded strain signals are visualized in *Catman* software. In the software, the required measurement channels are activated and sensor parameters are set in accordance with table 1. The sensors are initially zero-balanced to start measurements from zero strain. The time series of strain from both strain gauges are measured simultaneously with a duration of 2 seconds and sampling frequency of 2400 Hz giving 4800 samples plus 1 sample at time $t = 0$ s. A total of 3 measurements are recorded for each zone of the plate. The positions 5 selected query points for validation of classification model are shown in figure 3 (b). The event of application of actual mass on one of the zones is shown in figure 3 (c).

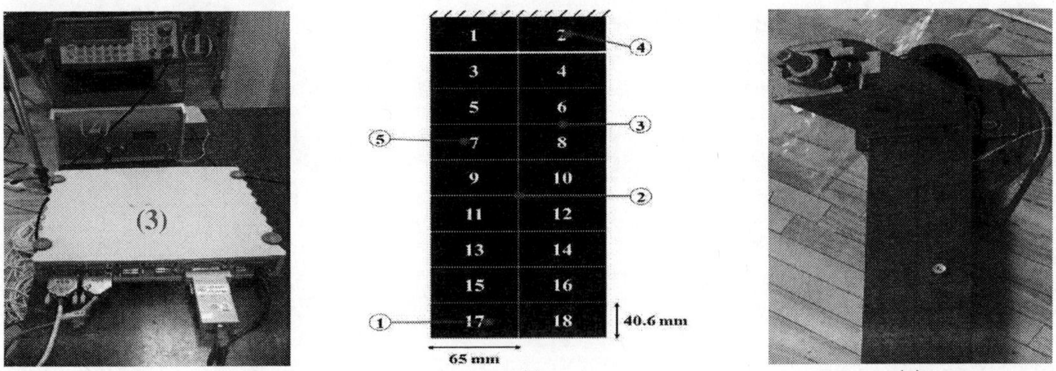

(a) (b) (c)

Figure 3. Preparation for dynamic strain measurement: (a) strain measurement equipment with (1) waveform generator, (2) signal amplifier, (3) strain acquisition system Spider 8; (b) partition of the plate into zones and application of 5 query point masses; (c) photo showing application of mass on the cantilevered composite plate.

Table 1. Strain measurement parameters.

Sampling rate	Sensor	Sensor amplifier	Transducer type	Measuring range	Gage factor	Filter	Filter frequency	Bridge factor
2400 Hz (0.42 ms)	SG 3 wire, 350 Ω	SR30 600 Hz (base)	Quarter bridge	3mV/V (4000 με)	1.99	Bessel low pass	300 Hz	1

4. Results and discussion

4.1 Natural frequencies

The measured spectrum of vibration velocity averaged over all scanning grid of the plate is shown in figure 4 (a). In the bandwidth of 800 Hz a total of 8 peaks are identified. Peak picking method is applied to extract the natural frequencies from the spectrum. Due to the limits imposed on the paper length, only the results for plate excitation using natural frequency of the fundamental bending mode (figure 4 (b)) are presented.

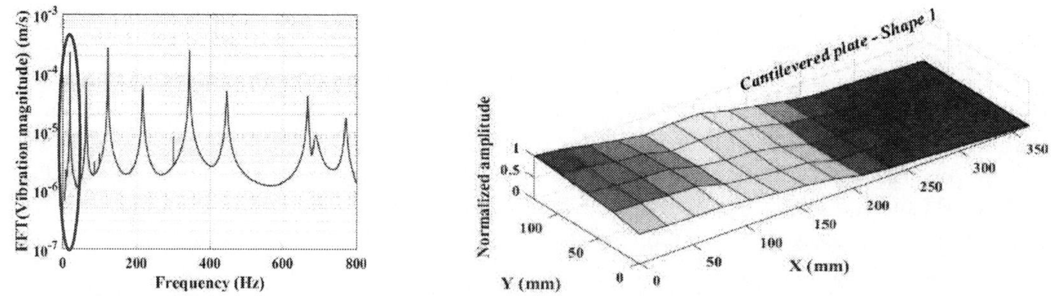

(a) (b)

Figure 4. Experimental modal analysis: (a) FFT of the vibration velocity with the first frequency marked; (b) fundamental bending mode at 19.5 Hz.

4.2 Strain features

The excitation signal with a driving frequency equal to the fundamental freuency of the cantilevered composite plate along with recorded strain timeseries for the 1st zone is shown in figure 5.

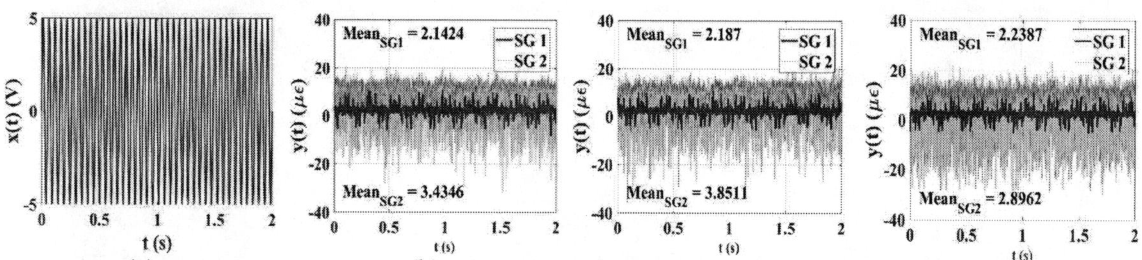

(a) (b) (c) (d)

Figure 5. Dynamic strain signals: (a) excitation signal (sine wave with an amplitude of 10 V peak-to-peak at 19.5 Hz); (b) strain timeseries #1; (c) strain timeseries #2; (c) strain timeseries #3.

As one can see, there is very small variation in mean values over the entire timeseries for 3 measurements. Also, the amplitude of strain for the strain gauge #2 is larger than for strain gauge #1. This is attributed to the fact that, firstly, strain gauge #2 is closer to the actuator, secondly, it lies in the direction of fiber orientation. These mean values are stored for the zone #1 and the experiment is repeted for the remaining zones. The final stage of the experiment involves the formation of classification matrix (refer to step 5 in Subsection 2.2).

4.3 Classification model and localization of added mass

The linear discriminant classifier is employed for training the classification model based on mean values of strains in every zone. The errors of the model are as follows: cross-validation error = 0.0926 (describes mean error of cross-validated model when predicting is based on data not used in training [14]) and resubstitution loss = 0.0741 (describes fraction of misclassifications over all the set of instances on the training data from the predictors of classification model). The number of cross-validation folds is set to the standard value of 10 folds [15]. A scatterplot of mean strain values from classification matrix, grouped by class labels (zones) is shown in figure 6 (a). Vertical dashed lines represent the class separation boundary according to the equation (1). The position of black stars in the plot represent the positions of 5 query points in the strain gauge #1 vs strain gauge #2 mean value plane.

The results of classification quality can be inspected using a confusion matrix with a row and the column for each class. Each matrix element shows the number of test samples for which the actual class is the row and the predicted class is the column. Good results correspond to large numbers down the main diagonal and, ideally, zero off-diagonal elements [15]. According to confusion matrix in figure 6 (b), 2 out of 3 observations (67 %) were classified to belong to the 2nd zone, while they actually belong to the 1st zone. 1 out of 3 observations (33 %) classified as 1st zone, while it is in 2nd zone. 1 out of 3 observations (33 %) classified as 14th zone, while it is in 10th zone. 1 out of 3 observations (33 %) classified as 10th zone, while it is in 14th zone.

The predicted probabilities for all 5 query points to belong to zones 1 to 18 are given in table 2. The query point #1 is correctly classified to lie in zone #1. According to figure 3 (b), query point #2 lies in the intersection of zones #9, 10, 11 and 12. Hence, there is an equal chance for it to be classified in one of these zones. The classifier classifies this point to belong to zone #10 (over 56 % probability) or to zone #12 (over 39 % probability). For zones #9 and 11 the probabilities are negligible. Misclassification is made for query point #3 – it is incorrectly classified to lie in zone #7, while it is on the boundary between zones #6 and 8. However, the mean strains in zones #6, 7 and 8 are, probably, very close to each other, hence this slight misclassification. For query point #4, the classifier predicts around 72 % probability for it to lie in zone #2 which is correct. The other 28 % of probability are attributed to zone #1 which is, of course, not correct but it is neighbouring zone and the strains vary just slightly. The query point #5 is correctly classified to belong to zone #5.

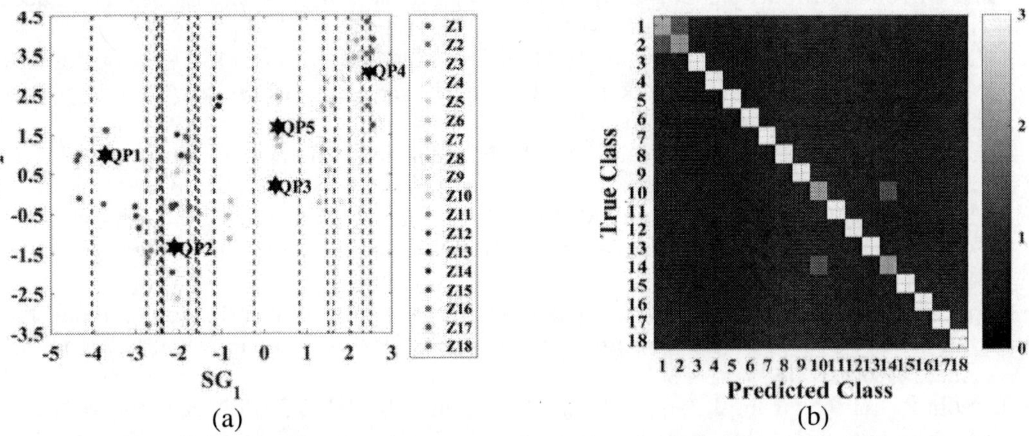

Figure 6. Linear discriminant classification results for the fundamental bending mode of the composite plate: (a) Scatterplot showing class separation; (b) confusion matrix.

Table 2. Probability of predicted labels for 5 query points (zone numbers in bold).

QP1	**1** (0)	**2** (0)	**3** (0)	**4** (0)	**5** (0)	**6** (0)	**7** (0)	**8** (0)	**9** (0)
	10 (0)	**11** (0)	**12** (0)	**13** (0)	**14** (0)	**15** (0)	**16** (0)	**17** (1)	**18** (0)
QP2	**1** (0)	**2** (0)	**3** (0)	**4** (0)	**5** (0)	**6** (0)	**7** (0)	**8** (0)	**9** (0)
	10 (0.564)	**11** (0)	**12** (0.391)	**13** (0)	**14** (0.045)	**15** (0)	**16** (0)	**17** (0)	**18** (0)
QP3	**1** (0)	**2** (0)	**3** (0)	**4** (0)	**5** (0)	**6** (0)	**7** (1)	**8** (0)	**9** (0)
	10 (0)	**11** (0)	**12** (0)	**13** (0)	**14** (0)	**15** (0)	**16** (0)	**17** (0)	**18** (0)
QP4	**1** (0.281)	**2** (0.719)	**3** (0)	**4** (0)	**5** (0)	**6** (0)	**7** (0)	**8** (0)	**9** (0)
	10 (0)	**11** (0)	**12** (0)	**13** (0)	**14** (0)	**15** (0)	**16** (0)	**17** (0)	**18** (0)
QP5	**1** (0)	**2** (0)	**3** (0)	**4** (0)	**5** (0)	**6** (0)	**7** (1)	**8** (0)	**9** (0)
	10 (0)	**11**(0)	**12**(0)	**13** (0)	**14** (0)	**15** (0)	**16** (0)	**17** (0)	**18** (0)

Conclusions

The localization of a pseudo defect on a cantilevered composite plate using strain data is demonstrated. The motivation behind this is that non-destructive approach allows for variations in defect size and other defect parameters without actually affecting the integrity of the structure. This problem is posed as one of classification – find the position of an added mass in terms of belonging of this position to one of zones of the plate.

At first, an experimental modal analysis using a scanning laser vibrometer is conducted and natural frequencies of the plate are identified. Secondly, the mass equal to 9.43 % of that of plate's mass is placed on each of plate's zones, the plate is excited harmonically with the frequency equal to the fundamental natural frequency of bending and strains measured by two strain gauges (one parallel to the orientation of fibres and other in the perpendicular direction) are recorded in the form of time series. The next step is to extract the position-sensitive strain features from the time series. In this case, simply the mean value over entire time series proved to be an efficient feature yielding classification accuracy of 90.7 %. Finally, the trained classifier is validated by classifying 5 different query points of mass application.

The proposed localization methodology is simple and gives promising results as all of the query points are localized with acceptable accuracy. Further research would deal with exploring the sensitivity of the approach with decreasing the added mass.

Acknowledgements
This work has been supported by the doctoral research grant at Riga Technical University.

References

[1] Mironov A and Mironovs D 2017 Experimental application of OMA solutions on the model of industrial structure *IOP Conf. Ser.-Mat. Sci.* **251** 012092.

[2] Tatarinov A, Barkanov E N, Davydov E and Mihovski M 2018 T- and L-Types of Long Range Guided Waves for Defect Detection – In: *Non-Destructive Testing and Repair of Pipelines*, Springer International Publishing 15-29.

[3] Ginzburg D, Pinto F, Iervolino O and Meo M 2017 Damage tolerance of bio-inspired helicoidal composites under low velocity impact *Compos. Struct.* **161** 187-203.

[4] Stelldinger E, Kühhorn A and Kober M 2016 Experimental evaluation of the low-velocity impact damage resistance of CFRP tubes with integrated rubber layer *Compos. Struct.* **139** 30-35.

[5] Topac T O, Gozluklu B, Gurses E and Coker D 2017 Experimental and computational study of the damage process in CFRP composite beams under low-velocity impact *Compos. Part A-Appl. S.* **92** 167-182.

[6] Toivola R, Lai P N, Yang J, Jang S H, Jen A K Y and Flinn B D 2017 Mechanochromic fluorescence in epoxy as a detection method for barely visible impact damage in CFRP composites *Compos. Sci. Technol.* **139** 74-82.

[7] Mechbal N, Uribe J S Uribe and Rébillat M 2015 A probabilistic multi-class classifier for structural health monitoring *Mech. Syst. Signal Pr.* **60-61** 106-123.

[8] Gaudenzi P, Nardi D, Chiapetta I, Atek S, Lampani L, Sarasini F, Tirillò J, Valente T 2015 Impact
damage detection in composite laminate plates using an integrated piezoelectric sensor and actuator couple combined with wavelet based features extraction approach *Proceedings of the 7th ECCOMAS thematic conference*, 2015.

[9] Vitola J, Pozo F, Tibaduiza D A and Anaya M 2017 A sensor data fusion system based on k-nearest neighbor pattern classification for structural health monitoring applications *Sensors* **17** 417-443.

[10] Vitola J, Pozo F, Tibaduiza D A and Anaya M 2017 Distributed piezoelectric sensor system for damage identification in structures subjected to temperature changes *Sensors* **17** 1252-1278.

[11] Janeliukstis R, Ruchevskis S and Chate A 2018 Classification model for damage localization in a plate structure *Mech. Compos. Mater.* **54**(6) 725-736.

[12] https://www.mathworks.com/help/stats/creating-discriminant-analysis-model.html

[13] https://www.mathworks.com/help/stats/fitcdiscr.html

[14] https://uk.mathworks.com/help/stats/classificationensemble.resubloss.html

[15] Witten I H, Frank E and Hall M A 2011 Data Mining: Practical Machine Learning Tools and Techniques, Third edition, Morgan Kaufmann Publishers.

IOP Conf. Series: Journal of Physics: Conf. Series **1106** (2018) 012023 doi:10.1088/1742-6596/1106/1/012023

Structural modification based analysis and design of tuned mass dampers

Andreas Kyprianou

Department of Mechanical and Manufacturing Engineering, University of Cyprus, , Cyprus

E-mail: akyp@ucy.ac.cy

Abstract. This article discusses the design of tuned vibration absorber for structures. A central notion in the identification of the parameters of a tuned mass damper (TMD) absorber for MDOF systems is that of modal mass of the original structure. The modal mass, as it is well known, is associated with the mode of vibration of the original structure whose frequency is required to be absorbed. The estimation of the modal mass can be achieved by either using a mass, stiffness, and damping model of the structure or, for an existing in service structure, from experimental modal analysis. In both of the above cases inaccuracies are introduced due to modeling errors and experimental noise. This article sets the theory of TMD in the context of structural modification. This enables the design of TMD by using only the frequency response function of the primary system without taking into consideration the original mass or modal mass. The theory developed herein is for one degree of freedom undamped primary vibrating system. The proposed structural modification based methodolgy of designing TMD is demonstrated on a specific single degree of freedom vibrating system.

1. Introduction

Vibration absorbers are one degree of freedom mass, spring, damper (MSD) systems that are attached on structures that vibrate excessively when subjected to external dynamic loads. The effective mitigation of excessive vibrations is normally achieved by tuning the vibration absorber to the frequency of these vibrations. The basic theory that underlies the vibration absorption of undamped one degree of freedom systems using MSD absorbers is well developed and reported in many textbooks of vibration theory [1].

The MSD absorber theory was subsequently extended to take into consideration (a) the effect of damping in the primary structure on the design of MSD absorbers [2, 3] and (b) the difficulties of using them to more complicated structures such as beams [4], plates, shells [5] and cylinders [6]. An outcome of the theory extension to more complicated structures is that the modal (and effective) mass are essential in designing vibration absorbers [7, 8] . However, estimation of the modal and effective mass for structures in service requires experimental modal analysis [9] which, especially for large and complicated structures, could be expensive and cumbersome [10].

This paper, through the theory of structural modification, extends the original theory of designing MSD absorber systems in order to overcome the difficulties involved when they are used on more complicated structures. The theory herein is developed for the undamped single degree of freedom system. The only requirement in designing the optimal MSD absorber is the frequency response function of the primary system. The actual values of the primary mass and spring constant are not required. It is expected that the extension of this theory to multi-degree of freedom systems will require only the measured frequency response function at the point of the attachment of the MSD absorber. Section 2 outlines the relevant structure modification theory whereas in section 3 this theory is exploited to find the optimal MSD. Finally section 4 summarises the conclusions of the work presented in this paper.

2. Structural modification theory in the context of spring mass damper absorber

This section develops a methodology of designing MSD absorbers that is based on the theory of structural modification. The advantage of the proposed methodology is that it requires only measured FRFs on the primary structure. The objective in mind is to design a MSD absorber to eliminate unwanted vibrations of a lightly damped structure excited at a single mode and, possibly, the frequency of excitation varies within a range around the resonance corresponding to the mode. Hence the primary structure can be thought as an undamped single degree of freedom system as shown in figure 1 of mass m and stiffness k. The MSD absorber of mass, m_a, spring, k_a and damper c_a is attached as shown in figure 2. In subsection 2.1 the relevant theory of structural modification in the context of MSD is presented and in subsection 2.2 the theory needed to develop the methodology is discussed.

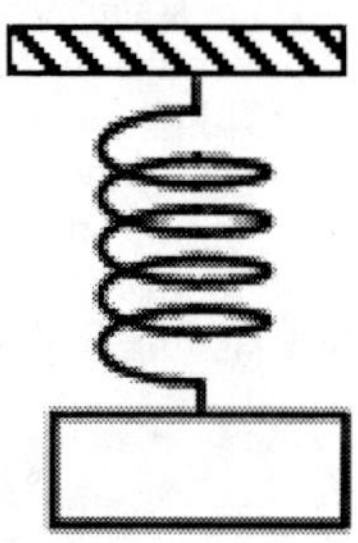

Figure 1. The single degree of freedom primary system consisting of spring of stiffness k and mass m.

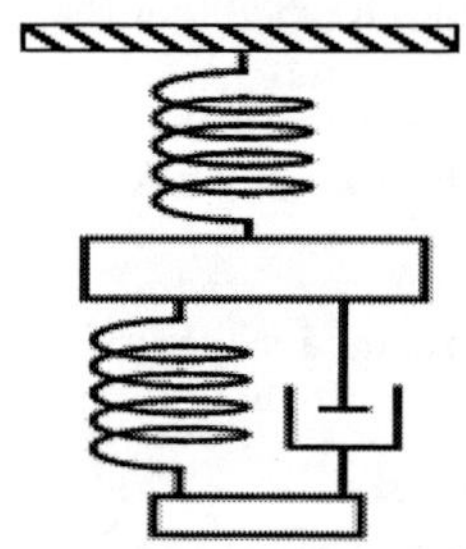

Figure 2. Mass, m_a, spring k_a, damper, c, absorber attached to the primary system of figure 1.

2.1. Theory of structural modification

The dynamics of a structure change due to modifications of its mass, stiffness and damping properties. The theory of structural modification deals with all aspects involved when such changes occur. In general, structural modification is separated to direct and inverse. In the direct modification, the structural modifications are known and the aim is to find the changes in the dynamic behaviour whereas in the inverse modification where the aim is to design the appropriate passive mass, damping, and stiffness modifications which bring about the desired dynamic behavior [11, 12, 13].

The vibration absorbtion problem can be viewed as a structural modification of the primary structure, shown in figure 1, by the the MSD absorber of figure 2. This modification involves the addition of a mass, m_a, at a new degree of freedom, and a spring of stiffness k_a and damper of damping coefficient c connecting the mass at the new degree of freedom with the original structure. The theory developed in [11, 13], by introducing the effect of damping c, can be readily extended to obtain the modified point frequency response function, $h^{mod}(\omega)$, at m from the original point FRF, $h^{or}(\omega)$, as follows

$$h^{mod}(\omega) = \frac{h^{or}(\omega)\left(-\omega^2 m_a + k_a + j\omega c\right)}{-\omega^2 m_a + k_a + j\omega c - \left(\omega^2 m_a k_a + j\omega^3 m_a c\right) h^{or}(\omega)} \quad (1)$$

where m_a, k_a and c are the mass, spring constant and the damping of the MSD absorber, ω the forcing frequency and j the imaginary unit. This equation demonstrates how the original frequency response function is affected by modifying the original structure by a mass, spring and damper absorber and the subsequent development of a design methodology for finding the optimal vibration absorber is totally relied on it.

2.2. Vibration absorber theory based on structural modifation

The vibration absorber theory states that the optimal m_a and k_a parameters of an undamped absorber, to be attached on a structure excited at a frequency ω close to one of its natural frequencies, are those that render the natural frequency of the vibration absorber $\omega_{n_a} = \sqrt{\frac{k_a}{m_a}}$ equal to the forcing frequency . This can be also seen by first setting the damping, c, in equation 1, to 0 and then by rendering the numerator zero. In practice when the mass (or the effective mass) of the original structure is known the mass of the absorber m_a is chosen to be a fraction of the original (effective) mass and then the stiffness k_a is selected to tune the absorber's natural frequency to the desired frequency.

In order to facilitate the development the mass and spring constant of the primary system, figure 1, are taken to be $m = 30$ Kg and $k = 300\ \frac{\text{kN}}{\text{m}}$ and hence its natural frequency is $100\ \frac{\text{rad}}{\text{s}}$. The magnitude of the actual frequency response function of this system is shown in figure 3. According to the classical theory of vibration absorbers the natural frequency of the undamped vibration absorber ω_{n_a} should be $100\ \frac{\text{rad}}{\text{s}}$. Choosing the absorber mass m_a to be 10 Kg the spring constant of the absorber becomes $100\ \frac{\text{kN}}{\text{m}}$. Setting $c = 0$ in equation 1 the modified FRF is computed and its magnitude is plotted in figure 4.

The two new resonance frequencies observed in figure 4 can be obtained from the original FRF $h^{or}(\omega)$ by setting the damping and the denominator of equation 1 to zero. By doing so, it is found that the new resonance frequencies should satisfy the following equalities,

$$\|h^{or}\|(\omega) = \pm\frac{-\omega^2 m + k}{\omega^2 mk} \tag{2}$$

where $\|.\|$ denotes the absolute if the argument is real or the magnitude if the argument is complex. A graphical solution is given in figure 3 in where the dashed and the dash-dotted curves correspond to the positive and the negative function respectively given in equation 2. The newly introduced natural frequencies are the abscissas of the intersections of these two curves with the magnitude, $\|h_{or}\|(\omega)$, of the original frequency response function. It can be seen from figure 3 that these values are $\Omega_1 = 75.21\ \frac{\text{rad}}{\text{s}}$ and $\Omega_2 = 135.46\ \frac{\text{rad}}{\text{s}}$

Damping is introduced in the absorber in order to reduce the amplitude vibrations that could occur when the frequency band of the excitation contains frequencies that are close to the newly introduced resonances. Figure 4 depicts the effect of various values of damping, namely $c = 250$, 500 and 750 $\frac{Nm}{s}$ on the magnitude of the modified frequency response function. It can be concluded that, even though damping reduces the resonance, high vibration aplitudes that extend over a frequency band might occur. As a consequence an optimum value of damping should be sought. In the classical literature this is done by observing that all curves of the amplitudes of the modified frequency response functions, figure 4, intersect at two specific frequency values, say ω_k and ω_l. Hence the FRF magnitudes at ω_k and ω_l are independent of the damping.

The values of these frequencies can be obtained from the original frequency response function $h^{or}(\omega)$ utilizing the structural modification theory. Since the FRF values at ω_k and ω_l are independent of the damping c the magnitudes of the modified frequency response functions for $c = 0$ given by,

IOP Conf. Series: Journal of Physics: Conf. Series **1106** (2018) 012023 doi:10.1088/1742-6596/1106/1/012023

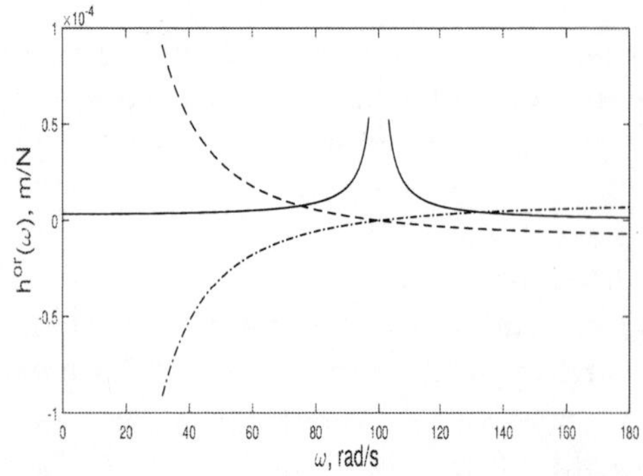

Figure 3. ——Frequency response function of the primary system for $k = 300\ \frac{\mathrm{kN}}{\mathrm{m}}$ and m=30 Kg. - - -The positive function of equation 2 — · —The negative function of equation 2

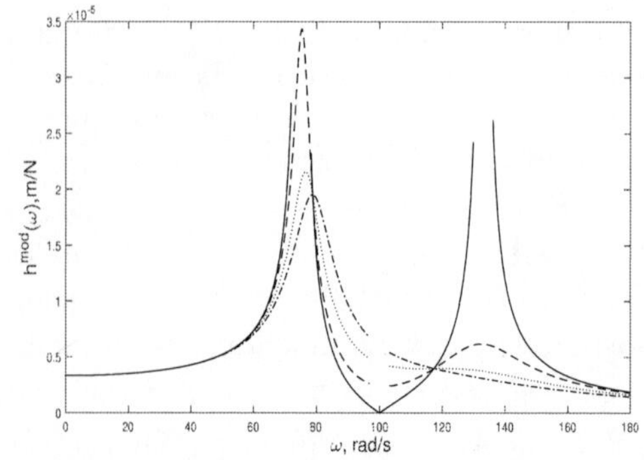

Figure 4. Modified frequency response function ——Undamped vibration absorber, $m_a = 10$ Kg and k_a 100 $\frac{\mathrm{kN}}{\mathrm{m}}$ and damped for c - - -250 ······, 500 and — · —750 $\frac{Nm}{s}$.

$$h^{mod}_{c=0}(\omega) = \frac{h^{or}(\omega)\left(-\omega^2 m_a + k_a\right)}{-\omega^2 m_a + k_a - \left(\omega^2 m_a k_a\right) h^{or}(\omega)} \tag{3}$$

and $c = \infty$ given by,

$$h^{mod}_{c=\infty}(\omega) = \frac{h^{or}(\omega)}{1 - \omega^2 m_a h^{or}(\omega)} \tag{4}$$

can be set equal, i.e. at ω_1 and ω_2 $\|h^{mod}_{c=0}\|(\omega) = \|h^{mod}_{c=\infty}\|(\omega)$. The magnitudes of the modified FRFs are depicted in figure 5 and as it can be seen they intesect at ω_k and ω_l. Expanding the equality $\|h^{mod}_{c=0}\|(\omega) = \|h^{mod}_{c=\infty}\|(\omega)$ and carrying out the subsequent algebraic manipulations it can be shown that at ω_k and ω_l are obtained by solving,

$$h^{or}(\omega) = \pm\frac{2\left(\omega^2 m_a + k_a\right)}{\omega^2\left(\omega^2 m_a + 2 m_a k_a\right)} \tag{5}$$

The graphical solution of this equation is shown in figure 6 where the intersections oocur at $\omega_k = 78.85\frac{\mathrm{rad}}{\mathrm{s}}$ and $\omega_l = 117.37\frac{\mathrm{rad}}{\mathrm{s}}$.

3. Optimal mass spring damper parameters: methodology and results

The optimal MSD absorber parameters are obtained by first finding the absorber mass, m_a, and stiffness, k_a that render the FRF maginitudes at ω_k and ω_l equal. Then the optimal damping is obtained by ensuring that the equal FRF magnitudes at ω_l and ω_k are also the maximum values over the excitation frequency range.

3.1. Optimal absorber mass and stiffness

It is assumed that absorber mass m_a is kept at $10Kg$ and thereby the absrorber spring constant k_a that would render the FRF magnitudes at ω_k and ω_l equal must be found. This can be achieved by setting $\|h^{mod}_{c=0}\|(\omega_1) = \|h^{mod}_{c=0}\|(\omega_2)$ which is equivalent to $h^{mod}_{c=0}(\omega_1) = -h^{mod}_{c=0}(\omega_2)$. Setting Expanding this equality a quadratic polynomial , $ak_a^2 + bk_a^2 + c$, is obtained. Its coefficients and constant term are respectively given by,

$$a = h^{or}(\omega_k) - h^{or}(\omega_l) + \left(\omega_k^2 - \omega_l^2\right) m_a h^{or}(\omega_k)\, h^{or}(\omega_l) \tag{6}$$

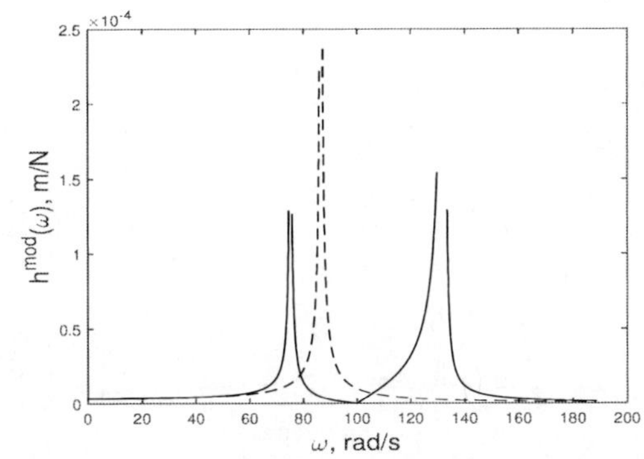

Figure 5. Modified frequency response functions: ——$c = 0$ and - - -$c = \infty$

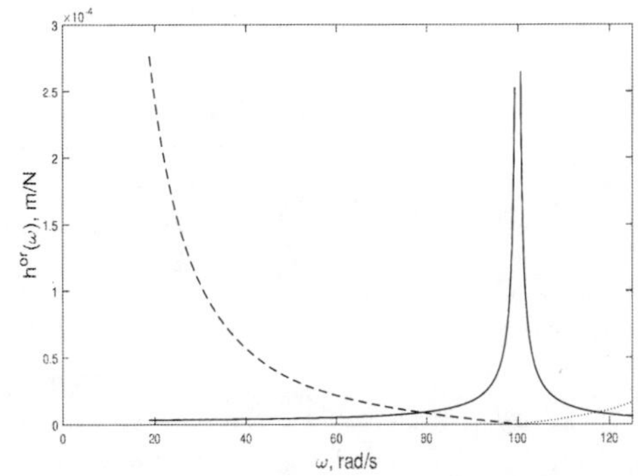

Figure 6. ——Original frequency response function - - -The positive function of equation 5 ······The negative function of equation 5.

$$b = m_a\left(\omega_l^2 - \omega_k^2\right)h^{or}(\omega_l) - \left(\omega_l^2 + \omega_k^2\right)h^{or}(\omega_k) \tag{7}$$

$$b = m_a\omega_k^2\omega_l^2\left(h^{or}(\omega_k) + h^{or}(\omega_l)\right) \tag{8}$$

Substituting the values $\omega_k = 78.85\frac{\text{rad}}{\text{s}}$, $\omega_l = 117.37\frac{\text{rad}}{\text{s}}$ and $m_a = 10$ Kg in 6, 7 and 8 and solving the quadratic polynomial the new absorber value is computed to be $k_a = 36007\frac{\text{N}}{\text{m}}$. The magnitude of the modified frequency response functions of the absorber with the optimal mass, m_a and spring constant k_a values for various damping values, namely, $c = 250$, 500 and 750 $\frac{Nm}{s}$ are shown in figure 7.

Since the optimal absorber mass and spring constant changed the values of the resonance frequencies, ω_1 and ω_2 and of the frequencies, ω_k and ω_l at which the FRF magnitude is idependent of the damping changed. Their values can be computed from the original FRF using the techniques described in section 2.2. The values obtained by doing this are $\omega_1 = 55.42$, $\omega_2 = 108.32$, $\omega_k = 57.24$ and $\omega_l = 96.95\frac{\text{rad}}{\text{s}}$. These values are also seen in figure 7.

3.2. Optimal absorber damping

The optimal damping value, as it is stated at the beginning of section 2.2 is the value that maximizes the magnitude of the modified FRF,$h^{mod}(\omega)$ given by equation 1 at the frequencies ω_k and ω_l. The magnitude of $\|h^{mod}\|(\omega)$ is given by,

$$\|h^{mod}\|(\omega) = \sqrt{\frac{N_R^2 + N_I^2}{D_R^2 + D_I^2}} \tag{9}$$

where N_R, D_R, N_I and D_I are the real and imaginary parts of the numerator and denominator of $h^{mod}(\omega)$, equation 1, and are given by,

$$N_R(\omega) = \left(-\omega^2 m_a + k_a\right)h^{or}(\omega) \tag{10}$$

$$N_I(\omega) = \omega c h^{or}(\omega) \tag{11}$$

$$D_R(\omega) = -\omega^2 m_a + k_a - \omega^2 m_a k_a h^{or}(\omega) \tag{12}$$

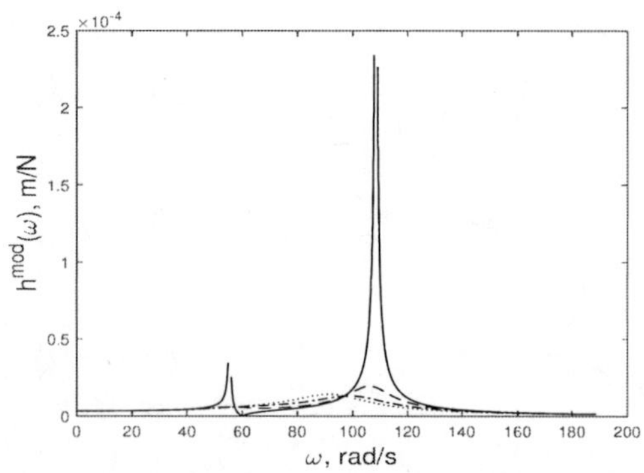

Figure 7. Modified frequency response function ——Undamped vibration absorber, $m_a = 10$ Kg and k_a 36007 $\frac{\text{N}}{\text{m}}$ and damped for c, – – –250, · · · · · ·500 and — · —750 $\frac{Nm}{s}$.

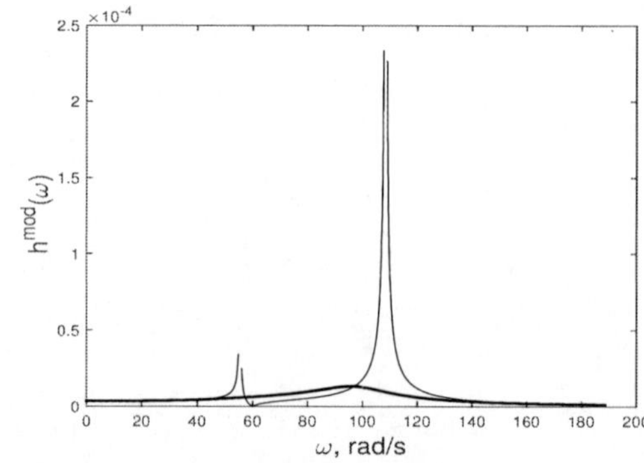

Figure 8. Overlay of the undamped and the damped by the optimum modified frequency response function.

$$D_I(\omega) = c\left(\omega^3 m_a h^{or}(\omega) + \omega\right) \tag{13}$$

Therefore the optimal damping renders the derivative, $\frac{d\|h^{mod}\|^2(\omega)}{d\omega}$ zero at ω_k and ω_l.

The derivative $\frac{d\|h^{mod}\|^2(\omega)}{d\omega}$ is zero when the following quantity becomes zero,

$$2\left(D_R^2(\omega) + D_I^2(\omega)\right)\left(N_R(\omega)\frac{dN_R(\omega)}{d\omega} + N_I(\omega)\frac{dN_I(\omega)}{d\omega}\right)$$

$$-2\left(N_R^2(\omega) + N_I^2(\omega)\right)\left(D_R(\omega)\frac{dD_R(\omega)}{d\omega} + D_I(\omega)\frac{dD_I(\omega)}{d\omega}\right) \tag{14}$$

where

$$\frac{dN_R(\omega)}{d\omega} = -2\omega m_a h^{or}(\omega) + \left(-\omega^2 m_a + k_a\right)\frac{dh^{or}(\omega)}{d\omega} \tag{15}$$

$$\frac{dN_I(\omega)}{d\omega} = c\left(h^{or}(\omega) + \omega\frac{dh^{or}(\omega)}{d\omega}\right) \tag{16}$$

$$\frac{dD_R(\omega)}{d\omega} = -2\omega m_a - 2\omega m_a k_a h^{or}(\omega) - \omega^2 m_a k_a \frac{dh^{or}(\omega)}{d\omega} \tag{17}$$

$$\frac{dD_I(\omega)}{d\omega} = c\left(3\omega^2 m_a h^{or}(\omega) + \omega^3 m_a \frac{dh^{or}(\omega)}{d\omega} + 1\right) \tag{18}$$

Evaluating expression 14 at ω_l and ω_k two quartic polynomials in c are obtained whose one of their factors yields the optimal damping value. The value obtained for the parameters of this example is 588.281 $\frac{\text{Ns}}{\text{m}}$ and occurs for the $\omega_l = 96.95\frac{\text{rad}}{\text{s}}$ value. This value along with $m_a = 20$ Kg and $k_a = 36007\frac{\text{N}}{\text{m}}$ constitute the optimal values of the MSD absorber. The curve of the modified frequency response function corresponding to the optimum damping value is shown in figure 8.

IOP Conf. Series: Journal of Physics: Conf. Series **1106** (2018) 012023 doi:10.1088/1742-6596/1106/1/012023

4. Conclusions

In this article the structural modification theory was employed to extend the theory of TMDs. It was shown that using this theory the design of damped dynamic absorbers can be achieved by using only the measured frequency response function on the original structure. Structural modification provided the means (a) to compute the newly created resonance frequencies and the invariant frequencies at once, (b) to retune the MSD absorber when damping in the absorber is introduced and (c) compute the optimal damping of the retuned MSD. The only information required to achieve this was the frequency response function of the primary structure. This work will be extented to lightly damped primary structures, multi degree of freedom systems and continuum systems.

References

[1] Den Hartog J P 1985 *Mechanical Vibrations* (New York: Dover Publications, Inc).
[2] loi T and Ikeda K 1978 On the dynamic damped absorber of the vibration system *Bull. JSME* **21** 64-71
[3] Bapat V A and Kumaraswamy H V 1979 Effect of primary system damping on the optimum design of an untuned viscous dynamic vibration absorber *J. Sound Vib.* **63** 469 - 74
[4] Jacquot R G 1978 Optimal vibration absorbers for general beam systems *J. Sound Vib.* **60** 535-42
[5] Warburton G B and Ayorindes E O 1980 Optimum absorber parameters for simple systems *Earthq. Eng. Struct. D* **8** 197-217.
[6] Warburton G B and Ayorindes E O 1980 Minimizing structural vibrations with absorbers *Earthq. Eng. Struct. D* **8** 219-36.
[7] Huang J-S, Lee S-H, Kareem A 2006 Estimation of the modal mass of a structure with a tuned-mass damper using H-infinity optimal model reduction *Eng. Struct.* **28** 149-56.
[8] Huang J-S, Lee S-H, Kareem A 2015 Estimation of modal masses of a structure with a mass-type damping device *Eng. Struct.* **99** 149-56.
[9] Brownjohn J M W and Pavic A 2007 Experimental methods for estimating modal mass in footbridges using human-induced dynamic excitation *Eng. Struct.* **29** 2833 - 43.
[10] Cara J 2016 Computing the modal mass from the state space model in combined experimental operational modal analysis *J. Sound Vib.* **370** 94-110.
[11] Kyprianou A, Mottershead J E and Ouyang H 2004 Assignment of natural frequencies by an added mass and one or more springs *Mech. Sys. Signal Pr.* **18** 263 - 89
[12] Mottershead J E, Kyprianou A and Ouyang H 2005 Structural Modification. Part 1: rotational receptances *J. Sound Vib.* **284** 249 - 65
[13] Kyprianou A, Mottershead J E and Ouyang H 2005 Structural Modification. Part 2: Assignment of natural frequencies and antiresonances by an added beam *J. Sound Vib.* **284** 267 - 81

IOP Conf. Series: Journal of Physics: Conf. Series **1106** (2018) 012024 doi:10.1088/1742-6596/1106/1/012024

Nonlinear model reduction for control of an active constrained layer damper

C H Meyer[1], K Holeczek[2], M Meschenmoser[1] and C Lerch[3]

[1] Technical University of Munich, Department of Mechanical Engineering, Chair of Applied Mechanics, Boltzmannstraße 15, 85748 Garching b. München, Germany

[2] Technische Universität Dresden, Institute of Lightweight Engineering and Polymer Technology, Holbeinstraße 3, 01307 Dresden, Germany

[3] Technical University of Munich, Department of Mechanical Engineering, Chair of Automatic Control, Boltzmannstraße 15, 85748 Garching b. München, Germany

E-mail: christian.meyer@tum.de, klaudiusz.holeczek@tu-dresden.de

Abstract. One technique to reduce vibrations of low-damped structures is the application of constrained layer damping (CLD) treatments. A unique extension of the CLD is a compressible constrained layer damping (CCLD) where the usually stiff viscoelastic layer is replaced by a compressible one. Due to the adaptable deformation kinematics, as well as compression-dependent material properties, the dynamic behavior of the whole considered structure can be optimized during the operation. One challenge for this application is to determine the optimal compression levels in order to adapt the dynamic behaviour of the component with CCLD to the excitation characteristics. Through the application of finite element model such optimal control strategies can be extensively studied and the dynamics of actuation principle can be included. Herein, low computation times are highly desired, especially for applications where the CCLD undergoes large deformations. This contribution shows how nonlinear model reduction can be applied to accelerate the solution of such finite element models.

1. Introduction and motivation

In order to avoid function impairing or even damage relevant vibrations, efficient solutions for vibration mitigation are required especially for lightweight applications [1]. Besides material-intrinsic design properties, conventional post-manufacturing measures such as supplementary applied noise-absorbent mats as well as different kinds of structure-attached fluidic dampers or tuned mass dampers are used. Such modifications are however often considered as counterproductive with regard to maintain lightweight properties of the manufactured component. Instead, a number of lightweight-compatible solutions for vibration mitigating have been developed [2, 3].

One technique to reduce vibrations of low-damped structures is the application of constrained layer damping (CLD) treatments [4]. Such treatments consist of overall three elements: the main component to be damped, a damping layer made of a material with high energy dissipation properties, and a constraining layer that forces the damping layer to undergo shear deformations rather than bending to induce higher stress levels and hence better energy dissipation. A unique extension of the constrained layer damping consists of the replacement of the incompressible viscoelastic layer of the classical CLD with a compressible one – usually open-cell foam – whose damping properties can be adjusted through thickness adjustment. The thickness changing actuation principle of such compressible CLD (CCLD) bases on structural cavities generating evanescent deformations when supplied with fluidic medium.

However, since the damping properties of the constrained layer damper depend on the excitation frequency and compression stage of the material, the CCLD has to be actuated on-line in order to achieve improvement of the dynamic behaviour. One challenge to set up the optimal compression during operation is the determination of the optimal pressure in function of excitation characteristics like frequency or amplitude. This can be achieved either based on results of experimental investigations or numerical ones by using a finite element model. Experiments can be very cost-intensive because they have to be done for each new CCLD design. A computation of the optimal compression by using a finite element model is usually cheaper but still can take long simulation time due to high dimension of the model and nonlinearities considering potentially large deformations of the structure. Hence, for such applications, it is crucial to significantly reduce the computation time. This contribution shows how nonlinear model reduction can be used in order to speed up design and allow to infer appropriate design parameters for the structure and the control system.

2. Operating principle of compressible constrained layer damping treatment

The general overview of the proposed Compressible Constrained Layer Damping (CCLD) treatment is presented in Figure 1. The analyzed CCLD object is configured as a three layered beam consisting of the load-bearing structure as well as constraining and compressible viscoelastic layer. The unique actuating principle is based on structural cavities generating evanescent deformations when supplied with hydraulic fluid, compressed air or vacuum. The last possibility was used in combination with an open-cell foam architecture ensuring a low bulk modulus and thus a broad deformation range under atmospheric pressure. The evanescent morphing is used to deliberately alter the geometrical and material properties of the viscoelastic elements through compression in order to achieve a damping capacity adaptation. With the decreasing viscoelastic layer thickness, an increase of its damping and stiffness is expected. At the same time, the shear strain becomes higher at a given vibration amplitude. The change in geometry can also affect the structure's stiffness. The combination of these effects should broaden the range of CCLD-adaptation through the compression level setting. The studies regarding experimental characterization of the compression-dependent viscoelastic shear properties are not the main part of these studies and will be presented elsewhere.

In order to maintain optimal cavity pressure for any realistic configuration of excitation frequency and amplitude, a closed-loop control system is necessary. The structure and parameters of such control systems should be matched to the dynamics of the CCLD treatment for any potentially suitable CCLD configuration. Since the application of a finite-element-model with appropriate real-time capabilities could be used for later hardware-in-the-loop tests of the control system a suitable simulation model was developed.

3. Modeling

A finite element model for simulation of the dynamic response and damping properties of the adaptive constrained layer damper is created. Its geometry is a square plate with a side length of 550 mm. The plate consists of 5 layers as illustrated in Figure 2 in order to introduce some kind of symmetric dynamic behavior. Their material properties are summarized in Figure 2 on the right.

The geometry is meshed with quadratic Hexaeder elements with 20 nodes each. The boundaries of the constrained layer damper are fixed in each translational direction. The structure is assumed to be excited at one node of the outer-layer in out-of-plane direction.

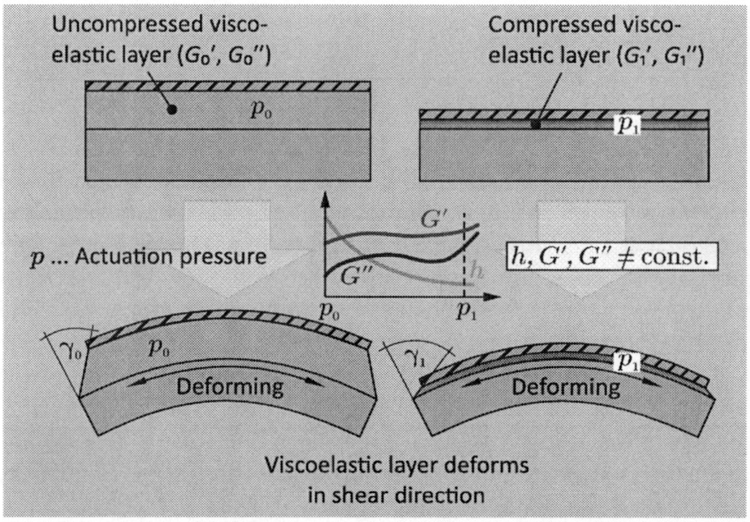

Figure 1. Cross section view of CCLD. The storage and loss modulus of the viscoelastic layer as well as the thickness of the viscolelastic layer depend on the applied pressure.

The full model has 556 308 degrees of freedom. A time-integration requires high computation times due to the geometric nonlinearity and the high number of degrees of freedom. We apply model reduction to reduce computation time.

4. Nonlinear model reduction

The outcoming equation of motion of the Finite Element model can be described by

$$\mathbf{M}\ddot{\mathbf{u}}(t) + \mathbf{D}\dot{\mathbf{u}}(t) + \mathbf{f}_{\mathrm{nl}}(\mathbf{u}(t)) = \mathbf{B}\,\mathbf{f}_{\mathrm{ext}}(t) \ , \tag{1}$$

where $\mathbf{u}(t)$ are the displacements of the nodes, $\mathbf{M}$ is the mass matrix, $\mathbf{D}$ the viscous damping matrix, $\mathbf{f}_{\mathrm{nl}}$ the nonlinear restoring force and $\mathbf{B}$ the matrix for distributing the external forces $\mathbf{f}_{\mathrm{ext}}$. The viscous damping matrix is modeled as

$$\mathbf{D} = a\mathbf{M} + b\left.\frac{\partial \mathbf{f}_{\mathrm{nl}}}{\partial \mathbf{u}}\right|_{\mathbf{u}=0} \ , \tag{2}$$

where the coefficients a and b are chosen in such way that the resonance peaks of the first and fourth eigenfrequencies of the linearized frequency response function at the excitation point match the measured damping ratios.

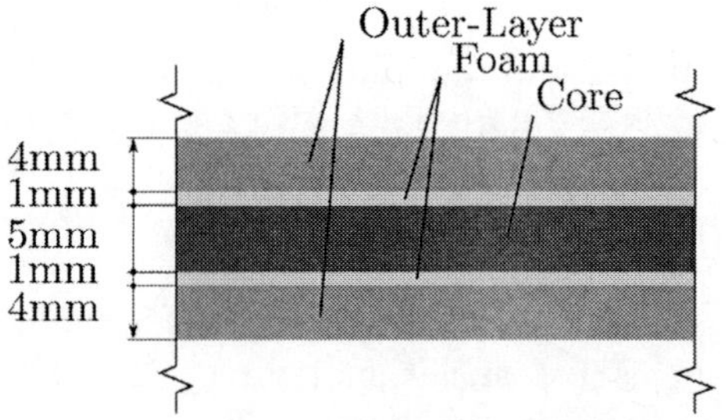

Layer	E	ν
Core	85 MPa	0.1
Foam	0.1 MPa	0.0
Outer-Layer	23 GPa[1]	0.3

[1] Orthotropic material, Young's modulus is given in direction of fibers

Figure 2. Left: Lay-up, Right: Material properties.

To reduce computation time for solving equation (1), the displacements $\mathbf{u}(t)$ are approximated by $\mathbf{u} \approx \mathbf{V}\mathbf{q}(t)$ with $\mathbf{V} \in \mathbb{R}^{N\times n}$ and $n \ll N$. Inserting this approximation into (1) and applying a Galerkin projection leads to the reduced equation of motion

$$\mathbf{V}^\mathsf{T}\mathbf{M}\mathbf{V}\ddot{\mathbf{q}} + \mathbf{V}^\mathsf{T}\mathbf{D}\mathbf{V}\dot{\mathbf{q}} + \mathbf{V}^\mathsf{T}\mathbf{f}_{\text{nl}}(\mathbf{V}\mathbf{q}) = \mathbf{V}^\mathsf{T}\mathbf{B}\,\mathbf{f}_{\text{ext}}\ . \tag{3}$$

This projection is similar to linear model reduction techniques (cf. [5, 6]). It reduces the dimension of the problem from N to n. However, no matter how small one chooses the reduced dimension n, the mere dimensional reduction does not reduce computation time because the nonlinear restoring force term $\mathbf{f}_{\text{nl}}$ still needs to be evaluated in the full element domain. It is computed by evaluating the element forces $\mathbf{f}_e$ for each element of the mesh and assembling

$$\mathbf{V}^\mathsf{T}\mathbf{f}_{\text{nl}} = \sum_{e\in E} \mathbf{V}^\mathsf{T}\mathbf{L}_e^\mathsf{T}\mathbf{f}_e(\mathbf{L}_e\mathbf{V}\mathbf{q})\ , \tag{4}$$

where $\mathbf{L}_e$ is a Boolean operator that maps the local degrees of freedom of element e to the global ones.

To accelerate the evaluation of the nonlinear term, a hyperreduction must be performed. Within this study, we use the Energy Conserving Sampling and Weighting method (ECSW) [7] for hyperreduction. This method only evaluates a subset of all element forces and assembles $\mathbf{V}^T\mathbf{f}_{\text{nl}}$ by weighting their contributions with a positive scalar $\xi_e^\star$, such that

$$\mathbf{V}^\mathsf{T}\mathbf{f}_{\text{nl}}(\mathbf{V}\mathbf{q}(t)) \approx \sum_{e\in\tilde{E}\subset E} \xi_e^\star \mathbf{V}^\mathsf{T}\mathbf{L}_e^\mathsf{T}\mathbf{f}_e(\mathbf{L}_e\mathbf{V}\mathbf{q})\ . \tag{5}$$

The weights $\xi_e^\star$ and element subset $\tilde{E}$ are chosen such that the hyperreduced nonlinear force term matches the projected nonlinear force vector $\mathbf{V}^\mathsf{T}\mathbf{f}_{\text{nl}}(\mathbf{V}\mathbf{q})$ within a certain error bound for N_τ training displacements $\mathbf{u}_\tau = \mathbf{V}\mathbf{q}_\tau$. This relation can be written as follows: Let

$$\mathbf{g}_{le}(\mathbf{V}\mathbf{q}_{\tau_l}) = \mathbf{V}^\mathsf{T}\mathbf{L}_e^\mathsf{T}\mathbf{f}_e(\mathbf{L}_e\mathbf{V}\mathbf{q}_{\tau_l})\ \in \mathbb{R}^n \qquad \mathbf{b}_l = \mathbf{V}^\mathsf{T}\mathbf{f}_{\text{nl}}(\mathbf{V}\mathbf{q}_{\tau_l}) = \sum_{e=1}^{N_e} \mathbf{g}_{le}(\mathbf{V}\mathbf{q}_{\tau_l})\ \in \mathbb{R}^n$$

$$\mathbf{G} = \begin{bmatrix} \mathbf{g}_{11} & \cdots & \mathbf{g}_{1N_e} \\ \vdots & \ddots & \vdots \\ \mathbf{g}_{N_\tau 1} & \cdots & \mathbf{g}_{N_\tau N_e} \end{bmatrix} \in \mathbb{R}^{nN_\tau\times N_e} \qquad \mathbf{b} = \begin{bmatrix} \mathbf{b}_1 \\ \vdots \\ \mathbf{b}_{N_\tau} \end{bmatrix} \in \mathbb{R}^{nN_\tau}\ .$$

The minimization problem

$$\xi^\star = \arg\min_{\xi\in\Phi} \|\xi\|_0 \qquad \Phi = \{\xi \in \mathbb{R}^{N_e} : \|\mathbf{G}\,\xi - \mathbf{b}\|_2 \le \varepsilon\|\mathbf{b}\|_2,\ \xi \ge 0\}\ ,$$

where $\|\xi\|_0$ is the number of all non-zero coordinates, returns the weights $\xi_e^\star$. The element subset $\tilde{E}$, whose element forces $\mathbf{f}_e$ are evaluated, consists of those elements that have a non-zero $\xi_e^\star$. It is noteworthy that this kind of problem is NP-hard. Therefore, it has been replaced by a similar problem that approximates the formulation above as proposed in [7].

Two entities that highly depend on the investigated system need to be determined to perform the nonlinear model reduction described above: first, a reduction basis $\mathbf{V}$ for Galerkin projection and, second, a training set $\mathbf{u}_\tau = \mathbf{V}\mathbf{q}_\tau$ for hyperreduction.

IOP Conf. Series: Journal of Physics: Conf. Series **1106** (2018) 012024 doi:10.1088/1742-6596/1106/1/012024

4.1. Determination of reduction bases $\mathbf{V}$

One option to determine a reduction basis is using the first n left singular vectors of a matrix with column-stacked displacement snapshots coming from a full simulation. This method is called Proper Orthogonal Decomposition [8]. Its characteristic is the dependence on full simulation results. These are typically very costly to compute. In this contribution, we make use of simulation-free methods that do not depend on solutions to the full model.

We use a combination of static solutions, vibration modes of the linearized system and their static modal derivatives. Thus, the reduction basis consists of three parts:

$$\mathbf{V} = [\mathbf{V}_{\mathrm{S}}\, \mathbf{V}_{\mathrm{VM}}\, \mathbf{V}_{\mathrm{SD}}] \; . \tag{6}$$

$\mathbf{V}_{\mathrm{S}}$ contains the static solutions of the model for different magnitudes of the external force. $\mathbf{V}_{\mathrm{VM}}$ consists of mode shapes of the linearized system that are computed by solving the eigenproblem

$$(\mathbf{K}_{\mathrm{T}} - \omega_i^2 \mathbf{M})\phi_i = \mathbf{0} \qquad \mathbf{V}_{\mathrm{VM}} = [\phi_1, \phi_2, \ldots, \phi_k] \; , \tag{7}$$

with tangent stiffness matrix $\mathbf{K}_{\mathrm{T}} = \left.\frac{\partial \mathbf{f}_{\mathrm{nl}}}{\partial \mathbf{u}}\right|_{\mathbf{u}_0}$ where $\mathbf{u}_0$ is a static solution to an actuated state $\mathbf{u}_0$ where a constant pressure has been applied to the surfaces that are connected with the foam. $\mathbf{V}_{\mathrm{SD}}$ contains static modal derivatives [9, 10]

$$\mathbf{K}_{\mathrm{T}}\mathbf{v}_{ij} = \frac{\partial \mathbf{K}_{\mathrm{T}}}{\partial \phi_i}\phi_j \qquad \mathbf{V}_{\mathrm{SD}} = [\mathbf{v}_{ij_1}, \mathbf{v}_{ij_2}, \ldots, \mathbf{v}_{ij_m}] \tag{8}$$

that insert nonlinear information into the reduction basis $\mathbf{V}$.

4.2. Determination of training sets for hyperreduction

Usually, some displacement vectors from the solution of the full model are used as training displacements. But we want to circumvent their high computation costs and use a simulation-free method. We use a method called Nonlinear Stochastic Krylov Training Sets (NSKTS) that has been proposed by Rutzmoser in [11].

The first step is to build a subspace that is able to approximate the nonlinear force vector $\mathbf{f}_{\mathrm{nl}}$. If the viscous damping term is neglected, one finds

$$\mathbf{M}\ddot{\mathbf{u}}(t) + \mathbf{f}_{\mathrm{nl}}(\mathbf{u}(t)) = \mathbf{B}\,\mathbf{f}_{\mathrm{ext}}(t) \quad \Rightarrow \quad \mathbf{f}_{\mathrm{nl}} \in \mathrm{span}(\mathbf{B},\, \mathbf{M}\ddot{\mathbf{u}}(t_1),\, \mathbf{M}\ddot{\mathbf{u}}(t_2),\, \ldots,\, \mathbf{M}\ddot{\mathbf{u}}(t_n)) \; . \tag{9}$$

As the accelerations $\ddot{\mathbf{u}}(t_i)$ are unknown, we need an approximation for the subspace described above. According to [11], it is approximated by the Krylov subspace

$$\mathbf{F}_{\mathrm{kry}} = \mathrm{span}\{\mathbf{B},\; \mathbf{M}\mathbf{K}^{-1}\mathbf{B},\; (\mathbf{M}\mathbf{K}^{-1})^2\mathbf{B}, \ldots\} = \mathcal{K}(\mathbf{M}\mathbf{K}^{-1},\, \mathbf{B}) \tag{10}$$

that is orthogonalized and normalized such that $\mathbf{F}_{\mathrm{kry}}^{\mathsf{T}}\mathbf{K}^{-1}\mathbf{F}_{\mathrm{kry}} = \mathbf{I}$.

Afterwards, some random vectors $\mathbf{f}_{\mathrm{NSKTS}}^{\tau}$ living in this Krylov subspace are generated and applied as external force to the nonlinear static problem $\mathbf{f}_{\mathrm{nl}}(\mathbf{u}_\tau) = \mathbf{f}_{\mathrm{NSKTS}}^{\tau}$. This equation must be solved by a nonlinear solver such as Newton-Raphson. The training set is then built by saving the solution $\mathbf{u}_\tau^{(k)}$ of each Newton iteration k for each applied external force $\mathbf{f}_{\mathrm{NSKTS}}^{\tau}$ as training vector.

5. Results

The finite element model is reduced with different hyperreduction tolerances ε and different reduction bases $\mathbf{V}$ that differ in size and combination of vectors to compare the performance of the reduction. We consider four different bases:

Figure 3. Reduced mesh of the plate with $\varepsilon = 0.001$ (left), $\varepsilon = 0.01$ (middle), $\varepsilon = 0.1$ (right)

$\mathbf{V}_{\text{VM}} + \mathbf{V}_{\text{SD}}$ full: 50 vibration modes (with smallest eigenfrequencies) and 1274 static derivatives

$\mathbf{V}_{\text{VM}} + \mathbf{V}_{\text{SD}}$ 600: 50 vibration modes and 550 static derivatives (truncated by using a SVD)

$\mathbf{V}_{\text{S}} + \mathbf{V}_{\text{VM}} + \mathbf{V}_{\text{SD}}$ full: Augmentation of **$\mathbf{V}_{\text{VM}} + \mathbf{V}_{\text{SD}}$ full** by 20 Newton iterations of the static solution

$\mathbf{V}_{\text{S}} + \mathbf{V}_{\text{VM}} + \mathbf{V}_{\text{SD}}$ 600 Augmentation of **$\mathbf{V}_{\text{VM}} + \mathbf{V}_{\text{SD}}$ 600** augmented by 20 Newton iterations of the static solution

The model also is hyperreduced with ECSW method using different values for the tolerance ε. 160 Nonlinear Stochastic Training Vectors are used as training set for all hyperreductions.

A constant pressure of -0.1 bar has been applied to the surfaces that are connected with the foam as prestress. This simulates an adaption of the CCLD through compression. All necessary linearizations (e.g. for determining vibration modes) are made relative to the static deformation due to the constant pressure.

Figure 3 shows the reduced meshes of the hyperreduced model for different values ε. One can see that higher values for ε lead to less elements that have to be evaluated. The number of evaluated elements are listed in Table 1.

Table 1. Amount of elements in element set computed by ECSW method depending on reduction basis **V** and tolerance ε. The full model has 45 375 elements

reduction basis **V**	tolerance ε			
	0.01	0.005	0.001	0.0005
$\mathbf{V}_{\text{VM}} + \mathbf{V}_{\text{SD}}$ full	1290	1622	2385	2670
$\mathbf{V}_{\text{VM}} + \mathbf{V}_{\text{SD}}$ 600	1073	1286	1840	2067
$\mathbf{V}_{\text{S}} + \mathbf{V}_{\text{VM}} + \mathbf{V}_{\text{SD}}$ full	1211	1550	2293	2625
$\mathbf{V}_{\text{S}} + \mathbf{V}_{\text{VM}} + \mathbf{V}_{\text{SD}}$ 600	1090	1337	1895	2095

We define the relative error

$$RE = \frac{\sqrt{\sum_{i=0}^{N_t} \Delta\mathbf{u}(t_i)^T \Delta\mathbf{u}(t_i)}}{\sqrt{\sum_{t=0}^{N_t} \mathbf{u}_{\text{ref}}(t_i)^T \mathbf{u}_{\text{ref}}(t_i)}} \quad \text{with} \quad \Delta\mathbf{u}(t_i) = \mathbf{u}(t_i) - \mathbf{u}_{\text{ref}}(t_i) \tag{11}$$

as measure for the accuracy of the reduced models where N_t is the number of timesteps, $\mathbf{u}$ is solution to the reduced model and $\mathbf{u}_{\text{ref}}$ is the solution to the full model. Table 2 shows relative errors and simulation times for a simulation with 220 timesteps.

The simulation time can be reduced significantly by applying nonlinear model reduction. But acceptable error measures can only be achieved with small tolerance values ε. Interestingly,

Table 2. Relative errors *RE* and simulation times t_{simul} depending on reduction basis **V** and tolerance ε. A linear full solution leads to $RE = 10\,\%$ and $t_{\text{simul}} = 0.5\,\text{h}$

	reduction basis **V**	tolerance ε 0.01	0.005	0.001	0.0005
$RE[\%]$	$\mathbf{V}_{\text{VM}} + \mathbf{V}_{\text{SD}}$ **full**	115.4	34.1	9.4	3.5
	$\mathbf{V}_{\text{VM}} + \mathbf{V}_{\text{SD}}$ **600**	177.9	80.5	83.3	82.1
	$\mathbf{V}_{\text{S}} + \mathbf{V}_{\text{VM}} + \mathbf{V}_{\text{SD}}$ **full**	58.6	34.1	7.3	1.4
	$\mathbf{V}_{\text{S}} + \mathbf{V}_{\text{VM}} + \mathbf{V}_{\text{SD}}$ **600**	89.4	72.2	67.4	67.2
$t_{\text{simul}}[h]$	**full solution**	23.70			
	$\mathbf{V}_{\text{VM}} + \mathbf{V}_{\text{SD}}$ **full**	6.44	5.78	5.81	6.18
	$\mathbf{V}_{\text{VM}} + \mathbf{V}_{\text{SD}}$ **600**	5.17	4.43	5.06	4.45
	$\mathbf{V}_{\text{S}} + \mathbf{V}_{\text{VM}} + \mathbf{V}_{\text{SD}}$ **full**	5.85	5.90	6.15	6.40
	$\mathbf{V}_{\text{S}} + \mathbf{V}_{\text{VM}} + \mathbf{V}_{\text{SD}}$ **600**	4.64	4.92	4.62	4.47

smaller tolerances ε for hyperreduction do not increase the simulation time significantly. The reason is that the Newton Raphson solver that is called during time-integration needs fewer iterations for smaller ε which compensates the higher costs for evaluating more elements.

6. Conclusion

Nonlinear model reduction can reduce computation time to estimate best values for the actuated pressure in a compressible constrained layer damper treatment. However, the computation time is still high which prohibits to use the method for model predictive control of the CCLD. Further investigations are needed to figure out if bases with smaller dimensions and equal accuracies can be determined or if other nonlinear reduction methods could improve performance.

Acknowledgments

The presented work is part of the research within the context of the Priority Program 1897 "Calm, Smooth and Smart", subprojects "Model Order Reduction of Parametric Nonlinear Mechanical Systems for Influencing Vibrations" and "Lightweight Structures with Adaptive Dynamic Behaviour through Evanescent Morphing". The authors are grateful to the Deutsche Forschungsgemeinschaft (DFG, eng. German Research Foundation) for the financial support.

References

[1] Koschichow R, Holeczek K, Koschichow D, Rüdiger F, Kostka P, Langkamp A, Fröhlich J and Modler N 2015 *Composites Theory and Practice* **15** 146–151
[2] Kozlowska J, Boczkowska A, Czulak A, Przybyszewski B, Holeczek K, Stanik R and Gude M 2016 *Smart Materials and Structures* **25** 035025
[3] Pawel Kostka, Klaudiusz Holeczek and Werner Hufenbach 2013 *International Journal of Composite Materials* **6B** 53–58
[4] Holeczek K, Dannemann M and Modler N 2015 ed VDI Wissensforum GmbH (Düsseldorf: VDI Verlag GmbH) pp 155–170 ISBN 978-3-18-092261-4
[5] Bampton M C and Craig Jr R R 1968 *Aiaa Journal* **6** 1313–1319
[6] Antoulas A C 2005 *Approximation of large-scale dynamical systems* vol 6 (Siam)
[7] Farhat C, Chapman T and Avery P 2015 *International Journal for Numerical Methods in Engineering* **102** 1077–1110
[8] Kerschen G, Golinval J c, Vakakis A F and Bergman L A 2005 *Nonlinear dynamics* **41** 147–169
[9] Idelsohn S R and Cardona A 1985 *Computer Methods in Applied Mechanics and Engineering* **49** 253–279
[10] Slaats P, De Jongh J and Sauren A 1995 *Computers & structures* **54** 1155–1171
[11] Rutzmoser J and Rixen D 2017 *Computer Methods in Applied Mechanics and Engineering* **325** 330 – 349

IOP Conf. Series: Journal of Physics: Conf. Series **1106** (2018) 012025 doi:10.1088/1742-6596/1106/1/012025

Constrained Layer Damping Treatment: Analysis of multi-strip application on thin plates

A R Kudal and A Cicirello

Department of Engineering Science, University of Oxford, Oxford, UK

E-mail: alice.cicirello@eng.ox.ac.uk

Abstract. This paper focuses on analysing the effects of using multiple strips of Constrained Layer Damping (CLD) treatments to cover entirely a thin aluminium plate and assess the vibration performance for various strips orientation. In particular, for a fixed type of CLD treatment, three strips orientation are considered: (i) parallel to the long axis of the plate, (ii) perpendicular to the long axis, and (iii) 45° to the long axis. By using both experimental and Finite Element (FE) modelling, it is shown that depending on the mode shape curvature and corresponding nodal line positions, the strip orientation would strongly affect the amount of energy that can be dissipated at one particular resonance. An approximate topology optimisation technique is investigated to target the first bending mode of the structure and is validated experimentally. This technique involves optimising a uniform CLD topology for maximum stiffness (using a topology optimization feature of a FE commercial software) and then manually locating optimal strip placement by taking into account orientation effects.

1. Introduction

Constrained Layer Damping (CLD) is a type of passive damping treatment commonly used in the aerospace and automotive industries for dissipating vibrations in lightweight structures. It consists of a viscoelastic polymer layer, with a high loss factor, sandwiched between the structure and a constraining metal layer with a high modulus. The vibrational energy is dissipated by shear distortion of the damping material when out-of-plane vibrations of the structural element occur, especially when resonant behaviour is dominant [1]. The efficiency of the CLD treatments is dependent on frequency and temperature, given the strong dependence of the complex modulus of the viscoeleastic material to these two parameters. CLD treatments are commercially sold as rolls of varying widths, and can be custom-made for larger footprints. Often larger footprints are covered by using multiple strips. This paper focuses on assessing the performance of a plate fully covered with multi-strip CLD treatment using Finite Element (FE) models and its validation with experimental data, and on the selection of the strips orientation.

Previous research by Moreira and Rodrigues [2, 3] has shown that the effects on the response of CLD treatment applied uniformly to a thin panel can be investigated by using the FE method. The complex modulus approach was used to model viscoelastic properties, obtaining good agreement with experimental results. Furthermore, several papers have investigated the topology optimization of CLD patches on planar structures. In particular, Chia et al. [4] employed a Cellular Automaton algorithm to optimally place CLD on a plate assuming a fixed temperature and frequency. Their approach considered the highest strain modal energy of an untreated plate to locate one cell of damping treatment. Then the patch was extended to zones which give highest damping effectiveness [4].

To the best of the authors knowledge, no studies so far have explicitly investigated experimentally and numerically the effects of employing non-overlapping multiple strips to cover large footprints, nor the effects of the orientation of the multi-strips treatment.

In this paper, thin aluminium plates measuring 300 mm x 200 mm x 2 mm are covered uniformly with non-overlapping CLD strips of 50 mm width, and three different orientations are considered: (i) parallel to the long axis of the plate, (ii) perpendicular to the long axis, and (iii) 45° to the long axis. Experimental Modal Analysis is performed to yield the Frequency Response Function. It is shown that strip orientation strongly affects the amount of energy that can be dissipated at one particular resonance because of the mode shape curvature and nodal line positions. Also presented is a comparison of these experimental results with FE models developed using the commercial FE software ANSYS, showing good agreement.

Finally, an approximate topology optimisation technique is investigated to target the first bending mode of the plate and is validated experimentally. This technique involves optimising a uniform CLD topology for maximum stiffness (using ANSYS topology optimization feature) and then manually locating optimal strip placement by taking into account orientation effects from the findings above.

To this end, this paper is split into a number of sections; Section 2 is focused on the Finite Element Model. The experimental set up is presented in Section 3, followed by analysis of orientation effects and validation of the FE model with respect to experimental data in Section 4. Finally an approximate topology optimisation approach is presented in Section 5.

2. Finite Element Modelling

The system to be modelled consists of an aluminuim plate measuring 300 mm x 200 mm x 2 mm (density $\rho = 2800\,\mathrm{Kg/m^3}$, Young's modulus $E = 70\,\mathrm{GPa}$ and Poisson ratio 0.3). The damping layer is characterised by a top aluminium layer of 0.20 mm and 0.14 mm damping polymer [7], and has a width of 50 mm. A number of special considerations have to be taken into account when developing FE models involving Constrained Layer Damping treatments. The first of which is the choice of elements and mesh density. It is well known that the mesh density required to reasonably predict dynamic behaviour of structural components is at least 6 elements per wavelength of deformation obtained at twice the maximum frequency of interest [5]. For this study, a density of 60 elements along the long axis of the aluminium plate, and 40 along the short axis is used to provide good model accuracy. Regarding choice of elements; solid 20-node hexahedral elements are used to model both the aluminium and viscoelastic layers. This ensures no issues will arise due to shear locking. Contact between the viscoelastic polymer and constraining metal layers was assumed to be perfect and therefore bonded contact was used to model this.

Another important consideration is the modelling of the constrained layer. This involves selecting an appropriate method for modelling viscoelastic material properties, which are both frequency ω and temperature T dependent. The approach used in this paper follows Johnson et al. [8] and Moreira et al. [3], whereby viscoelastic properties are modelled using the complex modulus method. Specifically, the modulus of the viscoelastic material is expressed as:

$$G(\omega, T) = G'(\omega, T) + jG''(\omega, T) \tag{1}$$

Where G' represents the shear storage modulus and G'' represents the shear loss modulus. The storage and loss modulus for the viscoelastic material are provided by the manufacturer in the form of a reduced-frequency nomogram [7]. This data is hand read for a fixed temperature (shown in table 1) and implemented into the commercial FE software ANSYS [6] by fitting data points using a Prony model [6].

Table 1. Storage and Loss modulus data for 3M 435 polymer @ 20°C

Frequency (Hz)	G' (MPa)	G" (MPa)
10	0.11	0.23
100	0.22	0.30
200	0.28	0.31
300	0.30	0.32
400	0.33	0.32
500	0.38	0.32

With this approach, the frequency response of the plate with constrained damping layer subject to a deterministic point load F (to yield the Frequency Response Function) is obtained by solving:

$$-\omega^2[M]X + \{[K_a] + [K_v(\omega, T)]\}X = F \qquad (2)$$

Where M is the mass matrix, K_a represents the stiffness matrix for the aluminium plate and top aluminium constraining layer, $K_v(\omega, T)$ is the complex stiffness matrix of the viscoelastic layer, and X is the vector of the displacement response of the system. The Harmonic Response module of ANSYS [6] is used to evaluate the virtual Frequency Response Function.

3. Experimental analysis

Experimental Modal Analysis (EMA) [9, 10] was performed to evaluate the Frequency Response function. The tests were performed on 3 aluminium alloy plates with CLD layers applied in varying layouts: (i) parallel to the long axis of the plate, (ii) perpendicular to the long axis, and (iii) 45° to the long axis, as shown in Figure 2. Care was taken to ensure no overlap and contact between adjacent strips. To ensure these free-free boundary conditions, a suspension system was devised to hang the plates (Figure 3).

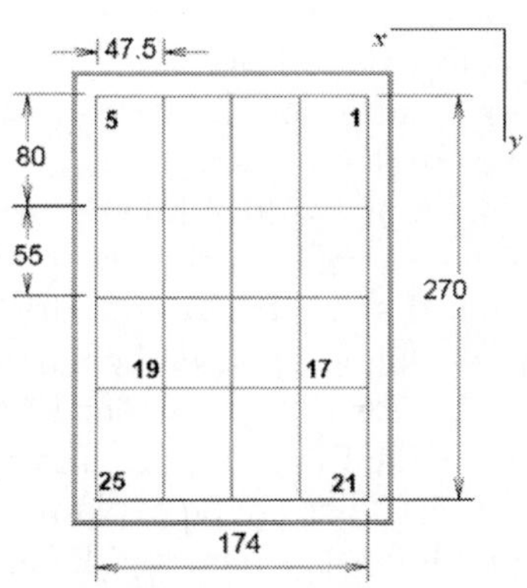

Figure 1. Measuring grid adapted from [2].

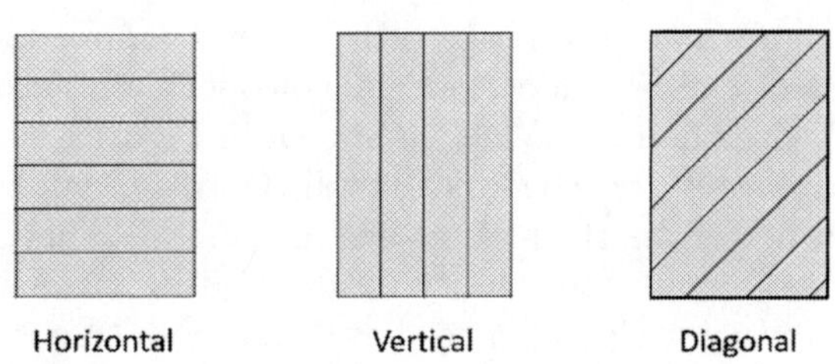

Figure 2. Orientations of CLD strips tested.

A measuring grid was employed for taking measurements on the plate (Figure 1). Each plate is excited by using a modal shaker (2004E mini-shaker) and the response is measured by using a single pointer Laser Doppler Vibrometer (Polytec NLV-2500). For this study, a sinusoidal chirp excitation is applied to the test structure through a modal shaker attached at point 17, with the force input to the system recorded by using an impedance head (PCB 288D01). The velocity response to this excitation is measured at point 19. A sampling frequency of 10kHz was implemented. The Frequency Response Function (in terms of acceleration divided by Force) and the modal parameters are evaluated by carrying out the appropriate signal processing strategies [9, 10].

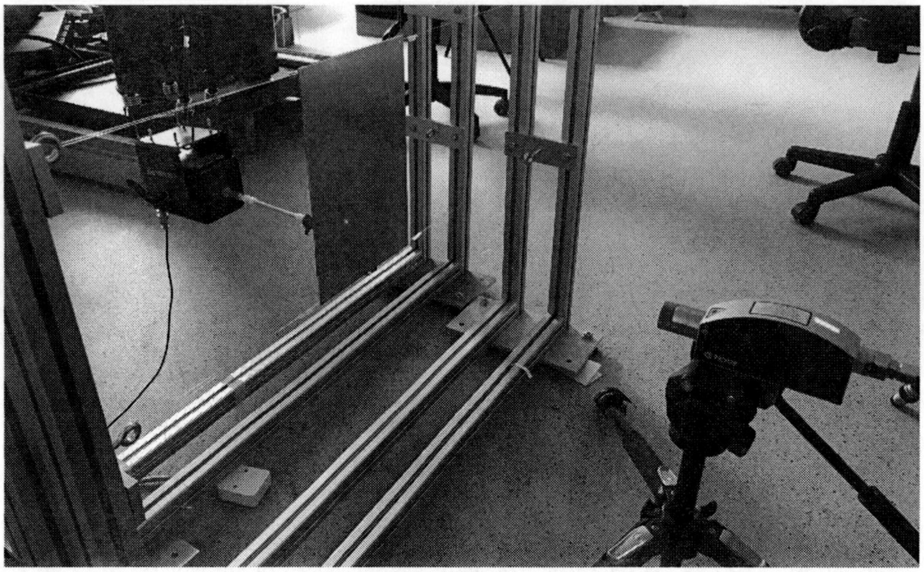

Figure 3. Experimental setup showing system used to suspend test items.

4. CLD Strip Analysis: experiments and modelling

In this section the FRF for the multi-strip orientations shown in Figure 2 are evaluated experimentally and compared with FE models.

The FRFs obtained experimentally for the three multi-strip orientations are compared in Figure 4. From this figure it is possible to observe that the orientation of the strips affects the damping performance in terms of FRF peak amplitude and resonant frequency shift.

This effect can be understood by considering for example the first two bending mode shapes (obtained with the FE model), as shown in Figures 5 and 6. The vertical multi-strip orientation performed better for the first bending mode, whilst the multi-strip horizontal orientation performed better for the second bending mode. The CLD strips aligned perpendicularly to the nodal lines show a better ability for dissipating vibrational energy for this particular resonance. On the contrary, strips aligned parallel to nodal lines show poor damping performance. In the case of torsional modes, the diagonal orientation was most effective. This can be explained by the fact that when the strip of damping layer is parallel to the nodal lines, the bending deformation, and therefore the consequent shear deformation of the viscoeleastic material, would be minimal.

To validate the effectiveness of FEA at modelling the orientation effects, equivalent FE models of test structures from Figure 2 were developed and compared with experimental data. During this study, it was found that the aluminium plates were manufactured using a rolling method, which leads to anisotropic material properties. To model this, orthotropic elastic modulus was implemented for the aluminium alloy. The values used were 67.5 GPa along the rolling direction and 72.5 GPa perpendicular to roll direction.

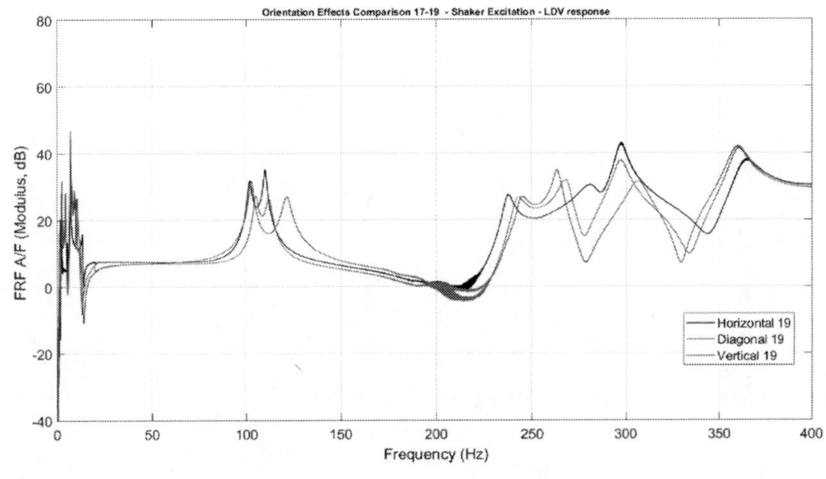

Figure 4. Graph showing comparison between the three orientations of CLD treatment tested experimentally.

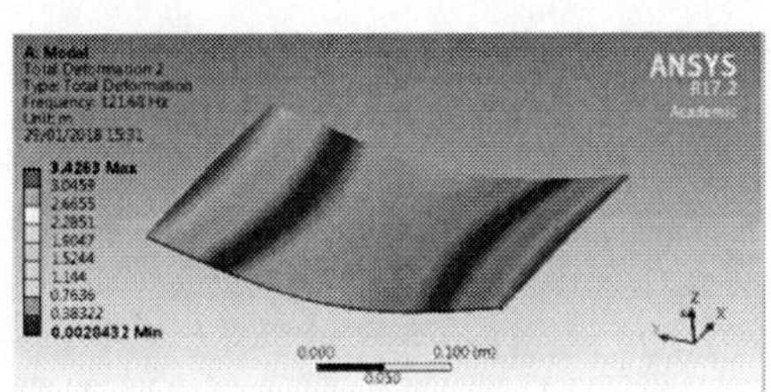

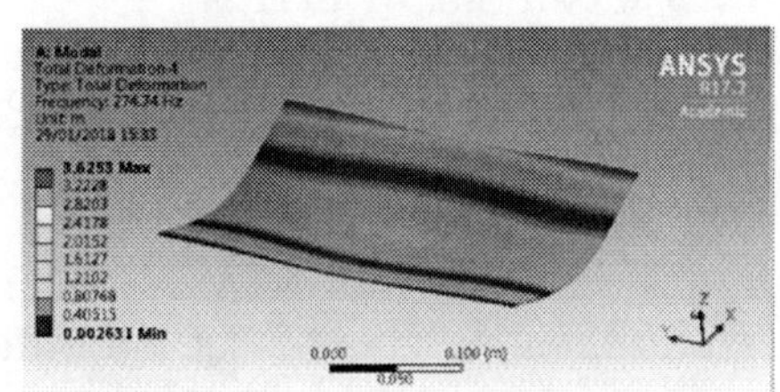

Figure 5. First bending mode **Figure 6.** Second bending mode

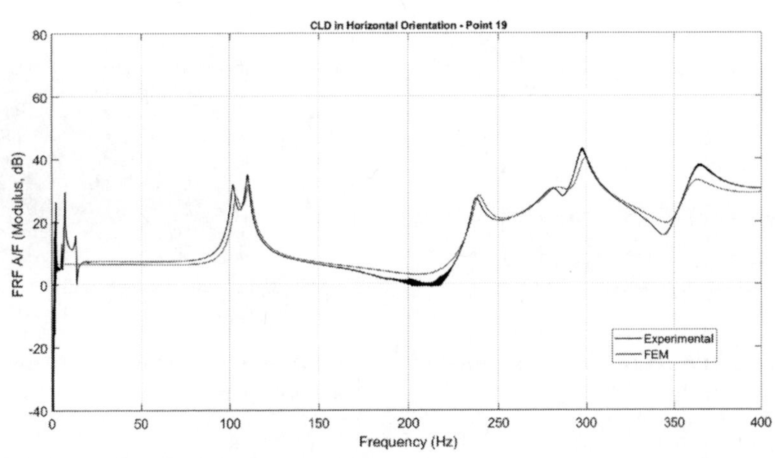

Figure 7. Comparison of FEM with experimental data for strips in horizontal orientation.

The Figures 7 and 8 show a comparison of the experimental and FE results for the horizontal and diagonal orientation. It is possible to observe in general a good agreement between the FEM and the experimental Frequency Response Functions computed.

IOP Conf. Series: Journal of Physics: Conf. Series **1106** (2018) 012025 doi:10.1088/1742-6596/1106/1/012025

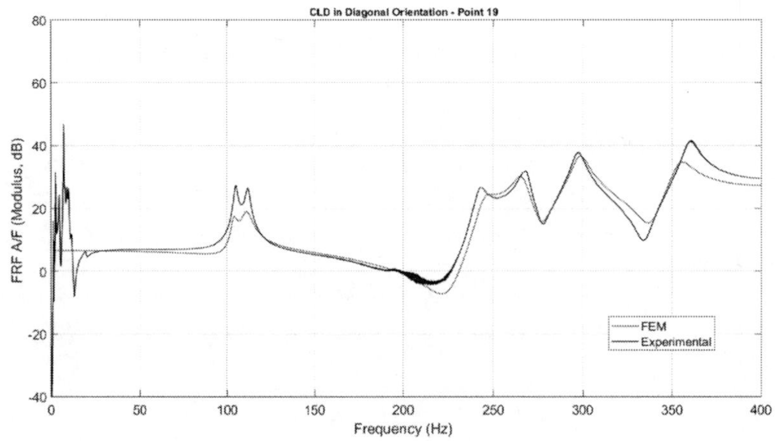

Figure 8. Comparison of FEM with experimental data for strips in diagonal orientation.

5. Optimal Placement of Strips

This section is focused on the description on an approximate approach for selection of optimal location of the CLD strips. The advantage of using CLD strips is that treatment can be selectively applied to target certain modes of vibration. As seen in the previous section, the orientation of strips has a large impact on the damping performance achieved, and this should be taken into account when deciding application of strips.

In this study, an approximate method is devised to locate the optimal placement of strips for suppressing the first bending mode of the structure (Figure 5). This mode of vibration occurs at around 121 Hz and is the second resonant peak on the FRF. The proposed approximate approach uses a combination of two metrics to decide on optimal placement.

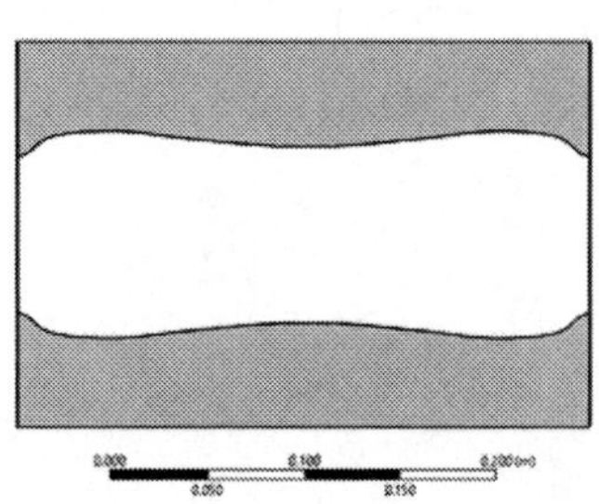

Figure 9. Stiffness optimised layout for first bending mode.

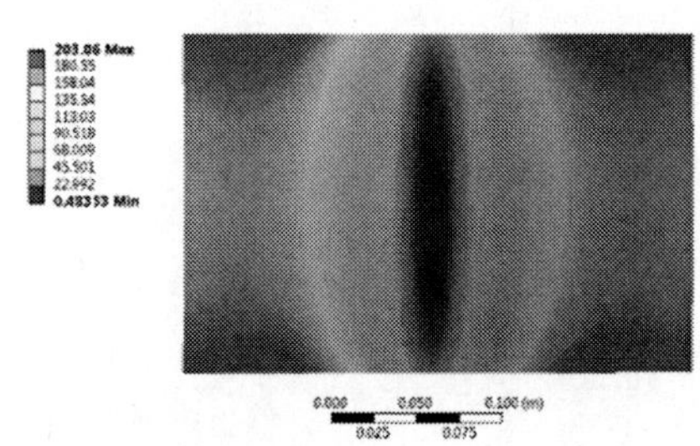

Figure 10. Shear strain distribution in viscoelastic layer for first bending mode.

The first metric is an optimisation for the CLD topology. The built-in ANSYS *Topology Optimization* [6] module is used to carry out this function. The objective of this optimisation is to evaluate the layout of CLD treatment on the plate that will maximise stiffness for the approximate modal deformation shape, with the constraint that 50% of the surface should be covered. The idea being that it is the stiffness provided by the constraining metal layer which is responsible for shearing within the viscoelastic layer. The results for this optimisation is shown in Figure 9.

The second metric is the shear strain distribution in the viscoelastic layer for the particular modal deformation. The rationale here being that areas of highest strain should ideally be covered with CLD as viscoelastic shearing is the main mechanism by which energy is dissipated by this treatment. Figure 10 shows the maximum shear strain distribution for the targeted mode shape.

Examining both metrics suggests the layout shown in Figure 9 should be ideal for targeting this mode of vibration as the areas of greatest shear strain lie within the optimised region. This is further reinforced by prior findings on orientation effect which suggest strips should be aligned perpendicular to the nodal line of this mode shape (to achieve highest curvature). To verify this, a plate configured with this layout of CLD is tested experimentally.

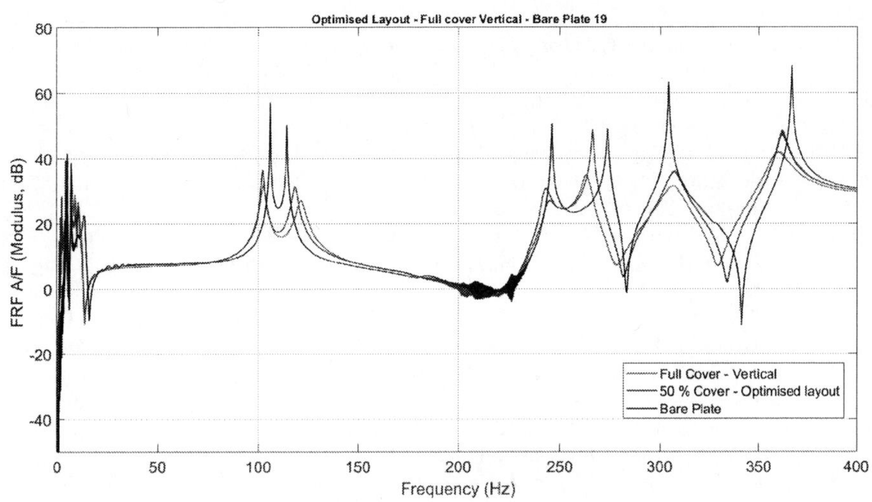

Figure 11. FRF of layout optimised for targeting the second mode of vibration. Bare plate and full cover vertical FRF included for comparison.

The results (Figure 11) show this layout was effective at damping the first bending mode of the structure, providing a reduction of almost 20dB form the bare plate with just 50% coverage. When compared to a fully covered plate (CLD in vertical orientation), the optimised layout was very close in terms of performance displaying a difference of about 4dB. These results further reinforce the findings on orientation effects as this layout provided almost no damping for the fourth mode of vibration (bending mode 2), for which the strips were aligned parallel to nodal lines.

6. Conclusions

Constrained Layer Damping (CLD) treatments provide an effective means of suppressing unwanted noise and vibrations in thin planar structures. Findings in this paper show that when a multi-strip CLD configuration has to be implemented, the orientation of the strips is an important parameter to take into account. The energy dissipated by the treatment was found to be strongly dependant upon alignment of strips with nodal lines of particular mode shapes. This was investigated both experimentally and with Finite Element models. In conclusion, findings suggest that strips should be placed to obtain the largest curvature for a specific modal shape, and in particular for bending mode shapes the strips should be placed perpendicular to the node lines in order to achieve best damping performance.

An approximate approach for optimal placement of strips was successfully devised for targeting a bending mode of vibration. The CLD topology was optimised for maximising stiffness

IOP Conf. Series: Journal of Physics: Conf. Series **1106** (2018) 012025 doi:10.1088/1742-6596/1106/1/012025

for a particular modal deformation. Placement of the CLD was then located manually by taking into account orientation effects.

Acknowledgments

AK is grateful to Balliol College for providing funding to participate to the MPSVA2018 Conference. The experiments were performed in the Dynamics, Vibration and Uncertainty Lab (University of Oxford), which has been established thanks to the John Fell Fund (163/029). AC gratefully acknowledges the financial support provided by Balliol College for her Career Development Fellowship in Engineering.

References

[1] Hambric S A, Sung S H, and Nefske D J 2016 *Engineering Vibroacoustic Analysis.* (*John Wiley and Sons, UK*).
[2] Moreira R A and Rodrigues J D 2002 The modellisation of Constrained Damping Layer treatments using the Finite Element Method: spatial model and viscoelastic behaviour *Proc.International Conference on Structural Dynamics Modelling, Test, Analysis, Correlation and Validation (Portugal).*
[3] Moreira R A and Rodrigues J D 2004 Constrained Damping Layer Treatments: Finite Element Modelling *Journal of Vibration and Control* **10** 575–95.
[4] Chia C M, Rongong J A and Worden K 2008 Evolution of constrained layer damping using a cellular automaton algorithm *Proceedings of the Institution of Mechanical Engineers, Part C: Journal of Mechanical Engineering Science* **222** 585–97
[5] Langer P, Maeder M, Guist C, Krause M and Marburg S 2017 More Than Six Elements Per Wavelength: The Practical Use of Structural Finite Element Models and Their Accuracy in Comparison with Experimental Results *Journal of Computational Acoustics* **25**
[6] ANSYS Mechanical User Guide *https://www.sharcnet.ca/Software/Ansys* (Accessed June 2018)
[7] 3M Vibration Damping Tapes 434-435-436 datasheet 2013 *https://multimedia.3m.com/mws/media/117317O/3mtm-scotchdamptm-vibration-434.pdf* (Accessed June 2018)
[8] C D Johnson and D A Kienholz 1982 Finite element prediction of damping in structures with constrained viscoelastic layers *AIAA Journal* **20** 1284–90
[9] P Avitabile 2001 Modal Testing: A Practitioner's Guide *Wiley*
[10] A Brandt, 2011 Noise and Vibration Analysis: Signal Analysis and Experimental Procedures, *John Wiley and Sons, New York*

IOP Conf. Series: Journal of Physics: Conf. Series **1106** (2018) 012026 doi:10.1088/1742-6596/1106/1/012026

Nonlinear Vibrational Analysis for Integrally Bladed Disk Using Frictional Ring Damper

Yekai Sun, Jie Yuan, Luca Pesaresi and Loïc Salles
Department of Mechanical Engineering, Imperial College London
London SW7 2AZ, UK

E-mail: ys5113@imperial.ac.uk

Abstract. The use of integrally bladed-disk is now very popular in turbomachinery industry since they feature significant aerodynamic and structural improvements along with a significant mass reduction. However, these integrated single structures can arise a major high cycle fatigue issue due to the lack of sufficient damping for dissipating the vibrational energy. This work describes a numerical investigation of the nonlinear dynamic behaviour and nonlinear normal mode for such a bladed-disk with frictional ring damper using the Harmonic Balanced Method (HBM) with alternating Fourier transformation. Jenkins element is used to model the nonlinear contact friction between the disc and ring damper. Using such a modeling strategy, the modal damping and resonance amplitude are directly and efficiently computed through nonlinear normal mode analysis. The initial results show the vibrational level on the blades can be effectively controlled by the parameters of the ring damper model. The effectiveness of ring damper and damping performance is evaluated. This study also indicates the nonlinear normal mode analysis based HBM may be an effective method to analyse the dynamic behaviour of the integrated bladed-disk with frictional ring damper.

1. Introduction

Blisk is one type of bladed-disk system used in turbomachinery analysis. It is key dynamic components in aircraft engine, which undergoes complicated operational environment with thermal loads, static fluid pressures and rotation-induced centrifugal loads [1]. The aeroelastic excitation on the blisk might cause severe vibration stress leading to high cycle fatigue (HCF). Consequently, HCF is regarded as a major reason for most of the failures in aero-engine. The flutter effect and mistuned blisk are also potential sources of vibration in the blisk structure [2]. In order to reduce vibration level of blisk, the frictional damper is widely used as main damping source. In a traditional manufacture techniques of blisk, the frictional joints and contact surfaces in assembly point between different components play an important role to dissipate the vibrational energy. Furthermore, friction between contact surfaces is regarded as major damping mechanism, such as friction from under-platform damper and shroud contact surfaces. Frictional joints and contact surfaces have been investigated for past twenty years by many researchers [3, 4, 5]. However, in integrated bladed-disk, the rotor is manufactured in a single piece of component, in stead of an assembled structure with a disk and individual blade. For this reason, the conventional friction joints are no longer favorable due to lack of contact surfaces. The alternative method to introduce contact friction is defined as a ring underneath the blisk wheel. This ring is used as a frictional damper to reduce the vibration of blisk [6].

In this paper, the dynamic behaviour of integrally blisk with frictional ring damper is analysed in two aspects, including forced frequency response and nonlinear modal analysis. There are several numerical approaches to directly compute periodic solution for nonlinear dynamic system in time domain or frequency domain. Time integration is traditional time domain method using various schemes, for example, explicit Runge-Kutta scheme, matrix exponential [7] and implicit Newmark scheme [7]. Generally, these methods can provide accurate results but with unaffordable computational costs because small time step is required to capture dynamics behaviour on contact surfaces [8]. Shooting method [9] is another popular method to solve nonlinear problem in time domain by solving boundary values. It provides monodromy matrix as a by-product which can be used for stability analysis. For frequency domain method, Harmonic Balance Method (HBM) is widely used and it discretises the nonlinear property directly with truncated Fourier series [10]. HBM has been extended and improved using various numerical algorithms. HBM with alternating Fourier transformation (AFT) is one of the improved HBM, which is proposed by Cameron in 1989 [11]. In this AFT scheme, nonlinear term is evaluated in time domain and transferred into frequency domain. It provides better approximation than analytical description of nonlinear term in frequency domain. In this work, HBM with AFT is used to solve nonlinear dynamic system.

The nonlinear normal mode (NNM) can be used to describe as an autonomous behaviour of the system. Rosenberg defined the concept of NNM as a vibration unison at the first place in 1960s [12]. For conservative system, the autonomous behaviour of the nonlinear dynamic system is similar to a linear system as a periodic solution. However, in non-conservative nonlinear system, the autonomous response is generally non-periodic and decays with time [1]. In the case of blisk with contact friction, the energy of system is dissipated during the vibration and system is non-conservative due to the friction. Shaw and Pierre proposed a definition of NNM as invariant manifold [13], which provides an accurate solution of NNM even in the case of internal resonance and modal localization. However, modal damping ratio and natural frequency ca not be computed directly. Laxalde et al used a concept of nonlinear complex mode in modal analysis of a non-conservative system [6].This nonlinear complex mode is obtained through solving complex eigen problems. Alternatively, Krack has proposed another NNM definition based on periodic solution by adding a negative artificial damping to compensate the frictional loss [14]. The modal analysis in this work is following Krack's strategy to obtain a periodic solution.

This paper is organised as follows. A detailed description of lumped parameter model and Jenkins element model is presented in Secition 2. Numerical formulations of HBM with AFT, computation of nonlinear normal mode and cyclic symmetric technique are explained in Section 3. Finally, the performance of ring damper is evaluated based on contact condition, damping effectiveness and modal characteristics from numerical results in Section 4.

2. Model Description

2.1. Lumped Parameter Model

Lumped parameter model has been widely used to investigate the dynamic system at the initial design stage, since it can provide fundamental understanding of the system with low computational costs [15]. In turbomachinery field, several lumped models representing blisk have been designed using a series of mass and spring. The lumped model with ring damper used in this work is referred to existing research by Laxalde et al [16] as shown in Figure 1. There are total four degrees of freedom (DOF) per sector is used to represent the blisk, with two DOFs representing the motion of blade, one DOF for the disk and the other one for the ring damper. There are overall 24 sectors within this cyclic symmetric structure, $N = 24$. The model parameters are listed in Table 1.

IOP Conf. Series: Journal of Physics: Conf. Series **1106** (2018) 012026 doi:10.1088/1742-6596/1106/1/012026

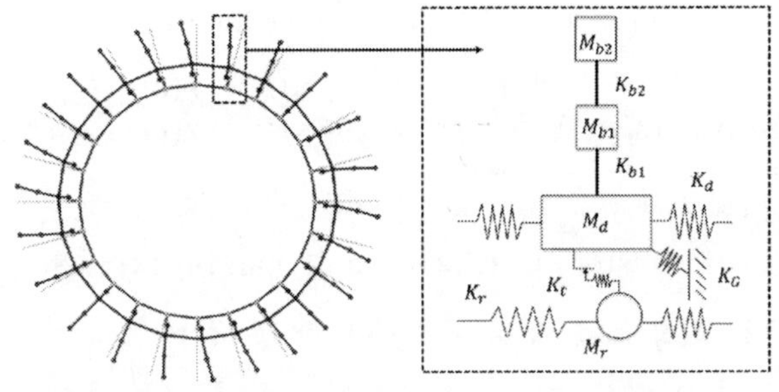

Figure 1. Lumped Parameter Model

Table 1. Lumped Model Parameter

M_{b2}	M_{b1}	M_d	K_{b2}	K_{b1}	K_d	K_G	K_r
0.2 kg	0.3 kg	1.2 kg	$2 \cdot 10^6$ N/m	10^6 N/m	$5 \cdot 10^7$ N/m	$6 \cdot 10^5$ N/m	$2 \cdot 10^7$ N/m

2.2. Jenkins Element Model

Only the friction in tangential direction is considered in this work. The nonlinear force due to friction between the disk and ring damper is computed through a Jenkins element based on Coulomb' law as shown in Figure 2. The relative displacement between disk and ring damper is x, z is internal variable for sliding position. k_t is tangential stiffness and N_0 is normal contact reaction force from rotation of engine. This reaction force is in radial direction about engine rotational axis. Because of normal contact reaction, the ring is assumed to constantly stay in contact with blisk wheel. $N_0 = m_r R \Omega^2$, where m_r is ring damper mass for each sector, Ω is engine rotational speed, R is distance between ring damper and rotational axis.

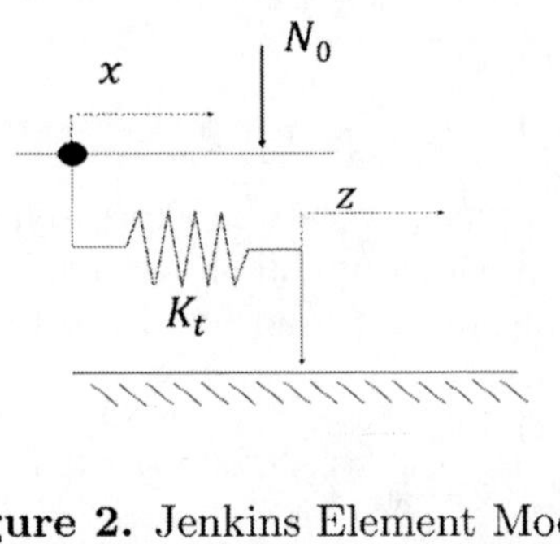

Figure 2. Jenkins Element Model

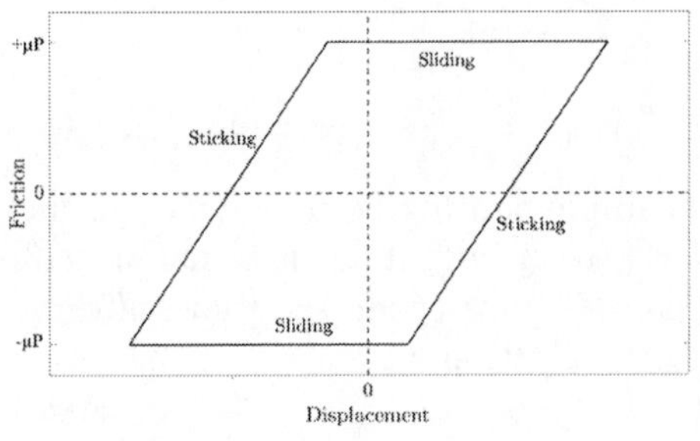

Figure 3. Hysteresis Loop

The Coulomb friction is calculated through Eq.(1), where μ is static friction coefficient 0.3 and spring stiffness of Jenkins element K_t is $5 \cdot 10^7 N/m$ referred to the parametric study by Schwingshackl et al [17]. There are two contact conditions in Jenkins element model, which are sticking and sliding, as shown in hysteresis loop. In sticking condition, the contact is elastic and no sliding happens. In sliding condition, contact point starts to slide and nonlinear force equals to maximum static friction μN_0 against the direction of sliding velocity.

$$\mathbf{f}_{nl}(t) = \begin{cases} K_t \times (\mathbf{x}(t) - \mathbf{z}(t)) & \text{if } |K_t(\mathbf{x}(t) - \mathbf{z}(t))| \leq \mu N_0 \text{ (Sticking)} \\ \mu N_0 \times \text{sign}(\dot{\mathbf{z}}(t)) & \text{if } |K_t(\mathbf{x}(t) - \mathbf{z}(t))| > \mu N_0 \text{ (Sliding)} \end{cases} \tag{1}$$

3. Numerical Methods

The dynamic equation of the system in time domain can be expressed as:

$$\mathbf{M} \cdot \ddot{\mathbf{x}}(t) + \mathbf{C} \cdot \dot{\mathbf{x}}(t) + \mathbf{K} \cdot \mathbf{x}(t) = \mathbf{f}_{ex}(t) - \mathbf{f}_{nl}(\dot{\mathbf{x}}, \mathbf{x}, t) \tag{2}$$

Where $\mathbf{M}$, $\mathbf{K}$, $\mathbf{C}$ are mass, stiffness and damping matrices respectively. $\mathbf{x}$ is displacement vector including the displacement of four DOFs for all sectors. $\mathbf{f}_{nl}$ is Coulomb friction described in Section 2.2. $\mathbf{f}_{ex}$ is external excitation force applied to blisk in forced response analysis.

To simulate the operating environment of the blisk, the engine order excitation is introduced, which is regarded as an effective traveling wave depending on the number of blades in previous stator. To clarify, a standing wave with constant phase angle on each blisk is used to model this traveling wave, where phase angle depends on the engine order number. The phase angle between the nth blisk and the first blisk is φ_n and EO is engine order applied. The excitation force vector is given by Eq.(3). Moreover, this engine order excitation is only applied to two DOFs of blade. The excitation frequency depends on engine order and engine rotational speed, $\omega = EO \times \Omega$.

$$\begin{aligned} \varphi_n &= 2\pi \cdot EO(n-1)/N \\ \mathbf{f}_{ex}^n(t) &= f_{ex}^{(1)} \times \exp[i(\omega t - \varphi_n))] \end{aligned} \tag{3}$$

3.1. *Harmonic Balance Method with Alternating Fourier Transformation*

Harmonic Balance Method with alternating Fourier transformation is used to solve the nonlinear dynamic equation Eq.(2) in frequency domain. HBM is also known as Fourier-Galerkin method, a continuous property is discretised by Fourier seires. Therefore, $\chi(t)$ representing periodic solution $\mathbf{x}(t)$, nonlinear force $\mathbf{f}_{nl}(t)$ and external excitation $\mathbf{f}_{ex}(t)$ is estimated by Fourier series truncated to N_H order of harmonics. $\tilde{\chi}$ is the collection of the all Fourier coefficients for both sine and cosine terms from 0th to N_H^{th} harmonics, where p is the order of harmonics.

$$\chi(t) = \sum_{p=0}^{N_H} \text{real}\left\{(\tilde{\chi}^{p,c} + i \cdot \tilde{\chi}^{p,s}) \times [\cos(p\omega t) + i \cdot \sin(p\omega t)]\right\} \tag{4}$$

$$\tilde{\chi} = [\tilde{\chi}^{0,c}, \tilde{\chi}^{1,c} + i\tilde{\chi}^{1,s}, \cdots, \tilde{\chi}^{N_H,c} + i\tilde{\chi}^{N_H,s}]^{\mathrm{T}}, \quad \chi \in [\mathbf{x}, \mathbf{f}_{ex}, \mathbf{f}_{nl}]$$

After implementing Fourier-Galerkin method, the dynamic equation can be expressed as Eq.(5) and Eq.(6), where $\mathbf{A}(\omega)$ is dynamic stiffness matrix in frequency domain shown in Eq.(7). $\mathbf{K}^p$, $\mathbf{M}^p$ and $\mathbf{C}^p$ are cyclic reduced stiffness, mass and damping matrices and will be discussed in details in Section 3.2.

$$\mathbf{A}(\omega) \times \tilde{\mathbf{x}} = \tilde{\mathbf{f}}_{ex} - \tilde{\mathbf{f}}_{nl}(\tilde{\mathbf{x}}) \tag{5}$$

$$\mathbf{R}(\tilde{\mathbf{x}}, \omega) \equiv \mathbf{A}(\omega) \times \tilde{\mathbf{x}} - \tilde{\mathbf{f}}_{ex} + \tilde{\mathbf{f}}_{nl}(\tilde{\mathbf{x}}) = 0 \tag{6}$$

$$\mathbf{A}(\omega) = \text{diag}(\mathbf{A}^0, \mathbf{A}^1, \mathbf{A}^2, \cdots, \mathbf{A}^{N_H}) \tag{7}$$

$$\mathbf{A}^p = \mathbf{K}^p - (p\omega)^2\mathbf{M}^p + i \times p\omega\mathbf{C}^p \tag{8}$$

$\mathbf{R}(\tilde{\mathbf{x}}, \omega)$ is residual of the HBM method expected to be solved iteratively based on Newton-Raphson method. In each iteration, the direct Fourier transformation(DFT) and inverse Fourier transformation(IFT) is alternated to compute $\tilde{\mathbf{f}}_{nl}$. The procedure is demonstrated in Eq.(9).

$$\tilde{\mathbf{x}} \overset{\text{IFT}}{\longrightarrow} \mathbf{x}(t), \dot{\mathbf{x}}(t) \overset{\text{Coulomb Friction}}{\longrightarrow} \mathbf{f}_{nl}(\dot{\mathbf{x}}, \mathbf{x}, t) \overset{\text{DFT}}{\longrightarrow} \tilde{\mathbf{f}}_{nl} \tag{9}$$

IOP Conf. Series: Journal of Physics: Conf. Series **1106** (2018) 012026 doi:10.1088/1742-6596/1106/1/012026

3.2. Cyclic Symmetric Techniques

In order to improve computational efficiency, the cyclic symmetric technique is applied based on the periodicity and cyclic symmetry of the blisk. The detailed description about cyclic symmetry has been demonstrated by Thomas [18]. In his work, he proposed $\mathbf{x}_R = \exp(\theta)\mathbf{x}_L$, where $\mathbf{x}_R$ and $\mathbf{x}_L$ are displacement of right and left boundaries. θ is purely imaginary value measuring the periodicity of mode shape. The periodicity of mode shape depends on harmonic order and engine order in this case. According to Wildheim's theory [19], θ is product of harmonic order and phase shift of each blisk, $\theta = \varphi_n p$. Therefore, the matrix for cyclic symmetric transformation is given in Eq.(10,11).

$$\Phi_p^{EO} = [\mathbf{I}, \mathbf{I} \times \exp(ip\varphi_1), \mathbf{I} \times \exp(ip\varphi_2), \cdots, \mathbf{I} \times \exp(ip\varphi_N)]^{\mathrm{T}} \tag{10}$$

$$\Phi^{EO} = \mathrm{diag}(\mathbf{I}, \Phi_1^{EO}, \Phi_2^{EO}, \cdots, \Phi_{N_H}^{EO}) \tag{11}$$

The Φ_p^{EO} for specific harmonic order p is used to obtained cyclic reduced mass, stiffness and damping matrices in Eq.(7). Φ^{EO} is globally transformation matrix including harmonic order from 0 to N_H and used to computing the cyclic reduced $\tilde{\chi}_{cyclic}$ and vice versa. $\mathbf{A}^+$ is Moore-Penrose pseudoinverse matrix of $\mathbf{A}$.

$$\Lambda_{cyclic}^p = (\Phi_p^{EO})^+ \Lambda \Phi_p^{EO}, \quad \Lambda \in [\mathbf{M},\ \mathbf{K},\ \mathbf{C}] \tag{12}$$

$$\tilde{\chi}_{cyclic} = (\Phi^{EO})^+ \times \tilde{\chi} \tag{13}$$

3.3. Nonlinear Normal Mode

According to Krack' work, a negative artificial damping is introduced into this system [20]. The autonomous system is expressed as Eq.(14), where $\zeta\mathbf{K}$ is an artificial stiffness proportional damping term. α is modal amplitude and ω_0 is natural frequency. The same Fourier-Galerkin process in Section 3.1 is applied and residual is given by Eq.(16). Compared to Eq.(6), there are two extra variables ζ and α. Hence, two additional constrians are expected which are mass normalization and phase normalization as shown in Eq.(17, 18). Firstly, mass normalization ensures a positive modal amplitude α. Secondly, a phase normalization is required, because absolute phase is arbitrary in autonomous system [14].

$$\mathbf{M} \cdot \ddot{\mathbf{x}}(t) - (\zeta \times \mathbf{K}) \cdot \dot{\mathbf{x}}(t) + \mathbf{K} \cdot \mathbf{x}(t) + \mathbf{f}_{nl}(\dot{\mathbf{x}}, \mathbf{x}, t) = 0 \tag{14}$$

$$\mathbf{x}(t) = \alpha \sum_{p=0}^{N_H} \mathrm{real}\left\{(\tilde{\mathbf{x}}^{p,c} + i \cdot \tilde{\mathbf{x}}^{p,s}) \times [\cos(p\omega_0 t) + i \cdot \sin(p\omega_0 t)]\right\} \tag{15}$$

$$\mathbf{R}_1(\tilde{\mathbf{x}}, \omega_0, \zeta, \alpha) = \mathbf{A}(\omega_0, \zeta) \times \alpha\tilde{\mathbf{x}} - \tilde{\mathbf{f}}_{nl}(\alpha\tilde{\mathbf{x}}) \tag{16}$$

$$\mathbf{R}_2(\tilde{\mathbf{x}}) = \tilde{\mathbf{x}}^{\mathrm{T}} \cdot \mathbf{M} \cdot \tilde{\mathbf{x}} - \mathbf{I} \tag{17}$$

$$\mathbf{R}_3(\tilde{\mathbf{x}}) = \mathrm{imag}(\tilde{\mathbf{x}}_{p=1}^1) \tag{18}$$

3.4. Continuation Method

For forced response analysis, system solution is expected to be computed within a range of excitation frequency ω. In the analysis of NNM, modal characteristics are required within a range of modal amplitude α. The continuation method is implemented to track the evolution of system behaviour as varying parameter, ω or α. A proper continuation process requires two steps. At the prediction step, the predicted point is found from previous solution to approximate accurate solution. Then, a new solution is updated from predicted point using Newton's method in correction step. The tangent prediction method and arc-length correction method is selected algorithms for predictor and corrector respectively. The reader can refer to [21] for a detailed description of continuation process.

IOP Conf. Series: Journal of Physics: Conf. Series **1106** (2018) 012026 doi:10.1088/1742-6596/1106/1/012026

4. Numerical Results

All natural frequencies (NF) of the lumped parameter model(without the ring damper) are plotted against nodal diameter(ND) in Figure 4. There are three families of mode, where each family of mode is associated with a nodal circle. The modal stiffness of disk increases dramatically with a growing ND [22]. Therefore, disk-dominated modes appear at the mode with relatively larger slop and strong blade-disk coupling, particularly at lower ND. The modes at higher ND tending to a horizontal line are blade-dominated modes with weaker blade-disk coupling. This blade-disk coupling is important because it gives access for communication of vibration energy between blade and disk. There is a veering region between 2nd and 3rd family of mode near ND 3. The modes in veering region tend to show high probability of mode interactions [23]. In order to evaluate the damping performance, the mode at 1st family ND 2 with strong coupling effect is excited. Due to strong bladed-disk coupling, both blade and disk show relatively large displacement when excitation is applied to blade.

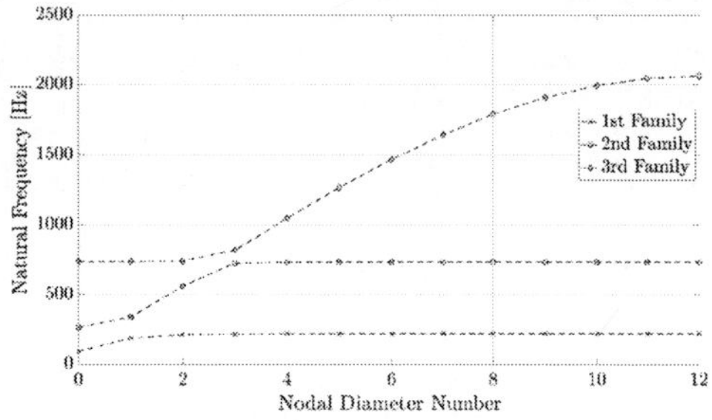

Figure 4. NF / ND Plot

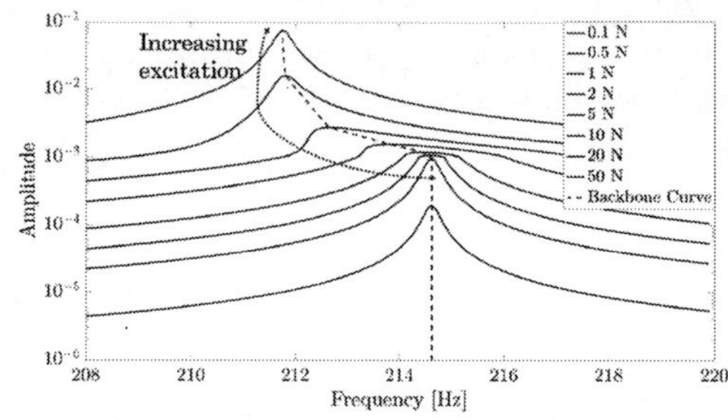

Figure 5. Different Excitation

4.1. Performance of Ring Damper

In this section, the damping ability of ring is discussed with the help of frequency response for a forced system. The amplitude of frequency response is recorded from $\mathbf{x}_{b2}$.

The frequency response with different excitation level is plotted in Figure 5. As can be seen, the magnitude of excitation force affects the contact condition resulting a softening effect. At excitation below 0.5 N, the contact is fully stuck and system shows linear behaviour. With greater excitation level, the displacement of system increases and sliding condition appears. The resonance frequency starts to shift left. When excitation is greater than 50 N, the contact is fully sliding and system becomes linear again with a lower resonance frequency.

The effect of damper mass and engine rotation speed on damping performance is analysed and results are shown in Figure 6 and Figure 7. This suggests that the optimal performance is reached with 30 g damper mass at rotational speed around 3000 rpm. As the increasing damper mass or rotation speed, the resonance peak shift right. The main reason contributing to this phenomenon is nonlinear softening effect as explained above. Normal contact force N_0 is proportional to the damper mass and squared of rotation speed. For this reason, N_0 is crucial factor in this case. The contact is fully stuck with larger normal contact force excited by same force level. When the normal contact force decreases, contact starts to slide and the resonance frequency is declining.

4.2. Modal Characteristics

During computation of nonlinear mode, the modal characteristics are obtained, including eigen-frequency, mode shape and damping ratio. Eigen-frequency and damping ratio are plotted against system energy shown in Figure 8 and Figure 9. As the system energy increases, natural

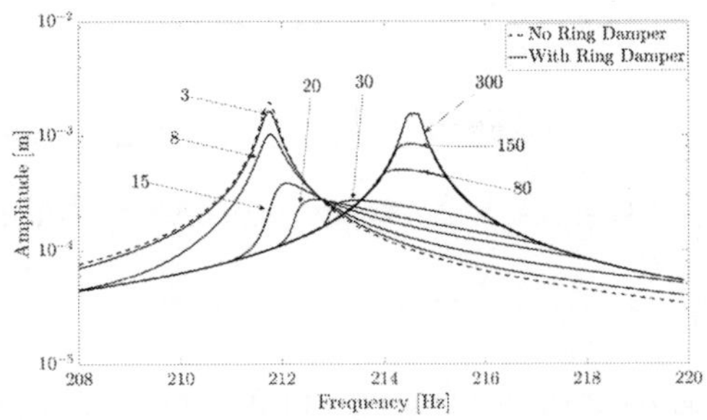

Figure 6. Different Damper Mass [g]

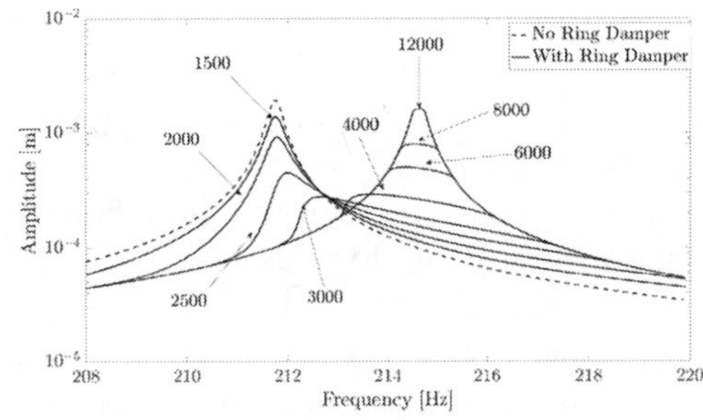

Figure 7. Different Rotation Speed [rpm]

frequency is decreasing due to softening effect caused by friction, which corresponds to the backbone curve in Figure 5. In terms of damping ratio, it stays at zero with lower system energy. At this point, the contact is fully stuck. It follows that there is no energy dissipation due to friction. At the transition from sticking to sliding, damping ratio reaches its maximum point and ring damper provides the optimal damping performance. When system energy is larger enough, damping ratio begins to drop because of fully sliding contact. The damping ratio will return to zero asymptotically with infinity system energy.

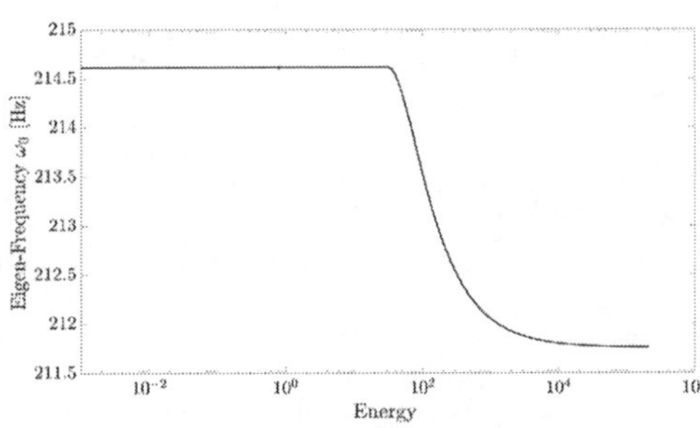

Figure 8. Frequency Energy Plot

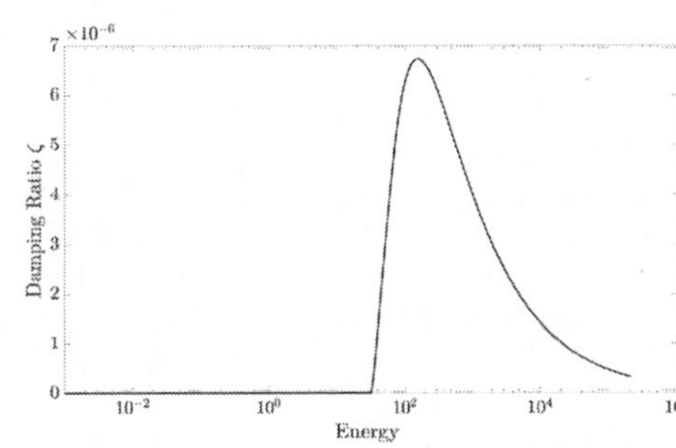

Figure 9. Damping Ratio / Energy

5. Conclusion

In summary, a lumped parameter model with Jenkins element model is used to simulate the nonlinear dynamic behavior of blisk with a frictional ring damper. This nonlinear dynamic system is analysed through forced frequency response and nonlinear modal analysis. The numerical results provided a qualitative analysis for nonlinear dynamic behaviour of the blisk with frictional ring damper.

Generally, both effectiveness and limitations of this ring damper are clearly pointed out from the analysis. The performance of the damper is highly depending on the relative motion between disk and ring. Ring damper is able to provide considerable damping when relative displacement is large. In other words, if a mode with weak bladed-disk coupling effect is excited, the motion of disk is relatively small and ring damper will show less effectiveness. In addition, the normal contact reaction force is also crucial for damping effect. In a sense, if the ring is well designed for blisk, this damping technique can be regarded as an alternative to classical friction joints.

In terms of numerical methodology, modal damping ratio and natural frequency are directly and efficiently computed within modal analysis. Forced response analysis provides corresponding results to the modal characteristics. Finally, this paper indicates that nonlinear normal mode is an efficient way to study dynamic behaviour of blisk with frictional ring damper.

Acknowledgement

The first author is grateful to China Scholarship Council (File NO. 201708060239) for providing the financial support for this project.

References

[1] Krack M, Salles L and Thouverez F, 2016, Vibration Prediction of Bladed Disks Coupled by Friction Joints, *Archives of Computational Methods in Engineering*, **24**, pp. 589–636
[2] Griffin J H and Sinha A, 1985, The Interaction between Mistuning and Friction in the Forced Response of Bladed Disk Assemblies, *Journal of Engineering for Gas Turbines and Power*, **107**, pp. 205–211.
[3] Griffin J H, 1990, A Review of Friction Damping of Turbine Blade Vibration, *International Journal of Turbo and Jet Engines*, **7**, pp. 297–307.
[4] Pesaresi L, Salles L, Jones A, Green J S and Schwingshackl C W, 2017, Modelling the Nonlinear Behaviour of an Underplatform Damper Test Rig for Turbine Applications, *Mechanical Systems and Signal Processing*, **85**, pp. 662–679.
[5] Armand J, Pesaresi L, Salles L and Schwingshackl C W, 2016, A Multiscale Approach for Nonlinear Dynamic Response Predictions With Fretting Wear, *Journal of Engineering for Gas Turbines and Power*, **139**(2), pp. 022505–022505–7
[6] Laxalde D, Salles L, Blanc L and Thouverez F, and Lombard J P, 2008, Non-Linear Modal Analysis for Bladed Disks With Friction Contact Interfaces, *Proceedings of ASME Turbo Expo 2008*, **5** (4), pp. 456–467.
[7] Cardona A, Lerusse A and Géradin M, 1998, Fast Fourier Nonlinear Vibration Analysis, *Computational Mechanics*, **22**, pp. 128–142
[8] Yuan J, El-Haddad F, Salles L and Wong C, 2018, Numerical Assessment of Reduced Order Modeling Techniques for Dynamic Analysis of Jointed Structures with Contact Nonlinearities, *Proceedings of ASME Turbo Expo 2018*, Accepted.
[9] Kerschen G, 2014, *Modal Analysis of Nonlinear Mechanical Systems*, (Vienna: Springer), pp. 215–250.
[10] Detroux T, Renson L, Massetand L and Kerschen G, 2015, The Harmonic Balance Method for Bifurcation Analysis of Large-scale Nonlinear Mechanical Systems, *Computer Methods in Applied Mechanics and Engineering*, **296**, pp. 18–38.
[11] Cameron T M and Griffin J H, 1989, An Alternating Frequency/Time Domain Method for Calculating the Steady-state Response of Nonlinear Dynamic Systems, *Journal of Applied Mechanics.*
[12] Rosenberg R M, 1960, Normal Modes of Nonlinear Dual-mode Systems, *Journal of Applied Mechanics*, **27**, pp. 263–268
[13] Shaw S W and Pierre C, 1991, Non-linear Normal Modes and Invariant Manifolds, *Journal of Sound and Vibration*, **150**(1), pp. 170–173
[14] Krack M, 2014, *The Concept of Nonlinear Modes Applied to Friction-Damped Systems* (Doctoral Thesis), Leibniz Universität Hannover
[15] Yuan J, Fabrizio S, Branislav T, Giuliano A, Sophoclis P and Ramesh R, 2017, Novel Frame Model for Mistuning Analysis of Bladed Disk Systems, *Journal of Vibration and Acoustics*, **139**, pp. 031016–031016–13
[16] Laxalde D, Sinou J J, Thouverez F, and Lombard J P, 2007, Qualitative Analysis of Forced Response of Blisks with Friction Ring Dampers, *European Journal of Mechanics A/Solids*, **26** (4), pp. 676–687.
[17] Schwingshackl C W, Petrov E P and Ewins D J, 2012, Effects of Contact Interface Parameters on Vibration of Turbine Bladed Disks With Underplatform Dampers, *Journal of Engineering for Gas Turbines and Power*, **134**(3), pp. 032507–032507–8
[18] Thomas D L, 1974, Standing Waves in Rotationally Periodic Structures, *Journal of Sound and Vibration*, **37**, pp. 288–290
[19] Wildheim S, 1979, Excitation of Rotationally Periodic Structures, *Transaction of ASME*, **46**, pp. 878–883
[20] Krack M, 2015, Nonlinear Modal Analysis of Nonconservative Systems: Extension of the Periodic Motion Concept, *Computers and Structures*, **154**, pp. 59–71
[21] Sarrouy E and Sinou J J, 2011, *Advances in Vibration Analysis Research* (Rijeka: nTech), pp. 419–434
[22] Castanier M P and Pierre C, 2006, Modeling and Analysis of Mistuned Bladed Disk Vibration: Current Status and Emerging Directions, *Journal of Propulsion and Power*, **22**(2), pp. 384–396
[23] Kenyon J, Griffin J and Kim N, 2004, Frequency Veering Effects on Mistuned Bladed Disk Forced Response, *Journal of Propulsion and Power*, **20**(5), pp. 863–870

IOP Conf. Series: Journal of Physics: Conf. Series **1106** (2018) 012027 doi:10.1088/1742-6596/1106/1/012027

Novelty Detection in a Cantilever Beam using Extreme Function Theory

Martucci D.[1], Civera M.[2], Surace C.[1], Worden K.[3]

[1] DISEG Department of Structural, Geothecnical and Building Engineering, Politecnico di Torino, Corso Duca degli Abruzzi 24, Turin, IT

[2] DIMEAS Department of Meachanical and Aerospace Engineering, Politecnico di Torino, Corso Duca degli Abruzzi 24, Turin, IT

[3] Dynamics Research Group, Department of Mechanical Engineering, University of Sheffield, Mappin Street, Sheffield S1 3JD, UK

E-mail: marco.civera@polito.it

Abstract. Damage detection and localisation in beam-like structures using mode shape features is well-established in the research community. It is known that by inserting a localised anomaly in a cantilever beam, such as a crack, its mode shapes diverge from the usual deflection path. These novelties can hence be detected by a machine-learner trained exclusively on the modal data taken from the pristine beam. Nevertheless, a major issue in current practices regards discerning between damage-related outliers and simple noise in observations, avoiding false alarms. Extreme functions are here introduced as a viable mean of comparison. By combining Extreme Value Theory (EVT) and Gaussian Process (GP) Regression, one can investigate functions as a whole rather than focusing on their constituent data points. Indeed, n discrete observations of a mode shape sampled at D points can be assumed as 1-dimensional sets of n randomly distributed observations. From any given point it is then possible to define its *Probability Density Function* (PDF) and the *Cumulative Density Function* (CDF), whose minima, according to the EVT, belong to one of three feasible extreme distributions - *Weibull, Frechet or Gumbel.* Thus, these functions - intended as vectors of sampled data - can be compared and classified. Anomalous displacement values that could indicate the presence of a crack are therefore identified and related to damage. In this paper, the effectiveness of the proposed methodology is verified on numerically-simulated noisy data, considering several crack locations, levels of damage severity (i.e., depths of the crack) and signal-to-noise ratios.

1. Introduction

A key factor in all probabilistic and semi-probabilistic approaches used in sciences and technology fields regards the assumptions on the statistical distributions of the available experimental data. On the basis of the well-known Central Limit Theorem (CLT), Gaussian or Gaussian-like probability distributions are mostly used. Nevertheless, when dealing with extreme deviations from the mean, hypothesising Gaussianity may misguide. On the other hand, Extreme Value Statistics (EVS), based on EVT [1], has been established specifically to handle the values of the independent random variables failing in the tails of the distribution of interest [2, 3, 4]. This is particularly interesting in the field of Structural Health Monitoring (SHM).

Indeed, most widely used damage detection techniques revolve around outlier detection (see, e.g., [5] and [6]), which in turn needs, in some sense, a sort of "thresholding" between data taken from damaged and undamaged conditions. This is due to the assumption that some features in the output recorded from the structure in an abnormal situation will be above the so-determined threshold. This can be intended as a statistical test performed on the experimental observations, which is regularly carried out considering some scalar values as the damage-sensitive features. Unfortunately, false positives are not uncommon in outlier analysis when performed on variables assumed as independent and identically distributed (*i.i.d.*) and/or Gaussian distributed. In fact, this assumption does not always hold true in practice.

EVT is an ideal statistical framework for evaluating the significance of extreme values departing from the normality model and the concept has already been exploited in this sense [7]. It must be remarked that *normality*, or *normal condition*, refers here to the pristine state of the structure; *Gaussian* is used instead to indicate the normal probability distribution function (PDF), to avoid any confusion.

In this work, the Generalised Extreme Value (GEV) distribution, which combines Gumbel, Frechet and Weibull ones in a unique model [9], has been used and its hyperparameters have been optimised through the Self-Adaptive Differential Evolution algorithm (SADE), [10]. Thanks to the adaptive nature of the adopted optimisation algorithm, the parameters of the genetic process have been deduced by the data themselves rather than being arbitrarily imposed. The novelty of this research, if compared to the previously mentioned papers and other works, resides in the application of GEV to the mode shapes as a whole. This represents a quite different approach from the more common pointwise comparison between models: the feature selected for the statistical test is not a univariate scalar, nor a multivariate collection of them, assembled in an array, but a function. This study relies on the EFT path initiated by [11] and [8]. In order to assess the mode shapes as continuous functions starting from a discrete amount of observations, GP regression has been applied. In fact, even if existing since more than a decade [12], GP Regression has become highly valued in the field of SHM only in recent years [13, 14, 15, 16]. Gaussian Processes return a Gaussian probability distribution, and the confidence intervals of the prediction; the principled Bayesian foundations make them exceptionally apt for statistical investigations. As will be explained in further detail later, combining GP Regression and the GEV distribution on modal data produces very promising results.

The proposed methodology has been here applied to a numerical case, based on a Finite Element (FE) model of a beam in various boundary conditions, where an edge crack has been simulated by deriving the stiffness matrix through the definition of the *stress intensity factor*, holding the hypotheses of invariance of crack in time, invariance of damaged element mass distribution and neglecting the *opening-closing* fracture mechanism [16, 17]. Several combinations of damage severity and noise levels (artificially included in simulations) have been investigated.

The paper is organised as follows: in Section two, some basic concepts of GP Regression, the EVT and the EFT, as well as some hints about the genetic algorithm used for this task, are briefly recalled; in Section three, results are reported and discussed; finally, a discussion is presented in Section four.

2. Gaussian Process and Extreme Function Theory

The theoretical FE mode shapes simulated using the Euler-Bernoulli dynamic elastic theory are noise free, hence to give them the aleatory behaviour that they would have in reality, Gaussian white noise has been added (at different levels) through the specification of the *Signal to Noise Ratio* (SNR), expressed in Decibels $[Db]$ as,

$$SNR = 20\log_{10}\left(\frac{A_{signal}}{A_{noise}}\right) \tag{1}$$

where A_{signal} is the noise-free signal amplitude and A_{noise} is the amplitude of the added noise. First, two large sets of data have been simulated with the FE Method, the first one made up from noisy undamaged transverse mode shapes, while the second one collects the noisy damaged ones. This step is operational to obtain a robust statistical pool of data from which different subsets will be randomly sampled later, avoiding intersections among them. Then three subsets of data are randomly sampled for each analysed mode shape from the parent sets as follows:

- *Training set (TR)*: composed only of the undamaged data.
- *Validation set (VA)*: composed only of the undamaged data.
- *Test set (TS)*: composed of the undamaged and the damaged data.

The training set TR is used to train the Gaussian Process, fitting the parameters of the variable distribution through the optimisation of the predictive joint conditional *Marginal Likelihood* by minimising its negative logarithmic form [12]. Combining the current parameter estimation with the validation set VA leads to the generation of the PDF standardised logarithmic values. As explained in [12], the noisy observations $\mathbf{y} = f(\mathbf{x}) + \varepsilon$ of a 1-dimensional (D) problem, that in this case are the beam transverse displacements at the node locations $\mathbf{x}$ disturbed by the i.i.d. noise ε, have a Gaussian joint distribution of the target values $\mathbf{f}^*$ and the function values at the test locations $\mathbf{x}^*$ defined as follows,

$$p\left(\mathbf{f}^* | \mathbf{x}, \mathbf{y}, \mathbf{x}^*\right) \sim \mathcal{N}(m^*, K^*) \tag{2}$$

where m^* is the predictive mean and K^* is the predictive covariance matrix, expressed as,

$$m^* = k\left(\mathbf{x}^*, \mathbf{x}\right)\left[k(\mathbf{x}, \mathbf{x}) + \sigma_n^2\right]^{-1}\mathbf{y} \tag{3}$$

$$K^* = cov(\mathbf{f}^*) = k\left(\mathbf{x}^*, \mathbf{x}^*\right) - k\left(\mathbf{x}^*, \mathbf{x}\right)\left(k(\mathbf{x}, \mathbf{x}) + \sigma_n^2\right)^{-1} k\left(\mathbf{x}, \mathbf{x}^*\right) \tag{4}$$

The resulting multivariate Gaussian probability distributions take the form,

$$p = \frac{1}{\sqrt{2\pi^D |K^*|}} \exp\left[-\frac{1}{2}(\mathbf{f}^* - m^*)^T K^{*-1}(\mathbf{f}^* - m^*)\right] \tag{5}$$

The prior information of the process is specified in the covariance matrix K^*, in this case assumed as a *Squared-Exponential.* In the one dimensional case it can be expressed in the form,

$$k_y(x_p, x_q) = \sigma_f^2 \exp\left[-\frac{1}{2l^2}(x_p - x_q)^2\right] + \sigma_n^2 \delta_{p,q} \tag{6}$$

where $\delta_{p,q}$ is the Kronecker delta and l, σ_f, σ_n are the *hyperparameters,* whose optimisation is based on the *marginalisation property* of the *marginal Likelihood* (ML). The variables x_p and x_q are the nodal locations of the training data data set corresponding to different position indices p, q. The ML is taken in the negative logarithmic form (NLML), since it is easier to perform a minimisation, resulting in,

$$NLML = \log p\left(\mathbf{y}|\mathbf{x}, l, \sigma_f, \sigma_n\right) = \frac{1}{2}\mathbf{y}^T L_{xx}^T \backslash (L_{xx} \backslash \mathbf{y}) + \sum \log\left(diag(L_{xx})\right) + \frac{n}{2}\log 2\pi \tag{7}$$

where L_{xx} is the lower Cholesky decomposition of the covariance matrix $k(\mathbf{x}, \mathbf{x})$ and the operator "\" indicates the left matrix division, required for a faster computation of the matrix inversion. At the end of the iterative minimisation process, when the difference in likelihood values are smaller than a given tolerance, the obtained hyperparameters are the best ones fitting the covariance function on the training data considered.

EVT is by definition a point-wise approach, generally applied to univariate data or extended to other low-dimensional spaces. It can be adapted for functional applications (EFT) [11] to identify extreme functions from a given n-dimensional multivariate Gaussian distribution; in the case proposed here, the functions are the mode shapes of the structure, interpolated from n discrete observations. Hence, defining a single value of PDF for each tested function as $\mathbf{z} = f_n(\mathbf{f}^*)$ allows a better distinction between the more extreme *normal* function (undamaged mode shapes) from the *abnormal* ones (damaged mode shapes) resulting in a reduction of wrong identifications.

The feasible extreme distributions for a sufficiently wide number of samples (tending to infinity), taken from an arbitrary parent distribution, are limited to three types: Gumbel, Frechèt and Weibull [1]. Alternatively, all the three mentioned cases can be collectively expressed using the generalised expression of the GEV, where the correspondence of their parameters is reported in equations (9).
A first posterior probability is calculated from equation (5) conditioning the validation data set VA on the training data set TR, in order to make statistically relevant the further results coming form this data combination and independent from the test data TS. The Gaussian probability z is taken in its logarithmic form ($lz = log(z)$) and processed through a genetic algorithm of the type *Differential Evolution* (DE) [18] to estimate the related CDF of the minima extreme distributions, since one could expect few occurrences of the anomalous values with respect to the normal ones. The best parameters are selected by minimising a cost function, which is the normalised mean-squared error (NMSE) in a fit of a parametric model to a given cumulative distribution function (CDF). Thanks to its heuristic approach of trying different random possible solutions and several restarts of the whole process, DE algorithms enhance robustness in the estimation while avoiding local minima, without requiring any *a-priori* knowledge on the defined range of parameters. In this study, the self-adaptive (SADE) modified version of the genetic differential algorithm proposed by [10] has been applied to fit the GEV cumulative function, with two relevant advantages: the adaptive feature of the optimisation does not require the specification of the initial parameters, such as the scaling factor or the crossover ratio, while the choice to use the generalised distribution bypasses the issue of the estimation of the domain of attraction. Regarding the GEV minima distribution (L), this has equation,

$$L(lz,\mu,\sigma,\gamma) = 1 - exp\left\{-\left[1+\gamma\left(\frac{\mu - lz}{\sigma}\right)\right]^{-\frac{1}{\gamma}}\right\} \tag{8}$$

where μ, σ, and γ denote, respectively location, scale, and shape of the GEV distribution. A direct correlation with the parameters of the 3 feasible limit minima distributions is reported in [9],

$$\begin{aligned}
GUMBEL: &\quad if \quad \gamma \to 0, \quad L_G(lz,\lambda,\delta,\beta) \\
&\qquad \mu = \lambda, \quad \sigma = \delta \\
WEIBULL: &\quad if \quad \gamma < 0, \quad L_W(lz,\lambda,\delta,\beta) \\
&\qquad \mu = \lambda + \delta, \quad \sigma = \frac{\delta}{\beta}, \quad \gamma = -\frac{1}{\beta} \\
FRECHET: &\quad if \quad \gamma > 0, \quad L_F(lz,\lambda,\delta,\beta) \\
&\qquad \mu = \lambda - \delta, \quad \sigma = \frac{\delta}{\beta}, \quad \gamma = \frac{1}{\beta}
\end{aligned} \tag{9}$$

To individuate the anomalous mode shapes that could indicate the presence of a damage in the beam structure, it is necessary to discern them from the mode shapes that fall in a normal range. After reconstructing the CDF it is possible to define a threshold lz_{lim} in correspondence of a given quantile α, in this case set to the lower 1%. For the test data set TS, a new logarithmic posterior probability lz_{test} is now calculated from equation (5) for each tested mode shapes. When a logarithmic probability falls in a quantile α' beyond the defined limit α, the value is recorded as an outlier and the entire mode shape identified as taken from a damaged structure. Since the TS data set contains undamaged and damaged mode shapes, when an undamaged mode shape is detected as damaged it is recorded as "false positive" and, on the other way, when a damaged mode is not recognised, a "false negative" misclassification occurs. The overall rate of success in correct classification is registered since it is of primary interest to determine the effectiveness of the used damage features, in this particular case the transverse displacement of the beam. Figure 1 shows an overview of the whole algorithm.

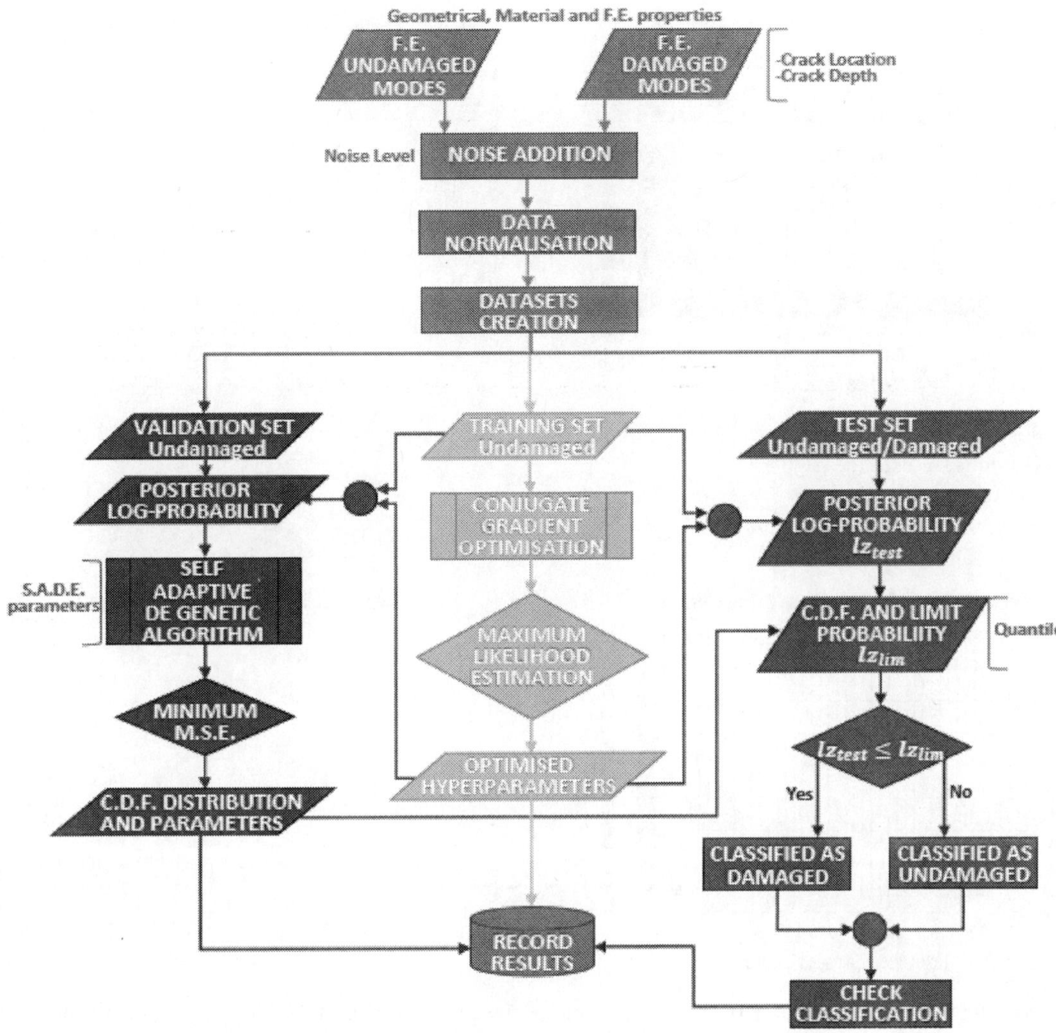

Figure 1: Flow chart showing with different colours data creation, training, validation and detection steps.

3. Results

As introduced previously, different boundary conditions were simulated; due to space limitations, only the results of the cantilever beam structure are here reported as reference. The results are displayed as comprehensive graphics where the x-axis and y-axis represent the crack location and the crack depth, respectively, while the cell colour in grey scale is the percentage of successful identifications of damaged modes. Figures from 2 to 4 below on the left side show resulting damage assessment using an additive noise with SNR = 80 dB for the first 3 modes. Instead, Figures from 2 to 4 on the right side show the theoretical (noise-free) inverse Euclidean distance in transverse displacements between undamaged and damaged modes, normalised to 1, in order to check the correctness of the obtained results. Initially, the different simulations were elaborated taking into account all the FE nodes as sensor locations in order to assess the correctness of the method. Successively, the number of input was reduced to verify the performance of the algorithm when dealing with fewer measurement points. In this case, among the 51 nodes, only 11 of them were considered as equally spaced sensor locations and used to train and validate the process (30 and 200 data copies, respectively). As one would expect, peaks of error occur near the free end, since the

IOP Conf. Series: Journal of Physics: Conf. Series **1106** (2018) 012027 doi:10.1088/1742-6596/1106/1/012027

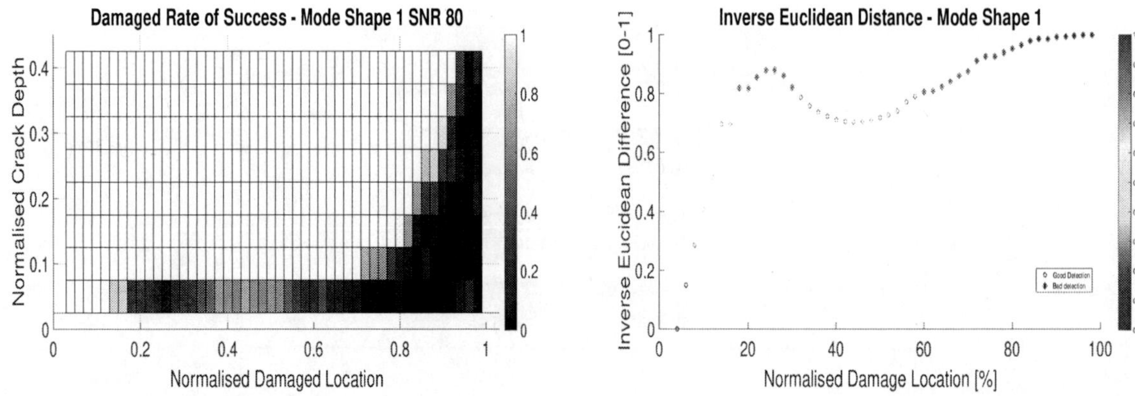

Figure 2: Results Cantilever Beam (CB) - Mode Shape (MS) 1, considering 11 sensors and SNR=80 dB (left) and its inverse Euclidean distance (right).

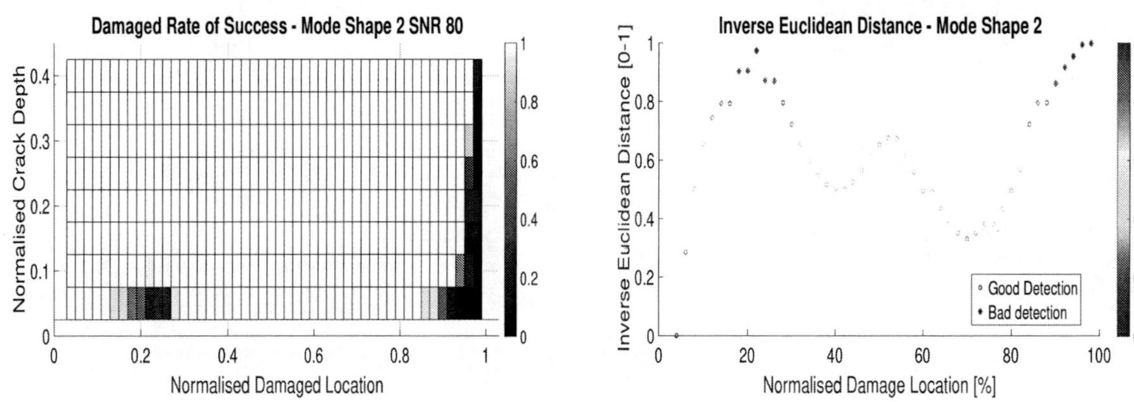

Figure 3: Results CB - MS 2 considering 11 sensors and SNR=80 dB (left) and its inverse Euclidean distance (right).

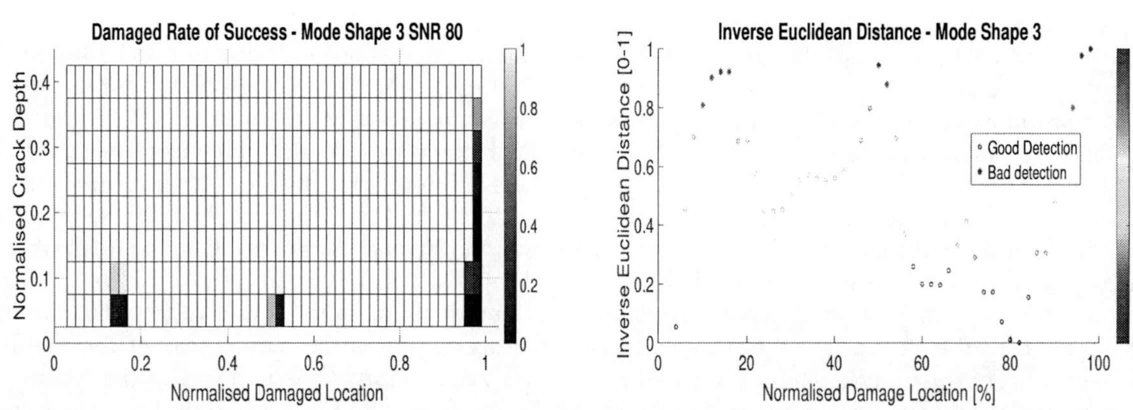

Figure 4: Results CB - MS 3 considering 11 sensors and SNR=80 dB (left) and its inverse Euclidean distance (right).

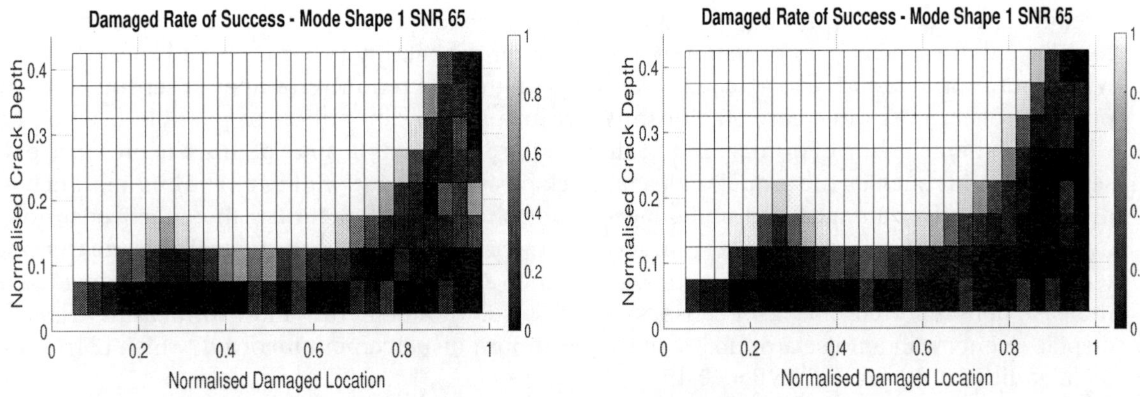

Figure 5: Results CB - MS 1 SNR=65 dB, considering 30 sensors (left) and 11 sensors (right).

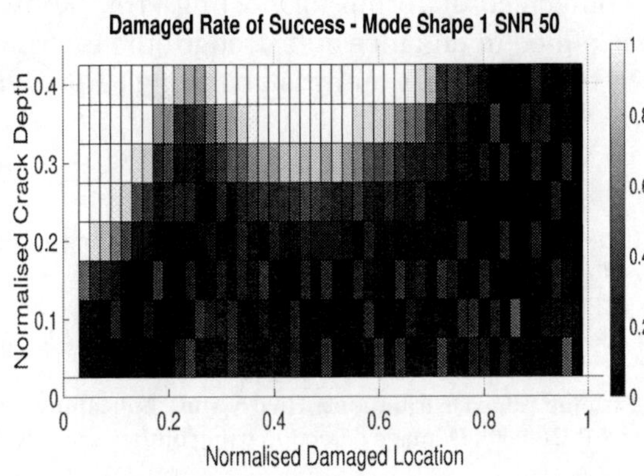

Figure 6: Results CB - MS 1 considering 11 sensors and SNR=50 dB.

crack influence decreases moving away from the clamped end, and in correspondence of the null-displacement points (i.e. modal nodes). Less intuitive is the peak of error located around the 20% of beam length for the particular boundary conditions of the cantilever beam: when a crack occurs in this location, the Euclidean distance between damaged and undamaged mode shapes is very small thus a correct detection result is difficult. The correspondence between peaks of the error in detection and peaks in the curve representing the inverse Euclidean distance is evident from the Figures 2 to 4. A comparison between results obtained with a different number of used sensors is shown for the first mode Figure 5: the left figure is obtained considering a noise level of 65 dB and it is evaluated with a coarser discretisation of 30 elements (instead of 50) and taking into account all of them as sensor locations. The right figure, instead, considers 11 sensors among the 31 nodes. As can be noted, the overall performance of the algorithm remains almost unvaried, while the computational time is definitely lower. Finally, Figure 6 shows the obtained results for a noise level of SNR = 50 dB, that can be assumed as limit for damage assessment beyond which the noisy mode shapes are too disturbed to be analysed.

4. Conclusions

The method proposed here shows some advantages, but also some weak points. A first point of interest surely is the capability to work without differentiating transverse displacement data, which always involves a manipulation of values that leads to an amplification of noise. Moreover, from a computational point of view, the algorithm elaborates sufficiently accurate results starting from a reduced number of input data or without a dense discretisation of the structure, remaining able to collect all the available information in the posterior probability in a few seconds per whole cycle with a non-optimised Matlab ® code. When working with a reduced number of sensors, the "false positive" occurrences decrease significantly because of the diminution of possible false outliers, which is a characteristic of primary importance in the SHM field. In spite of this, the presence of less sensitive points such as the modal nodes or crack position corresponding to a minimum Euclidean distance between damaged and undamaged mode shapes locally reduces the effectiveness of the method. Another issue regards the accuracy of the sensors used: the current available measurement technology, state-of-art devices able to consider the vast number of output channels considered in the case here reported, like laser sensors or high resolution cameras, allow to measure transverse displacements with a noise around 3% of the maximum displacement involved, which corresponds to an additive noise level of SNR = 40 dB.

Acknowledgements
The authors wish to thank Politecnico di Torino for hosting Prof. Keith Worden in the context of the *Visiting Professors Project*, aimed to enhance and strength the internationalisation process. The project has been supported by the European Commission through *Horizon 2020* and *HRS4R - Human Resources Strategy for Researchers* programs.

References

[1] Fisher R.A., Tippett L.H.C., (1928) Limiting forms of the frequency distributions of the largest or smallest members of a sample, *Proceedings of the Cambridge Phil. Soc.*, **24**:180-190.
[2] Gumbel E.J. (1958), Statistics of Extremes, *Columbia University Press*, New York.
[3] Castillo E. (1988), Extreme Value Theory in Engineering. *Academic Press Series in Statistical Modelling and Decision Science*, San Diego, CA.
[4] Kotz S., Nadarajah S. (2000), Extreme Value Distributions, Theory and Applications. *Imperial College Press.*
[5] Worden K., Manson G., Fieller N.R.J. (2000), Damage detection using outlier analysis. *Journal Of Sound And Vibration*, **229(3)**:647-667.
[6] Worden K., Manson G., Sohn H., Farrar C. (2005), Extreme value statistics from differential evolution for damage detection, *Proc. 23rd Int. Modal Analysis Conf.*
[7] Roberts S., Novelty detection using extreme value statistics (1998), *IEE Proc.in Vision, Image and Signal Processing* **146**:124-129.
[8] Papatheou E., Dervilis N., Maguire A.E., Campos C., Antoniadou I., Worden K. (2017), Performance monitoring of a wind turbine using extreme function theory, *Renewable Energy*, **113**:1490-1502.
[9] Park H.W., Sohn H., (2006), Parameter estimation of the generalized extreme value distribution for structural health monitoring, *Probabilistic Engineering Mechanics*, **21(4)**:366-376.
[10] Qin A.K., Suganthan P.N. (2005), Self-adaptive differential evolution algorithm for numerical optimization, *Proc. IEEE Congr. Evol. Comput.*, **2**:1785–1791.
[11] Clifton D.A., Clifton L., Hugueny S., Wong D., Tarassenko L. (2013), An extreme function theory for novelty detection. *IEEE Journal of Selected Topics in Signal Processing*, **7(1)**:28-37.
[12] Rasmussen C.E., Williams C. (2006) Gaussian Processes for Machine Learning, *The MIT Press*, Cambridge, MA, USA **38**:715-719.
[13] Worden K., Farrar C.R. (2013), Structural Health Monitoring: a Machine Learning Perspective, *Wiley and Sons.*
[14] Corrado N., Surace C., Montanari L., Spagnoli A. (2015), Comparing three derivative discontinuities detection methods for the localisation of cracks in beam-like structures. *Proc. Int. Workshop on Structural Health Monitoring*, Palo Alto, CA, USA.
[15] Civera M., Surace C., Worden K. (2017), Detection of cracks in beams using treed Gaussian processes, *Structural Health Monitoring and Damage Detection*, Springer, **7**:85-97.
[16] Corrado N., Durrande N. Gherlone M, Hensman J., Mattone M., Surace C. (2017), Single and multiple crack localization in beam-like structures using a Gaussian process regression approach, *Journal of Vibration and Control*, Sage, **1**:16.
[17] Qian G.L., Gu S.N., Jiang J.S. (1990), The dynamic behaviour and crack detection of a beam with crack, *Journal of Vibration and Control*, **138(2)**:233-243.
[18] Storn R., Price R. (1997), Differential evolution a simple and efficient heuristic for global optimisation over continuous spaces, *Journal of Global Optimization*, **11**:341-359.

IOP Conf. Series: Journal of Physics: Conf. Series **1106** (2018) 012028 doi:10.1088/1742-6596/1106/1/012028

Detection and Location of Nonlinearities using Reciprocity Breakdown

MG Wood[1,2], JET Penny[1] and MI Friswell[3]

[1] School of Engineering and Applied Science, Aston University, Birmingham B4 7ET, UK
[2] Mechanical Engineering, University of Sheffield, Sheffield S3 7QB, UK
[3] College of Engineering, Swansea University, Swansea SA1 8EN, UK

E-mail: j.e.t.penny@aston.ac.uk

Abstract. Damage in a structure often results in local stiffness nonlinearities and detecting these nonlinearities can be used to monitor the health of the structure. It is well-known that nonlinearities in structures lead to a breakdown in reciprocity, where the frequency response function between two points on the structure depends upon the forcing location. This paper proposes a measure to quantify the level of non-reciprocity in a structure and investigates the effect of the location and form of nonlinearity on this non-reciprocity measure. A simulated discrete mass-spring system was used to determine the effect of the excitation and response locations on the ability to detect the nonlinearity. Stepped-sine testing is commonly used to characterise a nonlinear system since harmonic excitation emphasises nonlinear phenomena and can, for example, allow the system to exhibit multiple solutions. Thus, a simulation of a stepped sine test was used as the benchmark to highlight reciprocity breakdown in the most favourable case. However, impact excitation is much easier and faster to implement in practice and consequently the effect of the type of excitation on the detection and location of nonlinearities was considered. Finally, the prospects for using a measure of reciprocity in a structural health monitoring system are discussed.

1. Introduction

Operational integrity is an essential requirement for any structure. Typically structures comprise an assembly of component parts, some of which may allow easy access to assess the robustness of those individual components through using either vision-based techniques or non-destructive examination methods [1, 2]. Conversely, some parts of a structure may not be accessible and a visual inspection or conventional non-destructive testing methods cannot be applied.

Safety critical structures range from composite aircraft skins and panels to concrete nuclear containments and so the aim of preventative maintenance is to identify damage within these structures as early as possible. In these cases, damage is most difficult to detect if it has occurred deep within a structure away from direct access for standard non-invasive assessment. Thus a great deal of research and innovation has centred upon identifying damage within continuous structures without the need for direct access [3].

Many approaches to damage location within structures involve the use of vibration in one form or another since its application is non-destructive [4]. In the case of safety critical structures, vibration based inspection is an approach that shows great promise [5]. Vibration testing provides a wealth of data and the interpretation of the data and the combination of outputs proves most useful in damage location [6]. A number of authors have addressed the problem of quantifying the magnitude and location of damage within vibrating structures [7]. One approach to damage location using vibration is to use non-reciprocity as a measure [8, 9]; Manson *et al.* [10] provided an interpretation of non-reciprocity with a sound mathematical basis.

This paper investigates the non-reciprocal vibration outputs from an idealised discrete system that simulates damage by a simple nonlinear spring. A scalar based on the measured data between reciprocal locations is proposed to quantify the change of state (i.e. damage).

2. Non-Reciprocity

The vibration data from a structure or system under investigation is usually obtained experimentally to validate a mathematical model that represents its behaviour. Such mathematical models may be linearised to describe characteristics such as the natural frequencies, damping ratios and mode shapes [11]. Thus, in the practical case of periodic monitoring, any changes in these parameters should identify damage within a structure.

The principal vibrational signature of any structure or system obtained through vibration testing is the Frequency Response Function (FRF). A common type of FRF is the Inertance, $\mathbf{A}(\omega)$, given by

$$\mathbf{A}(\omega) = -\omega^2\alpha(\omega) = -\omega^2(\mathbf{K} - \omega^2\mathbf{M} + \imath\omega\mathbf{C})^{-1} \tag{1}$$

where $\mathbf{M}$, $\mathbf{K}$ and $\mathbf{C}$ are the mass, stiffness and damping matrices of the linearised system, and α is the Receptance. $A_{jk}(\omega)$ is the acceleration response at point j due to a unit harmonic excitation of frequency ω at point k, with no other forces being applied to the system. $A_{jk}(\omega)$ may also be expressed in terms of the natural frequencies and mode shapes; for modal damping the Inertance is

$$A_{jk}(\omega) = -\omega^2 \sum_{i=1}^{n} \frac{\{\phi_i\}_j \{\phi_i\}_k}{\omega_i^2 - \omega^2 + 2\imath\zeta_i\omega_i\omega} \tag{2}$$

where ω_i and ζ_i are the i th natural frequency and damping ratio, $\{\phi_i\}_j$ and $\{\phi_i\}_k$ are the j th and k th elements of the i th mode shape, and $\imath = \sqrt{-1}$.

An important characteristic of any linear structure is its reciprocal behaviour. Thus, if a structure is linear, then $A_{jk}(\omega)$ must equal $A_{kj}(\omega)$. However, the FRF can be affected by non-linearities in the structure and by the testing regime. For example, a system with a cubic-stiffness characteristic can exhibit a nonlinear response, such as softening or hardening behaviour, or jumps in the response, due to the type and level of excitation. From a purely physical perspective, regardless of the levels of care employed in their manufacture, all structures contain imperfections or flaws ranging from microscopic voids to material irregularities and both result in local nonlinearity. Subject to loading even within their operational envelopes, these structural voids and material irregularities become precursors of physical damage whereby repeated and / or cyclic loading propagate and magnify that damage using the original voids and irregularities as its source [12].

One measure of non-reciprocity may be expressed as the Difference of Reciprocity between degrees of freedom j and k (DoR_{jk}), where the absolute difference is averaged over the frequency range $[\omega_1, \omega_2]$, as

$$\mathrm{DoR}_{jk} = \int_{\omega_1}^{\omega_2} |A_{jk} - A_{kj}| d\omega \tag{3}$$

In order enhance this difference, the reciprocity may be normalised to give the Non Reciprocity Index (NRI_{jk}) as

$$\mathrm{NRI}_{jk} = 2\int_{\omega_1}^{\omega_2} |A_{jk} - A_{kj}|\, d\omega \Big/ \int_{\omega_1}^{\omega_2} |A_{jk}|+|A_{kj}|\, d\omega = 2\mathrm{DoR}_{jk} \Big/ \int_{\omega_1}^{\omega_2} |A_{jk}|+|A_{kj}|\, d\omega \tag{4}$$

IOP Conf. Series: Journal of Physics: Conf. Series **1106** (2018) 012028 doi:10.1088/1742-6596/1106/1/012028

The frequency range $[\omega_1, \omega_2]$ can either cover a whole frequency range of a test, or focus on specific resonances. In the examples used in this paper the DoR and NRI will be given for each resonance, since the effect of the location of the force and response degrees of freedom, and the location of the nonlinearity, will affect the non-reciprocity of each resonance differently.

The definitions of DoR and NRI in Eqs. (3) and (4) are given in terms of Inertance, as often acceleration is measured experimentally. However, Receptance or Mobility may be used, which will give a slightly different weight to the FRFs at each frequency; however this will have a limited influence on the conclusions from the DoR and NRI for each resonance as the frequency range will typically be small. Indeed the NRI values for small frequency ranges will be very similar for all types of FRFs because of the inherent normalisation.

3. Modal responses for weak nonlinearities

Damage is often local and the resulting nonlinearities are relatively weak. In this case the dynamics of the multi-degree-of-freedom system can be understood by transforming to the modes of the underlying linear system. This also helps to understand the magnitude of the differences (i.e. the DoR and DRI values) caused by the nonlinearity, as a function of the forcing location and the location of the nonlinearity. This can be demonstrated explicitly using a single cubic spring, but a similar analysis and conclusions could be undertaken for other nonlinearities.

Suppose a cubic spring is located between degrees of freedom r and s, then the equations of motion are

$$\mathbf{M}\ddot{\mathbf{q}} + \mathbf{C}\dot{\mathbf{q}} + \mathbf{K}\mathbf{q} + h\mathbf{t}x_c^3 = \mathbf{f}_0 f(t) \tag{5}$$

where $\mathbf{q}$ is the vector of generalised coordinates, h is the coefficient of the cubic spring and x_c is the relative displacement of this spring. In Eq. (5), vector $\mathbf{t}$ consists of zeros, except for 1 in position r and -1 in position s, and the vector $\mathbf{f}_0$ is a vector of zeros, with a 1 to select the excitation location.

Suppose now that the system is excited close to the i th natural frequency of the linear system, with the corresponding mode shape $\boldsymbol{\phi}_i$. Note that here we also assume light damping (or at least modal damping) and so $\boldsymbol{\phi}_i$ represents the real mass normalised mode shape of the undamped system. We will assume that the response is of the form

$$\mathbf{q}(t) = \boldsymbol{\phi}_i p_i(t) \tag{6}$$

where p_i is the scalar modal coordinate. This implies that degrees of freedom vibrate spatially as the mode shape, which will be only approximately true, even for weak nonlinearities. Transforming the equation of motion to the modal coordinate p_i gives

$$\ddot{p}_i + 2\zeta_i\omega_i\dot{p}_i + \omega_i^2 p_i + h\left(\boldsymbol{\phi}_i^\top \mathbf{t}\right) x_c^3 = \left(\boldsymbol{\phi}_i^\top \mathbf{f}_0\right) f(t) \tag{7}$$

The relative displacement of the cubic spring, x_c, can be expressed in terms of the vectors $\mathbf{t}$ and $\mathbf{q}$ thus:

$$x_c = \mathbf{t}^\top \mathbf{q} \tag{8}$$

Hence

$$x_c = \left(\mathbf{t}^\top \boldsymbol{\phi}_i\right) p_i = \left(\boldsymbol{\phi}_i^\top \mathbf{t}\right) p_i, \tag{9}$$

and Eq. (7) becomes

$$\ddot{p}_i + 2\zeta_i\omega_i\dot{p}_i + \omega_i^2 p_i + h\left(\boldsymbol{\phi}_i^\top \mathbf{t}\right)^4 p_i^3 = \left(\boldsymbol{\phi}_i^\top \mathbf{f}_0\right) f(t) \tag{10}$$

It is now clear how the different mode, force location and nonlinear spring location affects the nonlinear response. The modal force is given by $\left(\boldsymbol{\phi}_i^\top \mathbf{f}_0\right)$, which is not affected by the nonlinear

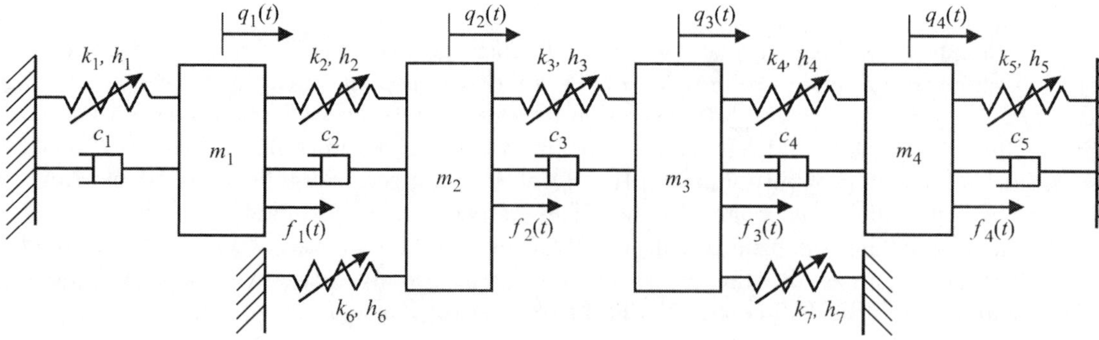

Figure 1. A schematic of the discrete four degrees of freedom system.

spring. The cubic spring coefficient is scaled by $(\phi_i^\top \mathbf{t})^4$, which depends on the mode shape of interest (i.e. the frequency of excitation) and the location of the nonlinearity. This will be explored further in the following examples.

4. Mass/spring example with stepped sine excitation

The lack of reciprocity, and its dependence on the locations of the excitation and nonlinearity, will now be demonstrated for a discrete mass-spring system with a cubic spring included selectively at a number of locations. The system is the four degree of freedom system shown in Figure 1, where k_i and h_i represent the linear and cubic coefficients for spring i. In all simulations the linear springs have stiffness $k_i = 1\,\text{kN/m}$ for $i = 1, \ldots, 5$ and $k_6 = k_7 = 0$. The discrete masses are $m_1 = 4\,\text{kg}$, $m_2 = 1\,\text{kg}$, $m_3 = 3\,\text{kg}$, and $m_4 = 2\,\text{kg}$. The natural frequencies of the linear system are 2.03, 3.39, 5.52 and 7.71 Hz, and discrete dampers are included to give damping ratios of 0.5% for all four modes. The mode shapes are

$$\phi_1 = \begin{Bmatrix} 0.268 \\ 0.361 \\ 0.396 \\ 0.236 \end{Bmatrix}, \quad \phi_2 = \begin{Bmatrix} 0.397 \\ 0.076 \\ -0.280 \\ -0.255 \end{Bmatrix}, \quad \phi_3 = \begin{Bmatrix} 0.076 \\ -0.214 \\ -0.247 \\ 0.611 \end{Bmatrix}, \quad \phi_4 = \begin{Bmatrix} -0.122 \\ 0.904 \\ -0.193 \\ 0.072 \end{Bmatrix}. \tag{11}$$

Only one cubic spring is simulated for each case, and hence only one of the h_i coefficients is non-zero. Table 1 shows the values of the equivalent modal constants for various locations of the nonlinear spring, and for the four modes. The values for the grounded springs are just the magnitude of the corresponding mode shapes element, and this also corresponds to the relative modal force for the excitation at this location. The table clearly shows that both the mode number and the location of the cubic spring will have a significant effect on the level of excitation of the nonlinearity.

4.1. Harmonic balance for sinusoidal excitation

The nonlinear system differential equations can be solved in the time domain using a standard numerical time stepping procedure such as the 4th order Runge-Kutta that simulates the experimental stepped sine test. However, here only the frequency domain information is required, and it is more efficient to use the harmonic balance (HB) method. In many cases, the HB method provides a faster method of solution although for a system producing a chaotic response this solution is not valid. The method assumes that the response of a system can be expressed as a truncated Fourier series, with a defined fundamental frequency ω_f. Should the response contain subharmonics then the excitation frequency, ω, will be an integer multiple of the response of the

IOP Conf. Series: Journal of Physics: Conf. Series **1106** (2018) 012028 doi:10.1088/1742-6596/1106/1/012028

Table 1. Modal constants for spring locations, $|\phi_i^\top \mathbf{t}|$. A zero for the location indicates a grounded spring. Some cases do not correspond to a spring location given in Figure 1.

Spring Location	h_i	Mode 1	Mode 2	Mode 3	Mode 4
(1,2)	h_2	0.093	0.321	0.291	1.027
(2,3)	h_3	0.034	0.355	0.033	1.098
(3,4)	h_4	0.159	0.024	0.859	0.265
(1,3)		0.128	0.676	0.323	0.071
(2,4)		0.125	0.331	0.826	0.833
(1,4)		0.032	0.652	0.535	0.194
(0,1)	h_1	0.268	0.397	0.076	0.122
(0,2)	h_6	0.361	0.076	0.214	0.904
(0,3)	h_7	0.396	0.280	0.247	0.193
(0,4)	h_5	0.236	0.255	0.611	0.072

fundamental frequency ω_f. To solve a system using the HB method we substitute the truncated Fourier series for the response into the equations of motion, and the harmonic coefficients (either cosines / sines or complex exponentials) of the fundamental and harmonic frequencies lead to a set of nonlinear algebraic equations. For simple nonlinearities the well known trigonometric identities can be used to simplify the equations. Often this is not possible, and the Discrete Fourier Transform (DFT) is used to obtain the harmonic coefficients using the Alternating Frequency/Time (AFT) method [13].

4.2. FRFs and reciprocity for the example system

The initial example places a cubic spring between degrees of freedom 1 and 2; thus the nonlinear coefficients are $h_2 = 100\,\mathrm{N/m^3}$, and $h_i = 0$ for $i \neq 2$. A stepped sine test is performed with a force amplitude of 20 N, and the frequency is swept up from 1 Hz to 9 Hz with a frequency increment of 0.001. The frequency is then swept back down to 1 Hz. Table 1 shows that mode 4 will show the strongest nonlinear effect, followed by modes 2 and 3 showing similar effects, and mode 1 showing a small effect. In terms of excitation, forcing at degree of freedom 2 will generate most force into mode 4, and forcing at degree of freedom 4 will produce the least force. Hence, Figure 2 compares the FRFs between degrees 2 and 4 (i.e. A_{24} and A_{42}) and both upward and downward frequency sweeps are plotted. It is clear that mode 4 shows a higher nonlinear response that the other modes, as expected. For A_{24}, the forcing level is low, and the response is near linear, with no jumps in the response. In contrast, or A_{42}, the forcing level is high, and the response shows a hardening behaviour and jumps in the response.

The DoR and NRI were calculated for these responses for each mode. The frequency bands used for each mode should be large enough to ensure that any nonlinear hardening or softening behaviour is captured, but small enough to ensure a single mode response is isolated. For this example a frequency bandwidth of 0.5 Hz is used, centred about the natural frequency of the linear system. The sweep up and sweep down are considered separately. Figures 3 and 4 show the calculated DoR and NRI and there is little difference between the up and down frequency sweeps. As expected, the DoR and NRI for mode 4 is most sensitive to the nonlinearity, and since the modal force amplitude for mode 4 is highest when the excitation is applied to degree of freedom 2, the DoR and NRI which include degree of freedom 2 are the highest. The conclusions from the DoR and NRI results are similar, but the NRI seems to give more emphasise to the nonlinear effects.

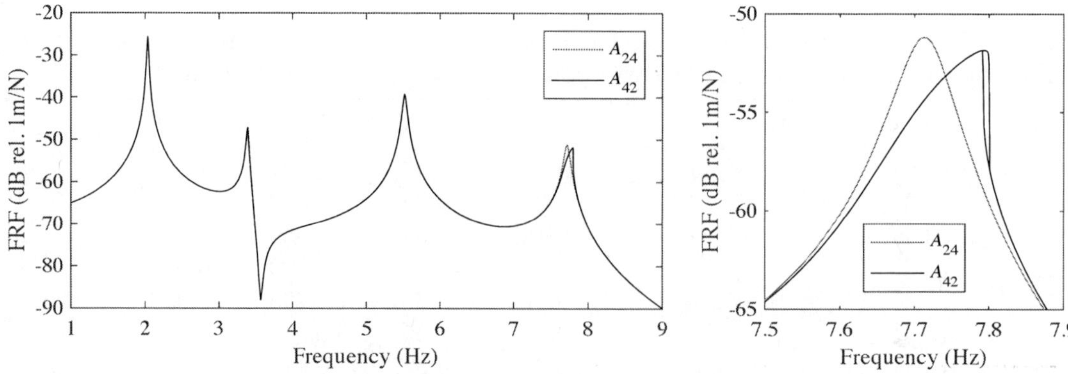

Figure 2. The FRFs between degrees of freedom 2 and 4, for both up and down frequency sweeps. The right plot is a zoom around mode 4.

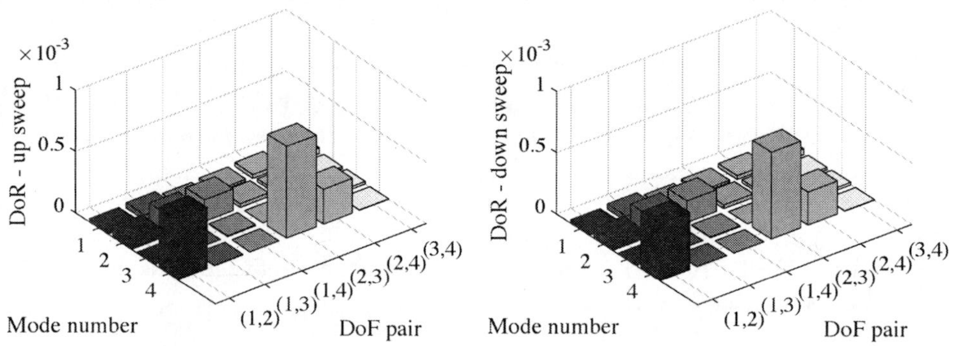

Figure 3. The Difference of Reciprocity for both up and down frequency sweeps.

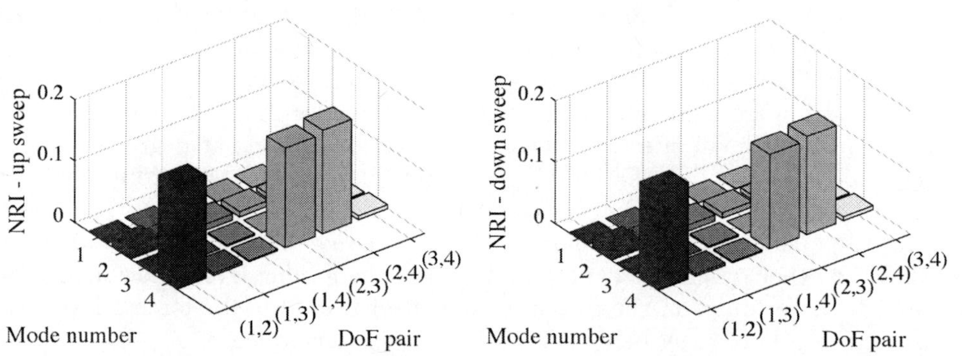

Figure 4. The Non Reciprocity Index for both up and down frequency sweeps.

The effect of force amplitude on $\mathrm{NRI}_{2,3}$ is shown in Figure 5 and, in this case, the NRI increases monotonically with the force amplitude. The amplitude of the NRI for mode 4 is highest, followed in order by modes 2, 1 and 3; from Table 1 it might be thought that mode 1 should have the lowest NRI, but the modal forcing and response obtained from the mode shapes, for degrees of freedom 2 and 3 are higher for mode 1 than mode 3.

A second example is now given for a grounded cubic spring at degree of freedom 2, i.e. $h_6 = 100\,\mathrm{N/m^3}$, and $h_i = 0$ for $i \neq 6$. Figure 6 shows the NRI for this case, and demonstrates

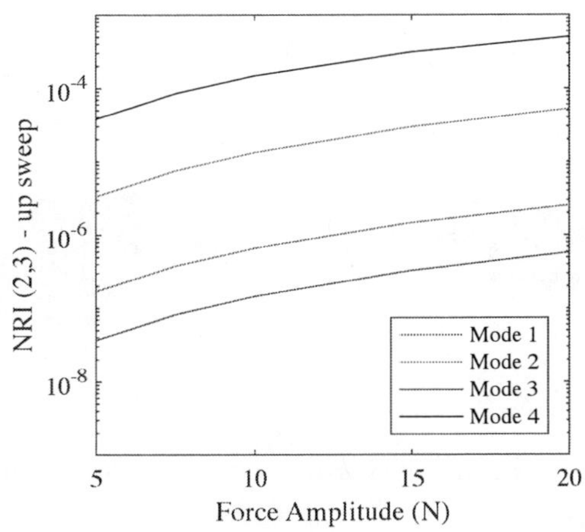

Figure 5. The Non Reciprocity Index $NRI_{2,3}$ for a frequency up sweep.

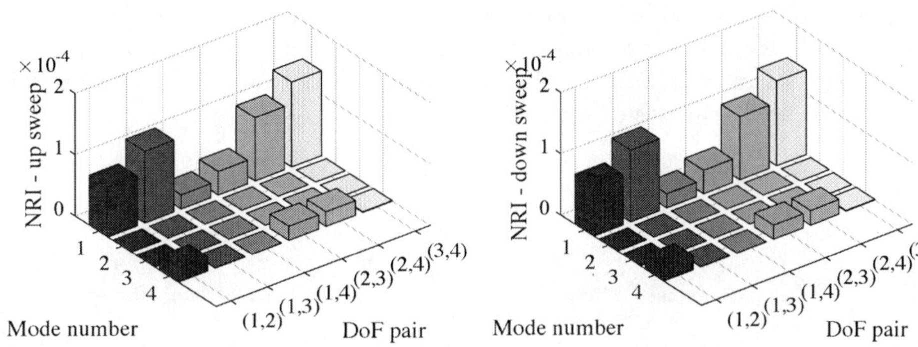

Figure 6. The Non Reciprocity Index for both up and down frequency sweeps for a grounded spring at DoF 2.

that modes 1 and 4 now have the highest NRI values. In this case mode 1 has the highest NRI values because of significantly higher modal excitation and response values.

5. Mass/spring example with impact excitation

A common excitation method is an impact force, and here this is modelled using a half sine pulse. The example used was the same 4 degrees of freedom mass-spring system used for the stepped sine testing with a cubic spring between degrees of freedom 1 and 2 with coefficient $h_2 = 100\,\mathrm{N/m^3}$, and $h_i = 0$ for $i \neq 2$. The impulse force has a peak amplitude of 2000 N and a duration of 25 ms. The response was simulated in the time domain using a 4th order Runge-Kutta algorithm and the FRF was obtained using the DFT. The DoR and NRI were obtained for each resonance, over a frequency range of 0.5 Hz centred about the linear natural frequencies, and the results are shown in Figure 7. The results are slightly different to those obtained from the stepped sine excitation, shown in Figures 3 and 4, but mode 4 is again the most sensitive to the nonlinearity and degree of freedom 2 also gives high NRI values.

6. Conclusions

Nonlinear multi-degree of freedom systems exhibit non-reciprocity in the measured FRFs, and the level of difference in the FRFs depends on the locations of the excitation, the response sensor,

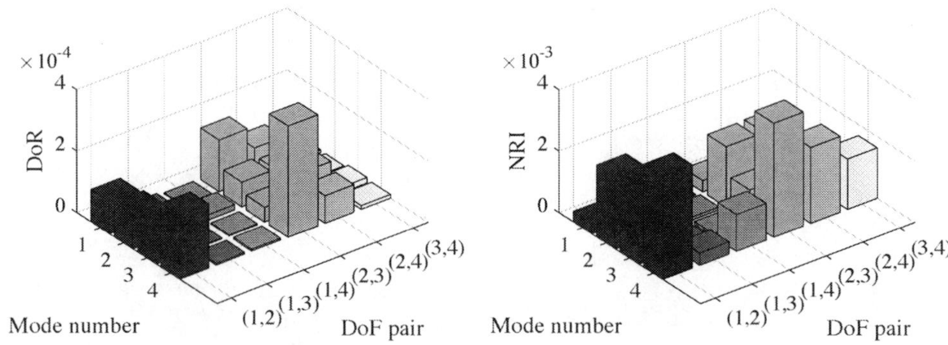

Figure 7. The Difference of Reciprocity and Non Reciprocity Index for impulse excitation.

and the localised nonlinearity. This paper has introduced two measures of non-reciprocity, the DoR and the NRI. A 4 degree of freedom mass-spring system has been used to demonstrate that the DoR and NRI are strongly dependent on the excitation of the nonlinearity, and for weak nonlinearities the linear mode shapes provide an efficient approach to understanding the resulting DoR and NRI values. In future this approach has the potential to efficiently determine the location of local nonlinearities in a structure.

7. References

[1] IAEA 2002 *Guidebook on Non-destructive Testing of Concrete Structures,* IAEA-TCS-17 Vienna, ISSN 10185518.

[2] Iliopoulos S, Aggelis D G, Pyl L, Vantomme J, Van Marcke P, Coppens E & Areias L 2015 Detection and evaluation of cracks in the concrete buffer of the Belgian nuclear waste container using combined NDT techniques, *Construction and Building Materials* **78** 369-78.

[3] Doebling S W, Farrar C R, Prime M B & Shevitz D W 1996 Damage identification and health monitoring of structural and mechanical systems from changes in their vibrational characteristics: a literature review, *Los Alamos National Laboratory,* New Mexico, Report LA-13070-MS

[4] Klepka A, Straczkiewicz M, Pieczonka L, Staszewski W J, Gelman L, Aymerich F & Uhl T 2015 Triple correlation for detection of damage-related nonlinearities in composite structures, *Nonlinear Dynamics* **81** 453-68.

[5] Zwink B R, Adams D E, Evans R D & Koester D J 2009 Wide-area damage detection in military composite helicopter structures using vibration-based reciprocity measurements, *Proc of the IMAC-XXVII,* 9-12 February 2009, Orlando, Florida USA.

[6] Rogers T, Manson G & Worden K 2016 On the behaviour of structures with many nonlinear elements, ed. De Clerck J & Epp D (eds) Rotating Machinery, Hybrid Test Methods, Vibro-Acoustics & Laser Vibrometry **8** *Proc of the Society for Experimental Mechanics Series* Springer, Cham.

[7] Chesne S & Deraemaeker A 2013 Damage localization using transmissibility functions: A critical review, *Mechanical Systems and Signal Processing* **38** 569-84.

[8] Wood M G, Penny J E T, Purkiss J A, Short N R & Owolawi O 2003 Measuring changes of state in concrete, *Materials Science Forum* **440-1**, 219-28

[9] Wood M G 1992 *Damage Analysis of Bridge Structures using Vibrational Techniques*, PhD Thesis, University of Aston, Birmingham, UK.

[10] Manson G, Worden K & Wood M G 2012 Analysis of reciprocity breakdown in nonlinear systems, *Proc Modern Practice in Stress and Vibration Analysis,* University of Glasgow, 31 August 2012, IoP Press.

[11] Gloth G & Sinapius M 2005 Analysis of swept-sine runs during modal survey and qualification tests, *Proc of the European Conference on Spacecraft Structures, Materials and Mechanical Testing* 10-12 May 2005, ESA SP-581 Noordwijk, The Netherlands.

[12] Patrício M 2008 *Crack Propagation on Highly Heterogeneous Composite Materials*, PhD Thesis, Eindhoven University of Technology, The Netherlands.

[13] Cameron T M & Griffin J H 1989 An alternating frequency/time domain method for calculating the steady-state response of nonlinear dynamic systems, *Journal of Applied Mechanics* **56** 149-154.

IOP Conf. Series: Journal of Physics: Conf. Series **1106** (2018) 012029 doi:10.1088/1742-6596/1106/1/012029

Microscopic investigation of subsurface initiated damage of wind turbine gearbox bearings

Jasim H AL-Bedhany and Hui Long*

Department of Mechanical Engineering. The University of Sheffield, Sheffield, UK.

* h.long@sheffield.ac.uk

Abstract. Wind turbine gearbox bearings experience premature failures by White Structure Flaking (WSF), often occurs much earlier than their designed life of 20 to 25 years. This results in increased operational and maintenance costs due to unplanned maintenance and early replacement. The main causes and damage initiation mechanism of this premature failure are not fully understood; despite extensive research and investigation in recent years. In this paper, two planetary bearings from a failed gearbox of a multi-megawatt wind turbine are destructively investigated to characterize the subsurface microstructural damage and to understand damage initiation mechanism leading to surface WSF. The results show that the non-metallic inclusions are not the only initiator of subsurface damage. The microcracks are also initiated in the subsurface to form macrocracks which then propagate or connect to other macrocracks to reach the rolling contact surface causing WSF. The characterization of different forms of subsurface microstructural damage shows a close correlation of the maximum shear stress with the damage initiation. Butterfly wings are found to initiate from the compound type of non-metallic inclusions with low aspect ratio and to be associated with inclusion internal cracking in a direction approximately parallel to the axis of the maximum wing length.

1. Introduction

The designed life of wind turbines (WTs) is 20 to 25 years however the premature failure of wind turbine gearbox (WTG) bearings by White Structure Flacking (WSF) has often been reported[1]. This failure is associated with microstructural alterations beneath the rolling contact surfaces of bearing raceways and rolling elements[1][2][3]. Despite extensive research effort investigating the bearing premature failure by WSF, the main causes and damage initiation and propagation mechanisms are still a scientific and debateable subject[4][5]. Microscopic investigation of samples obtained from failed bearings of field operating WTGs provides an insight into the microstructural alterations and various forms of damage occurred. Energy Dispersive X-ray Analysis (EDX) technique is commonly used to specify the chemical compositions of the bearing materials and other microstructural defects such as non-metallic inclusions[6].

Different forms of microstructural damage, especially butterfly wings, were investigated by a considerable number of studies[7][8][9]. The results found that the formation of microstructural alterations occurred at a specific depth range from the rolling contact surface. The butterfly wings approximately looked like shear stress distributions induced by contact pressure, which led to suggestions that the shear stress had an important effect on the initiation of this damage[10][11][12]. Non-metallic inclusions and material cleanliness had an important effect on damage initiation[13][14][15][16]. The inclusion type, size and distribution were the main parameters effecting the damage initiation[17][18]. Destructive investigation by sectioning the failed region of WTG bearings provided a two-dimensional view of investigated plane while serial sectioning technique allowed approximately a three-dimensional observation of cracking network[19].

In this study, a destructive investigation of two failed WTG planetary bearings is conducted to study the severely damaged regions of the bearing raceways by microscopic examination and to characterize different forms of damage such as butterfly wings, microcracks and damaged inclusions. The results show that the subsurface microcracks are another initiator of WSF initiation, in addition to the WSF resulted from non-metallic inclusions. Subsurface maximum shear stress has an important effect on the initiation of different forms of damage such as butterfly wings, inclusions with internal cracking and inclusions separations at inclusion-steel matrix boundaries.

2. Microscopic investigation

In this study, the severely damaged regions of two planetary bearings are destructively investigated. The samples for microscopic investigation are cut and examined in both axial and circumferential directions of the bearing raceway, as illustrated in figure 1. Samples are mounted using conductive resins to show the investigated surfaces. By using this cutting procedure, the entire axial plane located in the centreline of the severely damaged zone as well as six equally spaced circumferential sections of the raceway can be microscopically investigated. Samples are grinned, polished then etched with 2% Nital (2% Nitric acid and 98% ethanol). Optical microscope and Scanning Electron Microscope (SEM) are used in parallel with EDX technique to investigate the microstructure alterations and different forms of damage. Damage forms such as inclusion-initiated cracks and butterfly wings are characterised according to their damage features defined by dimensions, depth beneath the rolling contact surface, inclination angle relative to the rolling surface and inclusion Aspect Ratio (AR). The subsurface damage initiation is analysed by correlating the damage features with possible loading conditions experienced by the bearings during their operation. Subsurface stress distributions beneath the contact surface and the stress variation due to various loading levels are determined using Hertzian contact theory. Stresses distributions under the effect of surface traction are also calculated to analyse its expected role on damage initiation.

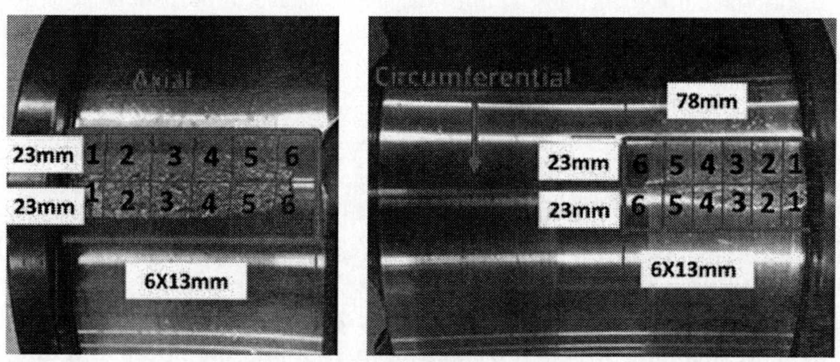

Figure 1. Investigated failed bearings and sample locations (upwind and downwind bearing raceways).

3. Results and discussions

In this study, 149 damaged inclusions are found, of which 55 inclusions (37%) are in the two axial sectioned samples and 94 inclusions (63%) are in the two circumferential-sectioned samples. These samples are chosen from the middle of axial and circumferential-sectioned samples, i.e. sample 3 and 4 as shown in Figure 1. The depths of theses samples investigated are within 1mm beneath the rolling contact surface, since the effect of maximum contact stresses do not expect to go beyond this depth[1][12].

Damaged inclusions either have separation damage, i.e. inclusions deboned at their boundaries from steel matrix, cracking damage or mixed damage of separation and cracking. table 1 illustrates these damage forms. Four different types of inclusion damage by separation are identified: upper separation, lower separation, upper and lower separation, and side separations.

Most distinctive microstructural alterations found are butterfly wings. Therefore, the investigation and analysis of this paper focus on this damage form. Butterfly wings are classified into single and double winged butterflies, and single winged butterflies are further classified into upper and lower single winged respectively. The upper winged butterflies have the butterfly wing above the damage initiating inclusion i.e. from the inclusion towards the rolling contact surface. The lower winged butterflies crack away from the rolling contact surface. The characterization parameters, including wing length, inclusion angle and depth of 49 butterflies, are analysed.

Table 1. Different forms of microstructural damage initiated by inclusions.

Separation at inclusion		**Cracking at inclusion**	
Damage type	**Examples**	**Damage type**	**Examples**
No separation	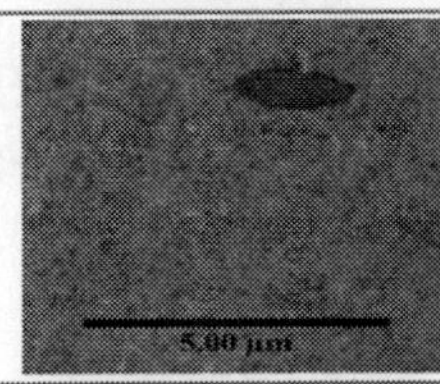	Upper cracking	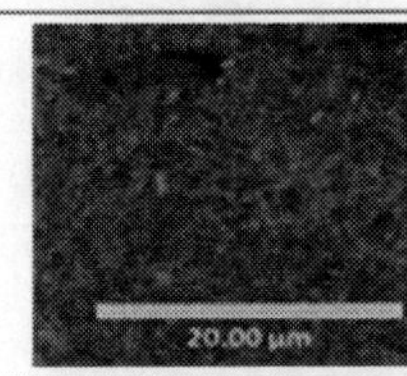
Upper separation	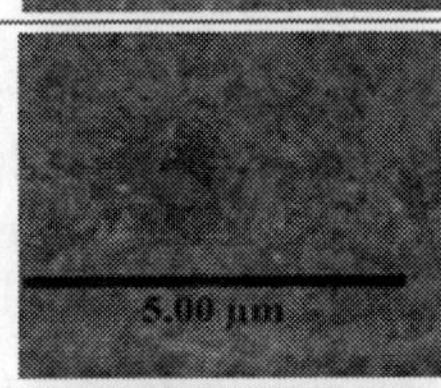	Lower cracking	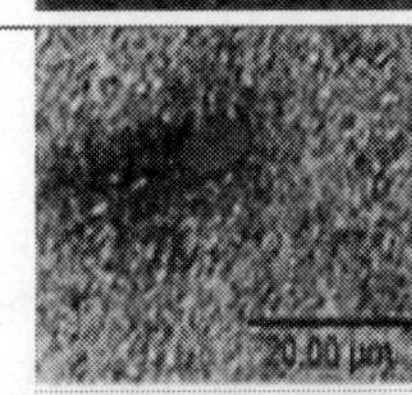
Lower separation	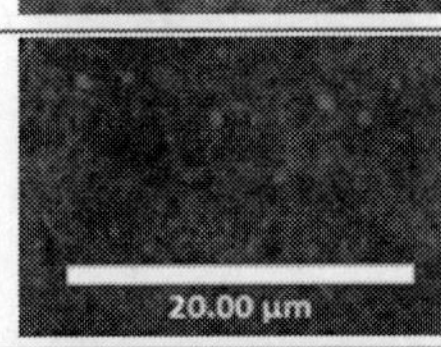	Upper and lower cracking	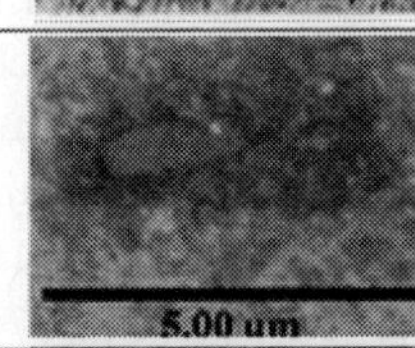
Upper and lower separation	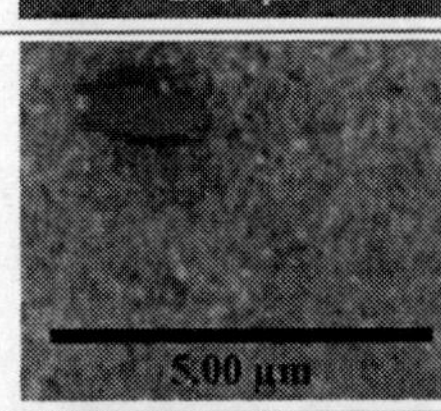	Cracked inclusion	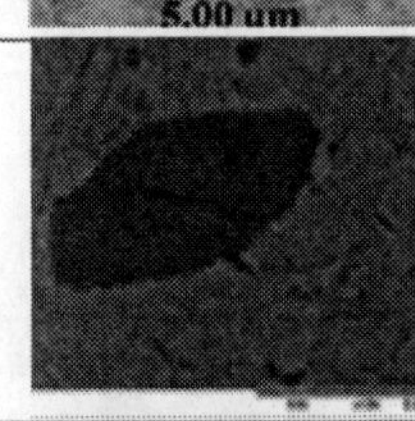
Side separations	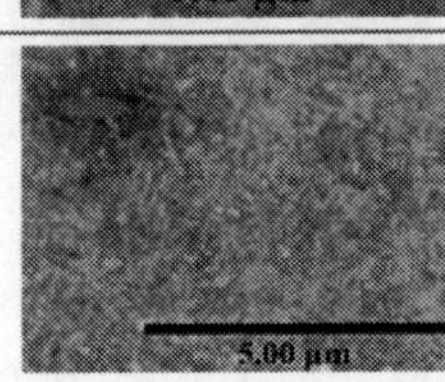	Mixed types of cracking	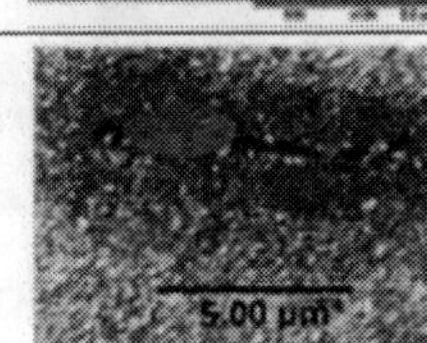

3.1. Characterization of damage initiating inclusions

Subsurface stress distributions beneath the contact surface are calculated using Hertzian contact theory by applying compression and traction loading according to the designed stress level specified by the international standards[20][21]. figure 2 shows the distributions of the maximum shear stress when under compression only and under combined compression and traction respectively. Increasing the compressive loading moves the maximum shear stress zone further away from the contact surface. Introducing surface traction into compression moves the maximum shear stress zone closer towards the contact surface. The maximum shear stress zone will be compared with the locations where various forms of damage are found in the following sections.

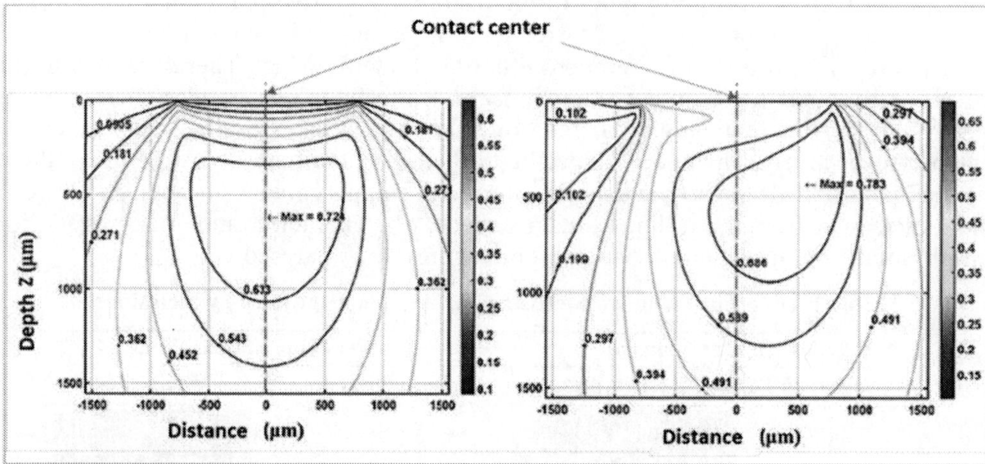

Figure 2. Distributions of maximum shear stress (in GPa), (left) under compression loading 1.8 GPa and (right) under combined compression loading 1.8 GPa and 0.2% traction.

Different forms of inclusion related damage are found in subsurface of the bearing rolling contact, including: inclusions separated from the matrix, cracks extended from inclusion into the matrix and butterfly wings. Percentages of different types of damaged inclusions with their depths are presented in figures 3 and 4. All separated inclusions are found to be located approximately within the depth of the maximum shear stress zone in the subsurface of the contact. For inclusions with separation in the centre of the upper side of the inclusion boundary, the damage may be due to shear stresses with surface traction, since the traction effect brings the maximum shear stress zone closer to the contact surface. The numbers of inclusions with separation at the lower side of the inclusion increase with the depth. By comparing the depths of damaged inclusions by separation in figure 3(a) with the depth of the maximum shear stress shown in figure 2, it is clear their depths are closely correlated.

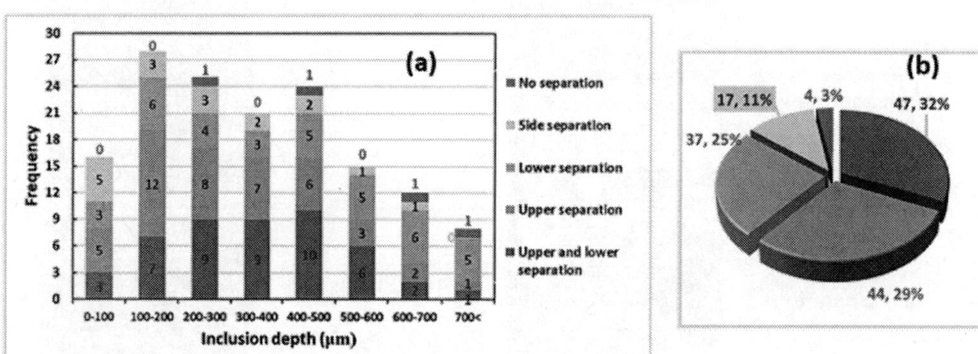

Figure 3. Damaged inclusions by separation: a) in relation to subsurface depth; b) numbers and percentages of separated inclusions.

Inclusions with separation are the most dominant damage form however damaged inclusions by cracking are also observed. The cracks appear in different forms: in the tip of the upper side of the inclusion (towards the rolling contact surface), in the tip of the lower side of the inclusion, internal cracking of the inclusion itself, or combined cracking of these different forms. Inclusions associated with cracks are approximately located within the depth of maximum shear stress in the subsurface of

IOP Conf. Series: Journal of Physics: Conf. Series **1106** (2018) 012029 doi:10.1088/1742-6596/1106/1/012029

the contact as shown in figure 4. The characterization of damaged inclusions supports the previous suggestion that the inclusion-initiated damage in subsurface may be resulted from effects of shear stresses and the cracking appears along the maximum shear stress plane.

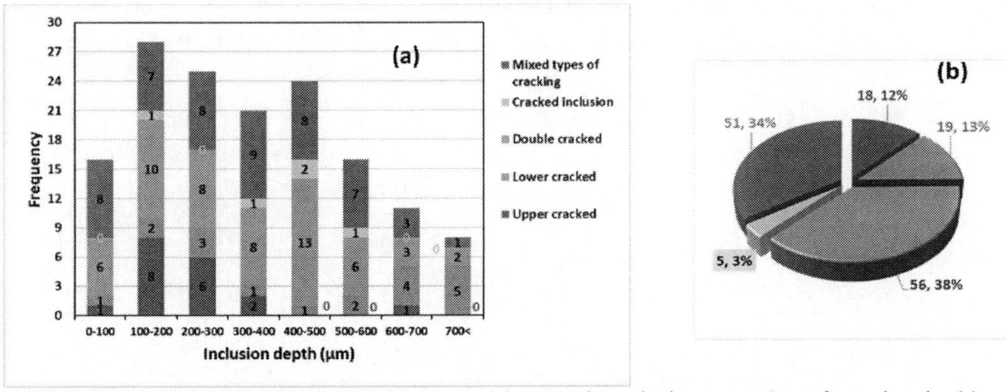

Figure 4. Damaged inclusions by cracking: a) in relation to subsurface depth; b) numbers and percentages of cracked inclusions.

3.2. *Butterfly wings*

The depth distribution of the 49 observed butterfly wings is shown in figure 5(a). All butterflies are in the circumferential-sectioned samples, approximately located at the depth of the maximum shear stress. Number of butterflies observed increases with increasing depth, however no butterfly is found in depth greater than 700 μm. Only one small butterfly wing is found parallel to the contact surface. Considerable butterfly wings have an inclined angle to the rolling contact surface at 25° or 40° respectively, counted as 29% and 21% of all butterflies, as shown in figure 5(b). figure 5(c) presents the distribution of butterfly wings according to the lengths of their wings. Because the investigated samples are cut from the severely damaged region of the bearing raceways where some parts of the contact surface are removed by severe spalling therefore no butterflies are found in shallow depths of the rolling contact surface.

Around 71% of the butterflies found have double wings while 57% of the remaining butterflies (29% of all butterflies found) have single wing at the upper side of the initiation inclusion and the rest butterflies have single wing at the lower side. Single winged butterflies have shorter lengths when compared with the double winged butterflies. This leads to a hypothesis that butterfly wings may have initiated as a single wing first and the other wing appeared later, then they propagated together to grow into longer wings. There is no evidence to support whether the upper or the lower wing has initiated first however the point of wing initiation may depend on the location of the damage initiating inclusion in relation to the location of maximum shear stress. These observations support the suggestion that the maximum shear stress is an influence factor on the initiation of butterfly microstructure damage [22]. It is observed that cracks associated with butterfly wings are probably not a part of macrocrack network that linked to rolling contact surfaces because they disappear after regrinding the sample surfaces where the butterfly wings are observed. However, the butterfly cracks may not play an important role on subsurface damage propagation. All forms of damage are probably produced due to the shear stress levels have exceeded a critical limit of the bearing material.

There is no clear correlation between the depths of damaged inclusions with the occurrence of butterflies with upper or lower wing. figure 6(a) shows two inclusions located approximately at the same depth (~320 μm); however, one inclusion has an upper single wing while another has a lower single wing. figure 6(b) shows two inclusions with butterfly wings, the inclusion with an upper wing is located at depth of ~250 μm; however, the inclusion with a lower wing is located deeper at ~148 μm.

Damaged inclusions, microcracks, and butterfly wings are located deeper than the region of the maximum shear stress calculated, which indicates that the contact surfaces may have been subjected to much higher loading levels than designed stress level specified by the international standards, possibly exceeded the yield strength of the bearing material. To confirm this, hardness of the bearing raceway contact surface inside and outside the loading zones are measured at 25 points then averaged. The surface hardness outside and inside the loading zones are 746 HV and 788 HV respectively and this indicates that the surface inside the loading zone has been hardened due to overloading.

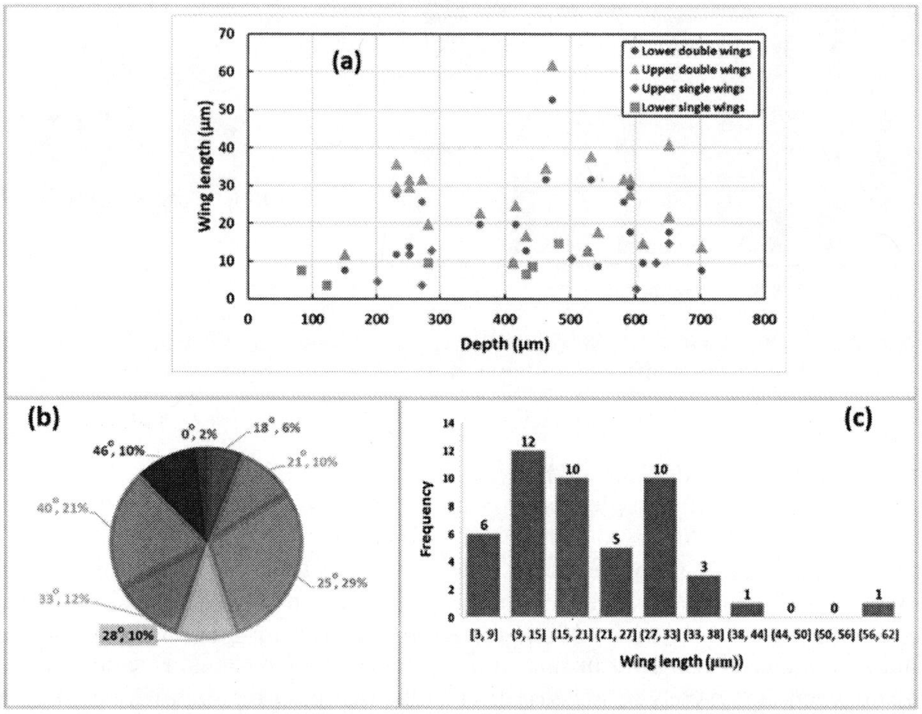

Figure 5. Butterflies: a) wing lengths in relation to subsurface depth; b) wing angles and percentages c) occurrences in relation to wing length.

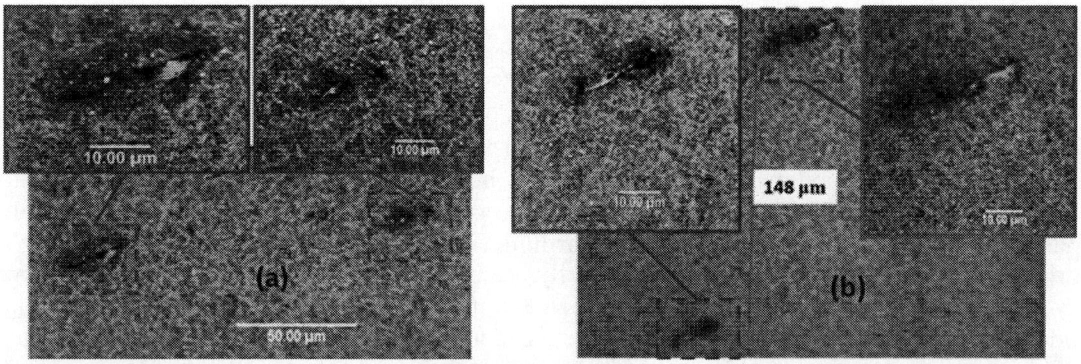

Figure 6. Depths of butterflies a) two butterflies with upper / lower wing located at the same depth; b) two butterflies with upper / lower wings located at different depths.

One of the largest butterflies found in this study is shown in figure 7(a). It is located at the end region of the severely spalling area away from rim of the downwind bearing (sample No.6 in figure 1). This butterfly has a depth of ~470 μm beneath the rolling contact surface, within the maximum shear stress zone as shown in figure 2. The inclusion in the centre of the butterfly wings shown in figure 7(b) has a darker colour compared to that of MnS inclusions which have light grey colour[13]. Energy Dispersive X-ray (EDX) analysis is used to confirm the inclusion's chemical compositions, as shown in figure 7(c). The analysis shows that it is a compound inclusion of MnS, aluminium oxide and silica in addition to other chemical compositions. Thus it is the inclusion type D_{Dup} according to the International Standard ISO-4967:2013[13]. Aspect Ratio (AR) is defined as the ratio of inclusion lengths along the major to minor axes and it is used to evaluate the inclusion shape. It is found that the majority of the butterflies are initiated at inclusions having low aspect ratio of around 2:1. It is also observed that the butterflies are likely associated with inclusions with internal cracks in a direction approximately parallel to the direction of maximum length of the butterfly wing.

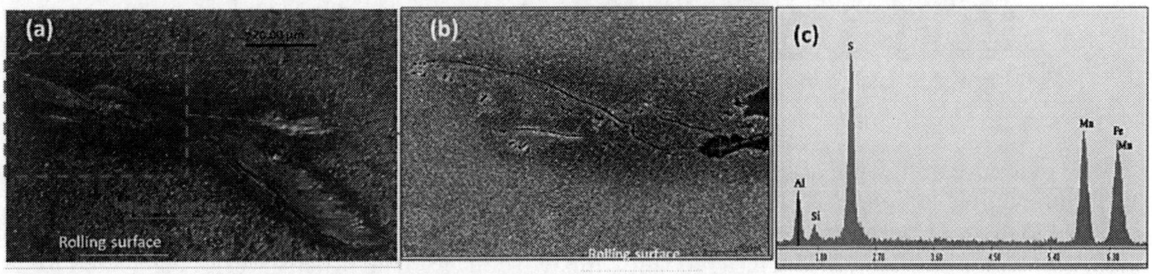

Figure 7. Inclusion initiated the largest butterfly: a) optical image; b) SEM image of one wing; c) EDX of central inclusion.

3.3. Subsurface microcracks
A considerable number of microcracks along a macrocrack are observed on a sample which is over etched as shown in figure 8(a) (marked with white arrows). Further inspections find that considerable microcracks exist on other samples in both axial and circumferential directions as shown in figure 8(b) and 8(c). The considerable number of subsurface microcracks and macrocracks have also been observed in the previous studies [2][22][23][24] however so far limited investigations exist to explain their role on WSF.

Microcracks are found near the sides and ends of the macrocracks where the number of microcracks is much higher than the number of the damaged inclusions. This leads to a postulation that both inclusion and microcracks are the initiators of subsurface microstructural damage leading to WSF. However, the weak boundaries and the residual stress around inclusions play an important role in subsurface damage initiation.

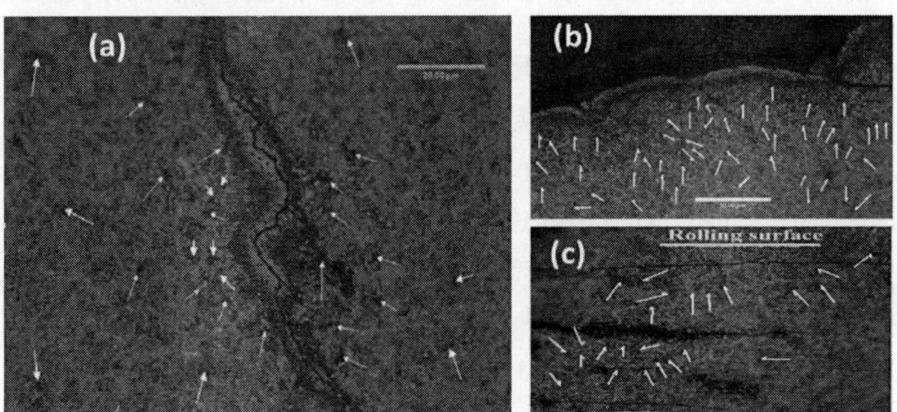

Figure 8. Over etched specimens showing microcracks: a) and b) circumferential sectioned samples c) axial sectioned sample.

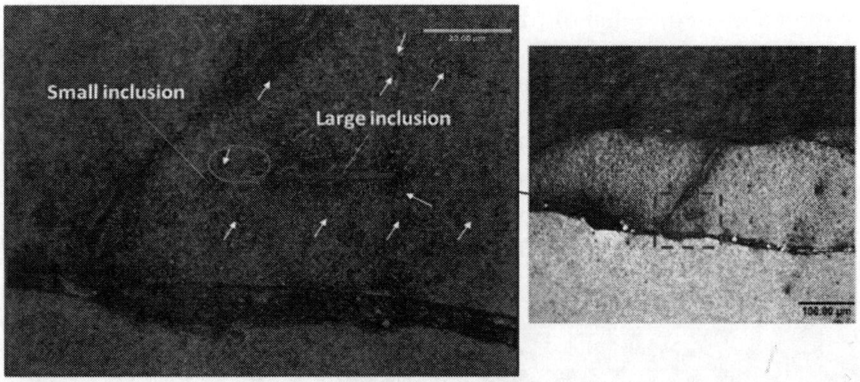

Figure 9. Large inclusion unconnected to surface crack network in axial sectioned sample.

The evidence of the role of subsurface microcracks on damage initiation is also shown in an axial sectioned sample in figure 9. A relatively large inclusion located close to the contact surface does not connect to the macrocracks network around it, despite the inclusion is connected to another small inclusion with a crack (marked with a red ring), the two connected inclusions are not connected to the macrocrack network. A considerable number of microcracks around these two inclusions can be seen (marking by white arrows). This leads to a postulation that subsurface microcracks and/or the cracks initiated from the separation of inclusion boundaries probably propagate towards each other, depending on the direction of the maximum shear stress in high stress locations, they then propagate towards the contact surface causing WSF.

4. Conclusions

This destructive investigation of two failed WTG planetary bearings by microscopic examination has found different forms of damage including butterfly wings, microcracks and damaged inclusions. The following conclusion may be drawn:

- Subsurface microcracks are another damage initiator in addition to non-metallic inclusions to produce subsurface microstructural damage leading to WSF.
- Butterfly wings are associated with the compound type of non-metallic inclusions with low aspect ratios. They are associated with inclusions with internal cracking in a direction approximately parallel to the axis of the maximum wing length.
- Butterfly wings and associated cracks may not be a part of the macrocrack network; the butterflies may firstly have a single wing and the other wing may appear later.
- Characterization of different microstructural damage forms confirms that the maximum shear stress is closely associated with the location of the subsurface microstructural damage.

Acknowledgment:

The authors would like to acknowledge the support of Vestas Wind System A/S by sponsoring this research. The first author would like to thank Iraqi Ministry of Higher Education and Scientific Research (MOHESR) and Engineering College of Misan University, Iraq for sponsoring his PhD study in the University of Sheffield, UK.

References

[1] M. Evans and M. Evans, "An updated review : white etching cracks (WECs) and axial cracks in wind turbine gearbox bearings," *Mater. Sci. Technology*, vol. **0836**, no. December, 2017.

[2] J. Lai and K. Stadler, "Investigation on the mechanisms of white etching crack (WEC) formation in rolling contact fatigue and identi fi cation of a root cause for bearing premature failure," *wear,* vol.

365, pp. 244–256, 2016.

[3] S. Sheng; McDade, M.; Errichello, R. Wind turbine gearbox failure modes. In Proceedings of the *International Joint Tribology Conference*, Los Angeles, CA, USA, 24–26 October 2011

[4] M.-H. Evans, White structure flaking (WSF) in wind turbine gearbox bearings: effects of butterflies and white etching cracks (WECs), *Materials Science and Technology*, vol. **28**, 3–22, 2012.

[5] Singh, H., et al.: Investigation of microstructural alterations in low- and high-speed intermediate-stage wind turbine gearbox bearings. *Tribol. Lett.* **65** (3), 81, (2017)

[6] Danielsen, H.K., et al.: Multiscale characterization of White Etching Cracks (WEC) in a 100Cr6 bearing from a thrust bearing test rig. *Wear,* vol. **370**, 73–82 (2017)

[7] M.-H. Evans, "White structure flaking (WSF) in wind turbine gearbox bearings: effects of 'butterflies' and white etching cracks (WECs)," *Mater. Sci. Technol.*, vol. **28**, no. 1, pp. 3–22, 2012.

[8] S. Mobasher *et al.*, "Effect of non-metallic inclusions on butterfly wing initiation , crack formation , and spall geometry in bearing steels," *International Journal of Fatigue,* vol. **80**, pp. 203–215, 2015.

[9] G. Fajdiga and M. Sraml, "Fatigue crack initiation and propagation under cyclic contact loading," *Eng. Fract. Mech.*, vol. **76**, no. 9, pp. 1320–1335, 2009.

[10] M. Evans, A. D. Richardson, L. Wang, R. J. K. Wood, and W. B. Anderson, "Confirming subsurface initiation at non-metallic inclusions as one mechanism for white etching crack (WEC) formation," *Tribology International,* vol. **75**, pp. 87–97, 2014.

[11] T. Bruce, E. Rounding, H. Long and R. S. Dwyer-Joyce: 'Characterisation of white etching crack damage in wind turbine gearbox bearings', *Wear*, **338–339**, 164–177, 2015.

[12] H. A. AL-Tameemi, H. Long and R. S. Dwyer-Joyce, "Initiation of sub-surface micro-cracks and white etching areas from debonding at non-metallic inclusions in wind turbine gearbox bearing," *Wear,* vols. **406-407**, pp. 22–32, 2018.

[13] ISO 4967-2013 — Determination of content of nonmetallic inclusions — Micrographic method using standard diagrams. 2013.

[14] P. Taylor, B. Jalalahmadi, F. Sadeghi, and V. Bakolas, "Material Inclusion Factors for Lundberg-Palmgren – Based RCF Life Equations Material Inclusion Factors for Lundberg-Palmgren – Based," *Tribology Transactions,* Vol. **54** (3), pp. 37–41, 2011.

[15] Y. Murakami and M. Endo, "Effects of defects , inclusions and inhomogeneities on fatigue strength," *International Journal of Fatigue,* vol. **16**, 1994.

[16] Y. Murakami, "Material defects as the basis of fatigue design," *International Journal of Fatigue*, vol. **41**, pp. 2–10, 2012.

[17] Atkinson HV, Shi G. Characterization of inclusions in clean steels: a review including the statistics of extremes methods. *Prog Mater Sci* 2003; vol. **48**: 457–520

[18] Lewis MN, Tomkins B. "A fracture mechanics interpretation of rolling bearing fatigue," Proceedings of the Institution of Mechanical Engineers Part J—*Journal of Engineering Tribology* 2012; vol. **226**: 389–485

[19] Evans M-H, Richardson AD, Wang L, Wood RJK. Serial sectioning investigation of butterfly and white etching crack (WEC) formation in wind turbine gearbox bearings. *Wear* 2013; vol. **302**: 1573–82.

[20] ISO 281: Rolling bearings — Dynamic load ratings and rating life," *British Standard*, 2008.

[21] BS EN 61400-1:2005, Wind turbines Part 1 : Design requirements, *British Standard*, 2010.

[22] A. Schwedt, L. Wang, W. Holweger, and J. Mayer, "Electron microscopy investigations of microstructural alterations due to classical Rolling Contact Fatigue (RCF) in martensitic AISI 52100 bearing steel," *International Journal of Fatigue*, vol. **98**, pp. 142–154, 2017.

[23] N. Shamsaei and A. Fatemi, "Small fatigue crack growth under multiaxial stresses," *International Journal of Fatigue*, vol. **58**, pp. 126–135, 2014.

[24] W. H. V Šmeļova, A Schwedt, L Wang, "A study of microstructure alteration in AISI bearings due to classic rolling contact fatigue and white etching crack," *STLE Annual Meeting & Exhibition*, L. Vegas, USA, 2016.

IOP Conf. Series: Journal of Physics: Conf. Series **1106** (2018) 012030 doi:10.1088/1742-6596/1106/1/012030

The effects of tribological factors and load sequence on surface pitting and cracks in bearing steel

Hamza A. Al-Tameemi[1,2], Hui Long[1] and Robert R. Dwyer-Joyce[1]

[1] Department of Mechanical Engineering, The University of Sheffield, Sheffield, UK.
[2] Department of Mechanical Engineering, The University of Baghdad, Baghdad, Iraq

E-mail: hamza.al-tameemi@coeng.uobaghdad.edu.iq

Abstract. This paper presents an investigation of the influence of various tribological parameters on surface initiated damage through Rolling Sliding Tests (RSTs) using bearing steel specimens. The RSTs were conducted on a benchtop twin-disc machine, consisting of a tribometer and a rolling contact fatigue testing system. The parameters investigated were contact pressure, slipping ratio, rotational speed, lubricant viscosity and load sequence, with each of them varying between two values. The first step was an investigation of the Coefficient of Traction (COT) under different testing conditions, followed by a set of RSTs to investigate surface damage initiation. It was found that the COT increased significantly under certain conditions of opposite rotational direction. The RST results showed that cracks and spalls on the surface were severer when higher slip ratio, higher contact pressure and higher rotational speed were applied first than that when lower levels of these parameters were applied first.

1. Introduction

Premature bearing failures have been frequently observed in Wind Turbine Gearboxes (WTGs) which makes the improvement of their reliability a top priority among other components [1]. The most common premature failure was due to White Etching Cracks (WECs) which caused flaking of material from the surface termed as White Structure Flaking (WSF) [2][3]. The root causes of the premature failure for WTG bearings are not fully understood, and further investigation is necessary in order to achieve the design life in field operation. Bearing failures may initiate either on the surface or under the surface of contact in the bearing raceways. The surface initiation hypothesis suggested that cracks could be caused by surface flaws which became worsened under loading conditions [4][5]. Another cause of the premature failure was material defects such as non-metallic inclusions which could serve as WEC initiators by forming butterflies in subsurface of the contact according to the sub-surface damage initiation hypothesis [6][7].

Premature bearing failure had been investigated experimentally in previous studies with different machines under various test parameters. Most of these earlier studies were carried out using discs tested on bench-top machines to represent the non-conformal contact between the inner or outer raceways and the rollers in the bearings. Since the WSF was considered as the premature failure mode of WTG bearings, most of the tests focused on replicating this failure mode and the related damage to WECs and WEAs, observed in the form of butterflies. Others investigated micro-pitting and the surface initiation of spalling. Both surface and sub-surface initiated damage were important, since they were both observed in the examined bearings retrieved from the field [6]. High numbers of cycles were commonly applied in testing, which reached tens to hundreds of millions [8]. Normally, to achieve this high number of cycles within a reasonable time, the tests were run at higher speeds, which may not be representative of those in-service operating conditions. The parameters investigated in the reported studies included the maximum contact pressure [9], the rolling-sliding ratio [10], lubricant additives [10], vibration and transient loading [11], sliding to an over-rolling direction [5], and the number of cycles [5][8]. It was found that the number of load cycles and friction between the contacted surface in a mixed lubrication affected the formation of WECs [8], in addition to the contact pressure and sliding ratio. One of the recent studies [10] found that a specific lubricant additive - metal sulphonate detergent - had a significant effect, triggering surface-initiation cracks that were similar to the WECs.

Despite a wide range of parameters were investigated under different testing levels of complexity, some issues still remained. For WTG planetary bearings the tangential speed is much lower than that tested in previous studies. Also, the values of the Coefficient of Traction (COT) have not been reported in some of the tests or lower values were reported, lower than that expected for roller bearings. Although the torque reversal is frequently reported in WTGs, no experimental investigation has been conducted to evaluate its effect due to the variation of the COT and reversed rotations. Some tests have investigated the effect of vibration or load variations, however the effect of the sequence in applying higher load levels or severe tribological conditions on damage initiation and propagation has not be investigated.

The paper reports the design of experimental tests to investigate the COT at three different operation conditions including dry and lubricated tests followed by nine RSTs conducted at different operational conditions. The research presented in this paper focuses on surface initiated damage by investigating the effect of surface traction and the loading sequence which may represent WTG operational conditions involving torque reversal and severe transient loading.

2. Experimental Testing

The machine used to perform rolling and sliding contact was the SUROS system (Sheffield University Rolling Sliding), shown in Figure 1 [12]. The instantaneous slip (Si) between the specimens is calculated by Equation 1, where R is the radius of the specimens and N is the rotational speed in RPM. During the test, the data is logged including load, rotational speed, number of revolutions, torque and COT. The machine is designed to run a dry or lubricated contact in an oil bath. One of the tested specimens has a crowned contact surface and the other has a flat contact surface, as show in Figure 2.

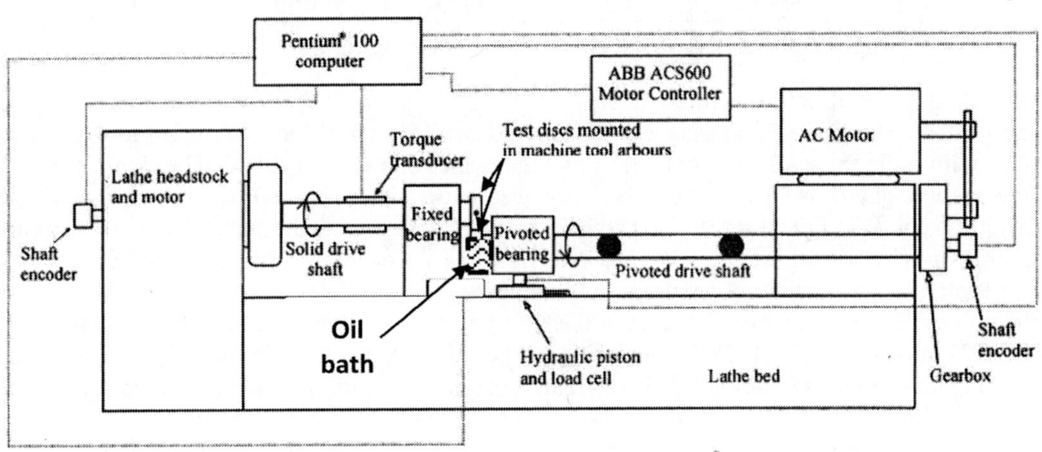

Figure 1: Schematic representation of SUROS machine adopted from [12]

$$S_i(\%) = 200 * \left(\frac{R_{Upper} * N_{Upper} - R_{Lower} * N_{Lower}}{R_{Upper} * N_{Upper} + R_{Lower} * N_{Lower}}\right) \quad 1$$

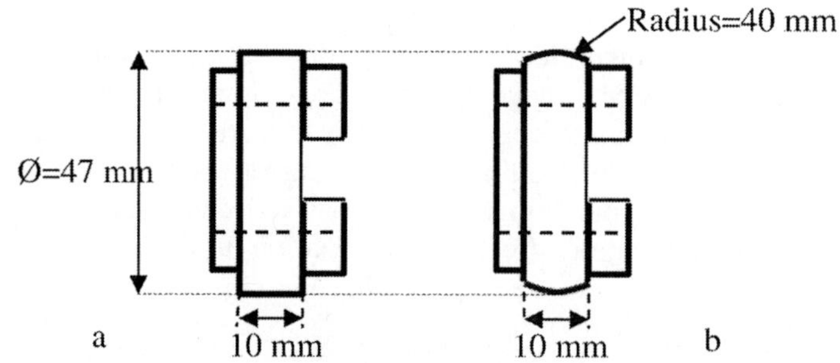

Figure 2: Specimens used for the tests (a) Flat (b) crowned surface of contact

The running track temperature close to the point of contact was measured using an Infrared thermocouple. The temperature was measured every 30 minutes during the tests and a record was made in which the average was applied for calculating the lubricant film thickness between the specimens at each test. In this research, two types of lubricants, Castrol Alpha SP 46 and 68, were used to investigate the effect of different lubricant viscosities. The average temperature measured close to the point of contact was considered for each RST to find the viscosity and calculate the minimum lubricant film thickness, according to Hamrock and Dowson in [13], and in turn to find the λ ratio, and thus the lubrication regime during the RSTs.

In each of the RSTs, two discs, crowned on a non-crowned, were rolling over each other under five controlled parameters. These were rotational speed of lower disc, contact pressure, slip ratio, lubricant viscosity and number of cycles. Since the surface traction, quantified by the COT, is one of the main test parameters to be investigated, a set of tests were conducted to gain an insight into the possible values of the COT under different operating conditions, as shown in Table 1. It was found that after a running in of 2000 cycles in a lubricated condition before the first measurement, the surface became very smooth and this reduced the effect of surface roughness.

Table 1: The settings for the COT measurement tests

Test	Lubrication	Load (kN)	P_{max} (MPa)	Speed (rpm)	Slip %
COT1	Sp 68	4.6	3052	20	-1
COT2	Sp 68	6.5	3425	400	-10
COT3	Dry	3.5	2786	400	-1

To investigate the effect of different tribological conditions on surface initiated damage, nine RSTs were performed. The first test consisted of three steps, while each of the other eight tests consisted of two steps with different settings of operating parameters, as shown in Table 2. The first test ran longer than the other eight tests, and it was the only test with a step of reversed rotations, which was designated by the (–) sign of speed in Table 2. In these tests, with the exception of the number of cycles, the variation of all parameters was investigated as well as a change in the sequence of test steps. In each of the eight tests, there was a step which was called the *high speed step* where the testing parameters were set to a rotation speed of 400 rpm, maximum contact pressure of 3425 MPa, 250,000 cycles, -10% slip and using SP 68 lubrication oil. The other step was called *low speed step* where the number of cycles was 50,000 and other investigated parameters varied between two values. In Table 2, *sequence 1* refers to the first step, step (a), as the low speed step and the second step, step (b), as the high speed step. *Sequence 2* refers to the first step, step (a), as the high speed step and the second step, step (b), as the low speed step.

The tests in Table 2 were designed to investigate the effect of each of the studied parameters by comparing certain tests with each other. The features of the contact surface used to evaluate the effect of the investigated parameters were the variation of surface roughness, the weight loss, and the characteristics of surface. The surface roughness variation was examined using the non-contact profilometer (INFINITEFOCUS Alicona). This system uses an optical focus variation method to create 3D measurements. The arithmetical mean roughness (Ra) was measured for the surface of each specimen outside and inside wear scars, in the axial and circumferential directions, with at least three lines at different locations. Then, the average and standard deviation were calculated to find the Ra inside and outside the wear scars in the axial and circumferential directions. The wear scar roughness for each RST was compared with the original roughness of the specimens in order to find the percentage difference.

Table 2: Specifications of RSTs

RSTs	Test step	Speed (rpm)	Slip %	Oil type SP	Sequence	Load (kN)	P_{max} (GPa)	Cycles	The radii of the elliptical contact area (mm)
1	1a	400	-10	68	-	6.5	3.42	900000	1.43 and 0.63
	1b	-20	0	68	-	1.5-3.5	2.1-2.78	36	(0.88 and 0.39) to (1.16 and 0.52)
	1c	400	-10	68		6.5	3.42	100000	1.43 and 0.63
2	2a	400	-10	68	2	6.5	3.42	250000	1.43 and 0.63
	2b	87	-10	68	2	6.5	3.42	50000	1.43 and 0.63
3	3a	87	-10	68	1	6.5	3.42	50000	1.43 and 0.63
	3b	400	-10	68	1	6.5	3.42	250000	1.43 and 0.63
4	4a	87	-20	46	1	6.5	3.42	50000	1.43 and 0.63
	4b	400	-10	68	1	6.5	3.42	250000	1.43 and 0.63
5	5a	48	-10	46	1	6.5	3.42	50000	1.43 and 0.63
	5b	400	-10	68	1	6.5	3.42	250000	1.43 and 0.63
6	6a	48	-20	68	1	6.5	3.42	50000	1.43 and 0.63
	6b	400	-10	68	1	6.5	3.42	250000	1.43 and 0.63
7	7a	87	-20	68	1	6.5	3.42	50000	1.43 and 0.63
	7b	400	-10	68	1	6.5	3.42	250000	1.43 and 0.63
8	8a	400	-10	68	2	6.5	3.42	250000	1.43 and 0.63
	8b	87	-20	68	2	9.75	3.92	50000	1.63 and 0.73
9	9a	400	-10	68	2	6.5	3.42	250000	1.43 and 0.63
	9b	87	-20	68	2	6.5	3.42	50000	1.43 and 0.63

3. Results and Discussion

In the following sections, the results of the COT at different operating conditions including dry contact and reverse rotation are presented first. Then the surface examination of the specimens is presented. Finally, the effects of the investigated parameters during the RSTs on the initiation of surface damage are compared.

3.1. Coefficient of Traction at Different Testing Conditions

The COT results were obtained according to the sequence and testing parameters in Table 1. For a lubricated test under certain testing conditions, the COT did not vary significantly after a specific number of cycles, as shown in Figure 3 (a). However, for dry contact, the COT reached a very high value as shown in Figure 3 (b). Although dry contact may not be expected in the WTG bearings, it shows the possible COT values during metal on metal contact. It is found that a low slip ratio, -1%, caused a rapid increase in the COT and considerable surface damage. The heat generated could be the main reason for the adhesive surface damage and the increase in COT. However, the accumulation of heat after a number of cycles, and increasing the flash temperature reduce the COT due to the oxide formations.

After the COT measurement tests, the surfaces examination of specimens used for lubricated tests showed a very mild wear track which only reduced the roughness of the surface, as shown in Figure 4 (a), while after the dry test, the surfaces were highly damaged, as shown in Figure 4 (b).

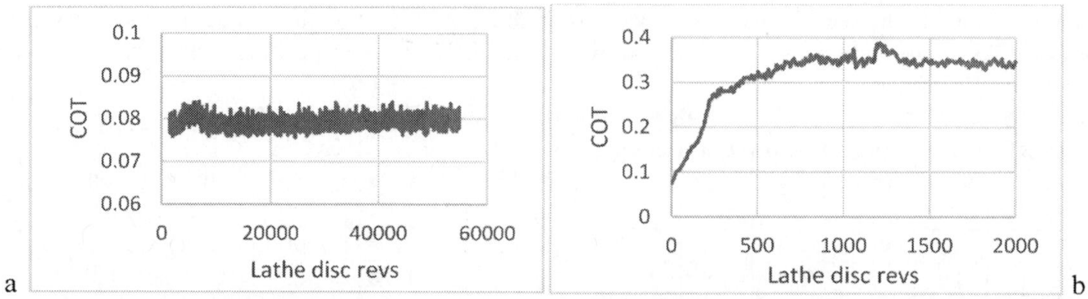

Figure 3: COT variation with number of cycles according to Table 1 (a) lubricated contact (COT2) (b) dry contact (COT3)

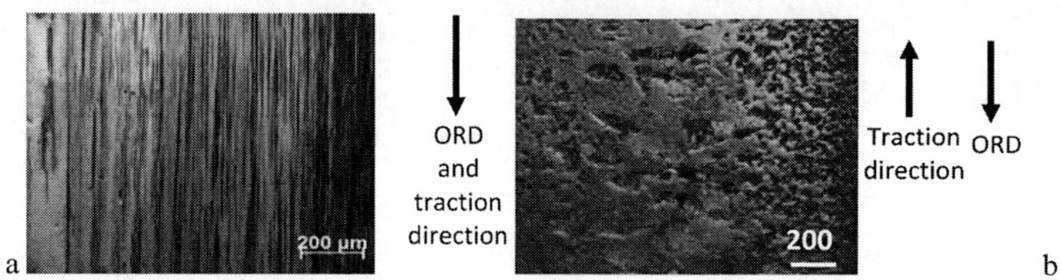

Figure 4: The surface after the COT tests (a) after lubricated tests; (b) after high speed dry test

To investigate the effect of reversed rotation on the COT, RST 1 (b), as specified in Table 1, was conducted. The results are shown in Figure 5. The variation of the COT shows that not only dry contact, such as in Figure 3 (b), can cause a high COT, but a very well lubricated contact surface can do so as well when the reversed rotation occurs for a brief time. In Figure 5, the COT jumped to a higher value after the maximum contact pressure reached 2.78 GPa (3.5 kN) as shown by the dashed line, although the load was being gradually increased. Accordingly, the instantly reversed rotation under this level of contact pressure can cause high values of COT. Reversed test can simulate the torque reversal when a roller moves in the reversed rotational direction. The accelerated damage observed on the surface supports the occurrence of premature failure of wind turbine bearings, which can be caused by changes in the direction of rotation. For the RSTs, the COT was in the range of 0.06 to 0.087 and the value at the low speed step of each test was higher than that during the high-speed step for each test. This indicates that the tests were running in the mixed lubrication regime where higher speed produces a thicker lubricant film thus lower COT.

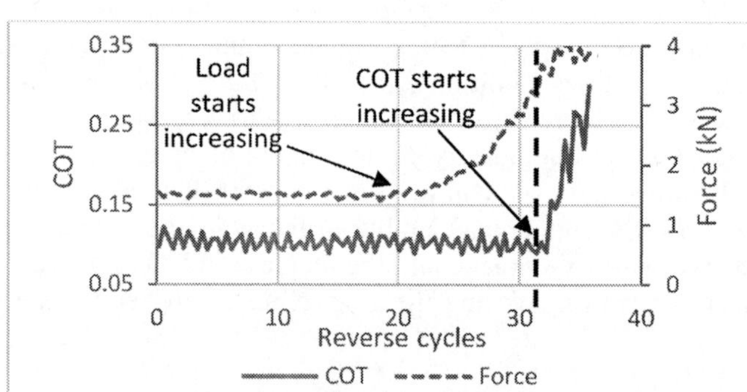

Figure 5: COT during reversed rotation according to testing conditions in Table 2

3.2. *Effects of Tested parameters on Surface Damage and Wear*

The optical microscope was used to take images, such as those in Figure 6, from four equidistant regions around the circumference and the maximum size of cracks and pits were measured. For all the tests that developed pits, it was found that this was mainly due to axial surface cracks that were close to each other.

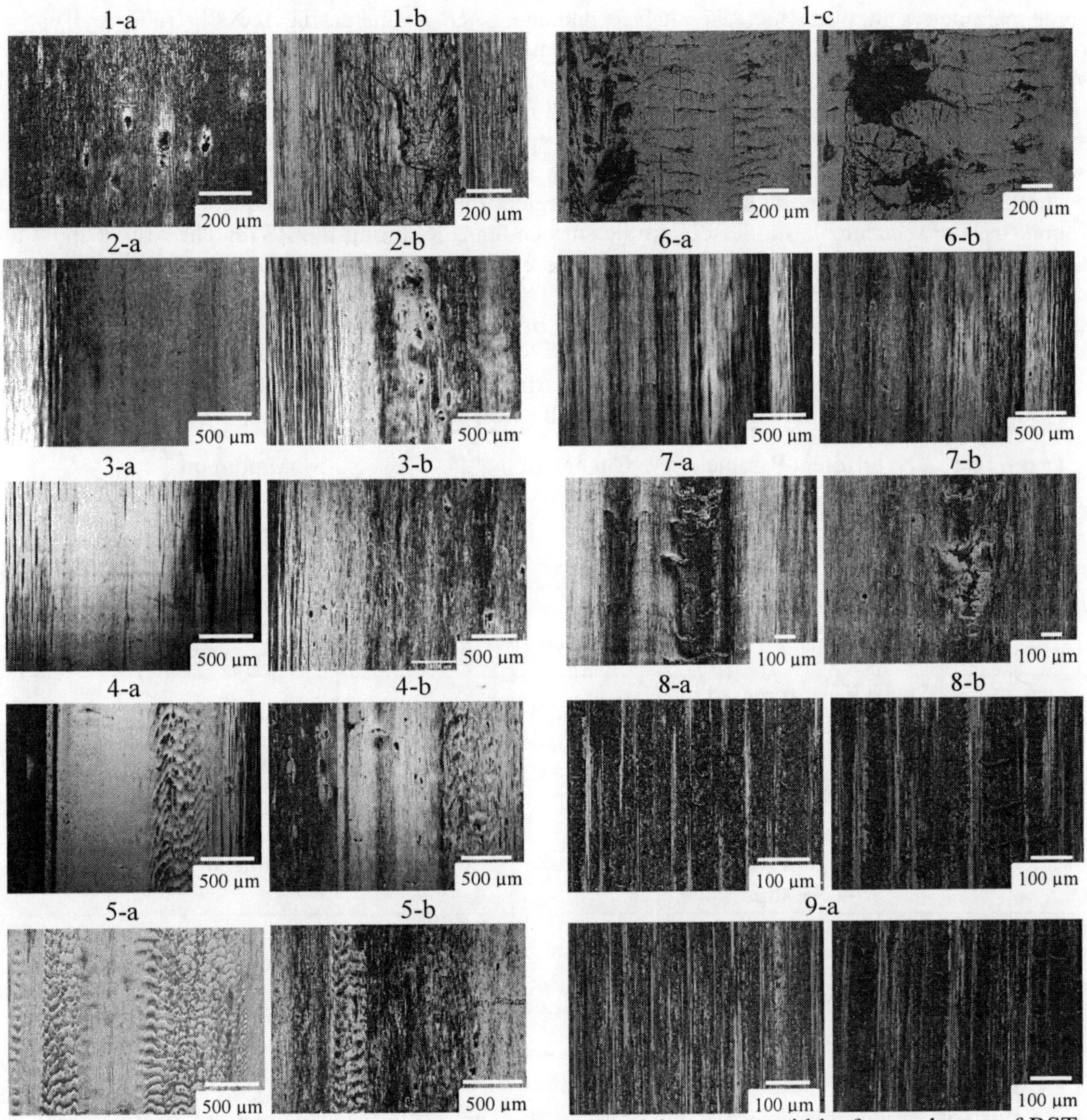

Figure 6: Non-crowned specimen surfaces at the middle of the contact width after each step of RSTs 1 to 9

For all the test steps conducted at 400 rpm, the λ values were almost one, which leads to the asperities on the surfaces being almost separated, whilst at low rotational speeds of 48 and 87 rpm, the h_{min} is considerably lower than the value of the composite roughness, and thus a mixed lubrication regime and partial contact between the surfaces are expected. No severe damage was observed on the surface, based on the observed optical images and the variation of the roughness. However, considerably more damage occurred during the reverse rotation. The observed variation in the initial roughness implied that the calculation of the lambda ratio should not be based on the initial surface roughness, which varies and reduces quickly after the running in time. The wear rate was measured by the weight loss from the flat and the crowned specimen after each step of the RSTs. In general, the wear rate is very limited and within the range of milligrams.

Table 3 shows pairs of tests carried out under similar conditions except one investigated parameter that varied in Test B. The observations from Test B were compared to that in test A. The results show some parameters that can increase surface damage, such as increasing the slip ratio and contact pressure (normal load), and other parameters that can mitigate this damage, such as reducing the speed, and changing the sequence of load steps to high speed first. A higher slipping ratio causes more relative sliding between the contacted bodies, and more surface strain that accumulates to fatigue the surface and onset cracking under mode II. Higher speed was found to increase the temperature and the surface damage, which could be due to the higher strain rate at a higher speed. Although damage accumulation theories such as the Miner rule, neglect the effect of load sequence, this research has found that the sequence has an effect on the surface damage during the RSTs. The results show that the propagation of surface damage was accelerated even under less severe conditions when the surface damage initiates first under low speed testing. This highlights the importance of avoiding severe tribological conditions or loading at the beginning of a bearings life.

Table 3: Effects of investigated parameters on surface observations from RSTs 1 to 9 where LS and HS are the low and high speed steps, respectively

Test A	Test B	The different Parameter in Test B	% Change of Roughness	Maximum surface Crack length	Maximum pit size	Wear rate
3	7	Slip ratio was increased in RST 7 at sequence 1	Reduced	Increased	Increased	Increased
7	4	Viscosity was decreased in RST 4	Increased	Increased (LS) Reduced (HS)	Reduced (LS) Reduced (HS)	Increased (LS) Reduced (HS)
7	6	Speed was decreased in Test 6	Increased	Reduced	Reduced	Reduced
9	8	Load was increased in RST 8	Increased	Increase (HS) Increase (LS)	Increase (HS) Reduced (LS)	Reduced (HS) Increase (LS)
7	9	Sequence was changed from 1 to 2 in RST 9 at -20% slip	Reduced	Reduced	Reduced	Reduced
3	2	Sequence was changed from 1 to 2 in RST 2 at -10% slip	Reduced	Reduced	Reduced	Increased

4. Conclusions

The effect of various tribological conditions on the surface initiated damage of bearing steel was investigated. Sets of tests were conducted and the observation of the COT and the topography of the specimens' surface were utilized to characterize the effect of contact pressure, slipping ratio, rotational speed, lubricant viscosity and load sequence. This study concludes that

- The COT did not vary significantly after a specific number of cycles for lubricated tests, and a high slip ratio did not necessarily correlate to high surface traction/COT under full lubrication. However, for dry contact or reversed rotation tests, the COT reached a very high value within a short time or few cycles.

- The severity of the damage observed during reversed rotation revealed more deteriorating effect of this condition compared with the other operating conditions. Accordingly, the severity of spalling correlated to the level of surface traction.
- The initial surface roughness was reduced significantly after the running-in cycles. This results in inaccuracies in the calculation of the lambda ratio based on the initial surface roughness, and invalidates the assumption of boundary lubrication made in many studies.
- It was also found that cracks and spalls on the surface were affected by sequence 1 (low speed step, then high speed step), higher slip ratio, higher contact pressure and higher speed at lambda ratio less than one.

Acknowledgment
The first author would like to thank the Iraqi Ministry of Higher Education and Scientific Research (http://www.mohesr.gov.iq/en/) and the University of Baghdad for sponsoring his PhD study in the University of Sheffield, UK.

References
[1] Faulstich S, Hahn B, Jung H and Rafik K 2009 Suitable failure statistics as a key for improving availability in *Proceedings of the European Wind Energy Conference EWEC* Online resource.
[2] Evans M H 2012 White structure flaking (WSF) in wind turbine gearbox bearings: effects of 'butterflies' and white etching cracks (WECs) *Mater. Sci. Technol.* **28** 1 3–22
[3] Bruce T, Rounding E, Long H and R. Dwyer-Joyce 2015 Characterisation of white etching crack damage in wind turbine gear- box bearings *Wear* **338–339** 164–177
[4] Stadler K and Stubenrauch A 2013 Premature bearing failures in industrial gearboxes *Annu. Rev. Mater. Sci.* **9** 1 283–311
[5] Gould B and Greco A 2015 The Influence of Sliding and Contact Severity on the Generation of White Etching Cracks *Tribol. Lett.* **60** 29 2 1–13,.
[6] Al-Tameemi H, Long H and Dwyer-Joyce R S 2016 Investigation of wind turbine gearbox bearing subsurface damage considering transient loading and the seperation of MnS inclusions *Wind Summit* 145
[7] Al-Tameemi H A, Long H, and Dwyer-Joyce R S 2018 Initiation of sub-surface micro-cracks and white etching areas from debonding at non-metallic inclusions in wind turbine gearbox bearing *Wear* **406–407** 22–32
[8] Kruhöffer W and Loos J 2017 WEC Formation in Rolling Bearings under Mixed Friction : Influences and ' Friction Energy Accumulation ' as Indicator *Tribol. Trans.* **60** 3 516–529
[9] Brückner M, Gegner J, Grabulov A, Nierlich W, and Slycke J 2011 Butterfly formation mechanisms in rolling contact fatigue *Proc VHCF-5* 101–106
[10] Ruellan A, Ville F, X Kleber, and Liatard B 2017 Understanding white etching cracks in rolling element bearings : State of art and multiple driver transposition on a twin-disc machine *Proc. Inst. Mech. Eng. Part J J. Eng. Tribol.* **231** 2 203–220
[11] Evans M H, Richardson A D, Wang L, and Wood R J K 2013 Serial sectioning investigation of butterfly and white etching crack (WEC) formation in wind turbine gearbox bearings *Wear* **302** 1–2 1573–1582
[12] Fletcher D I and Beynon J H 2000 Development of a Machine for Closely Controlled Rolling Contact Fatigue and Wear Testing *Test. Eval. J.* **28** 4 267–275
[13] Harris T A and Kotzalas M N 2007 *Advanced Concepts of Bearing Technology* 5th ed. CRC Press

IOP Conf. Series: Journal of Physics: Conf. Series **1106** (2018) 012031 doi:10.1088/1742-6596/1106/1/012031

A vibro-acoustic quality control approach for the elastic properties characterisation of thin orthotropic plates

F Igea and A Cicirello

Department of Engineering Science, University of Oxford, Oxford, UK

E-mail: alice.cicirello@eng.ox.ac.uk

Abstract. Many engineering structures are built by assembly of plates, and in recent years there has been a widespread use of composite plates, and in particular of orthotropic plates, to improve dynamic structural performance and reduce structural weight. Depending on the manufacturing process used, orthotropic plates might be affected by variability of the mechanical properties, which in turn will affect their performance. The four elastic constants of orthotropic plates can be obtained via direct static or dynamic tests, or via inverse methods which combine dynamic experiments with mathematical models. However these approaches may not be appropriate for establishing manufacturing variability across a number of manufactured plates, since they might require expensive testing equipment and they may be time consuming. In this paper an inverse procedure based on the Chladni patterns is investigated to analyse thin orthotropic rectangular plates. A detailed mathematical derivation of the equations needed for estimating the elastic constants based on the knowledge of five natural frequencies is presented. The repeatability and reproducibility of the results obtained with the proposed procedure is first assessed, and the evaluation of the manufacturing variability of an ensemble of "nominally identical" plywood plates with the proposed approach is then investigated.

1. Introduction

In many areas of aeronautical and mechanical engineering there has been a widespread use of composite plates to improve dynamic structural performance and reduce structural weight of built-up structures. To investigate the dynamic performance of built-up structures at the design stage using virtual prototypes, it is necessary to characterise the material and geometrical properties, as well as damping, of the structural components. However, the manufacturing of composites generally involves the moulding of several layers of dry fabrics together with resin to form a laminate that is cured in moulds, which can lead to manufacturing variability. As a result, in recent years extensive research has been devoted to the development of efficient analysis methods which can take into account manufacturing variability and are suitable for use in the iterative design process. Nonetheless, little work has been carried out on the development of efficient strategies for quantifying the variability of the mechanical properties of the manufactured structural components. In this paper we focus our attention on the analysis of thin orthotropic plates.

The main direct testing methods that are used for the determination of elastic constants can be classified in three main groups: static, ultrasonic and Dynamic Mechanical Analysis (DMA) [1-4]. Static tests enable the evaluation of the elastic properties from the stress-strain curves. Nonetheless, these tests are destructive and can be time consuming. Moreover, as significant levels of creep (time increasing strains under constant stress) are found on the composite materials, the results of this type of tests cannot be directly used for the resolution of dynamic problems. Ultrasonic tests are based on measurements of the propagation of ultrasonic waves in the material tested. Composite materials often dissipate a

significant amount of the energy carried by the ultrasonic waves transmitted through them. Additionally, the wavelength that this test requires is often similar to the size of some structural elements part of the composite materials to be tested. For these two reasons the ultrasonic test results are also inappropriate for dynamic modelling of vibrating structures. DMA tests measure the strains of a material under sinusoidal stress and it provides values for the complex forms of the Young's Modulus and Shear Modulus of the materials. However, due to the small size of the specimen that can be tested and the high cost of the equipment involved, the industrial use of this method is limited.
Alternative approaches are based on inverse methods which combine dynamic experiments with mathematical models [3-6]. These approaches can be mainly grouped into (i) vibrational methods based on the optimization of the mathematical model to fit with certain tolerances the Frequency Response Function obtained via Experimental Modal Analysis (EMA); and (ii) Chladni approaches based on the optimization of a mathematical model to yield nodal lines (the so-called Chladni patterns) corresponding to the ones obtained with a Chladni setup. The latter approaches do not require expensive testing equipment and are characterised by a significantly reduced testing time, and is the focus of this paper.
The Chladni patterns have been used by Woodhouse and co-workers [3-5] to develop a simple yet effective procedure for evaluating the four elastic constants that affect the vibration thin orthotropic rectangular plate, namely: Young's moduli in x and y directions, in-plane shear modulus and Poisson's ratio. Their approach is based on evaluating with the Chladni pattern the mode shapes of 5 modes and their corresponding natural frequencies, and to use closed-form solutions relating these frequencies with the four elastic constants were provided [3-5]. Nonetheless, these papers do not investigate the possible industrialisation of the procedure for investigating manufacturing variability, neither include a detailed mathematical derivation of the equations needed for estimating the elastic constants.

In this paper an inverse procedure for the evaluation of the manufacturing variability of an ensemble of "nominally identical" orthotropic thin rectangular plates based on the Chladni patterns is investigated. A detailed mathematical derivation of the equations needed for estimating the elastic constants based on the knowledge of the natural frequencies of five modes is presented. A more accurate estimate of each natural frequency, as well as damping, is obtained by evaluating a single driving Frequency Response Function. The Chladni setup is then used to quickly visualise the mode shapes. The repeatability and reproducibility of the results obtained with the proposed procedure is first assessed, and the evaluation of the manufacturing variability of an ensemble of "nominally identical" plywood plates with the proposed approach is then investigated.

The Chladni experimental setup is described in section 2. Section 3 focuses on the evaluation of approximate expressions for the estimation of the elastic constants starting from the resonant frequencies of five modes. Section 4 introduces a procedure for obtaining refined resonant frequencies by combining EMA, Chladni and a mathematical model. Section 5 shows the results obtained for a plywood plate. Section 6 addresses the acceptability of the proposed approach for assessing the manufacturing variability evaluation of 30 plywood plates.

2. Chladni Experimental Setup

The Chladni setup, shown in figure 1, is composed by a flat surface equipped with a loudspeaker (Visaton WS 17E) driven by a sinewave generator (HP 3325B) with an audio power amplifier Prism LA50B with a power supply GW GPC-3030D. The plate is driven by a sinusoidal acoustic input which produces an out-of-plane displacement of the plate. Fine powder (tea leaves or salt) is spread on the plate resting on a flat surface and being supported by small blocks of foams (located under the nodal lines of the modes to reproduce free-free boundary conditions). A broad range of frequency is explored, and when the frequency of the acoustic input coincides with one of the plate natural frequencies, the powder moves from the plate's vibrating areas to the nodal lines where no movement occurs, forming the so-called Chladni patterns which enable the identification of the corresponding mode shape. Examples of three mode shapes are shown in figure 2.

This is a fast test that requires relative inexpensive equipment to evaluate the mode shapes (no sensor is attached to the plate, nor laser is used to measure the response) compared to that required standard Experimental Modal Analysis [6]. Knowing the frequency at which a set of mode shapes

IOP Conf. Series: Journal of Physics: Conf. Series **1106** (2018) 012031 doi:10.1088/1742-6596/1106/1/012031

occurs, it is possible to determine the elastic constant of the thin orthotropic plates by using some closed-form expressions [3-5]. In the next section, these expressions are derived.

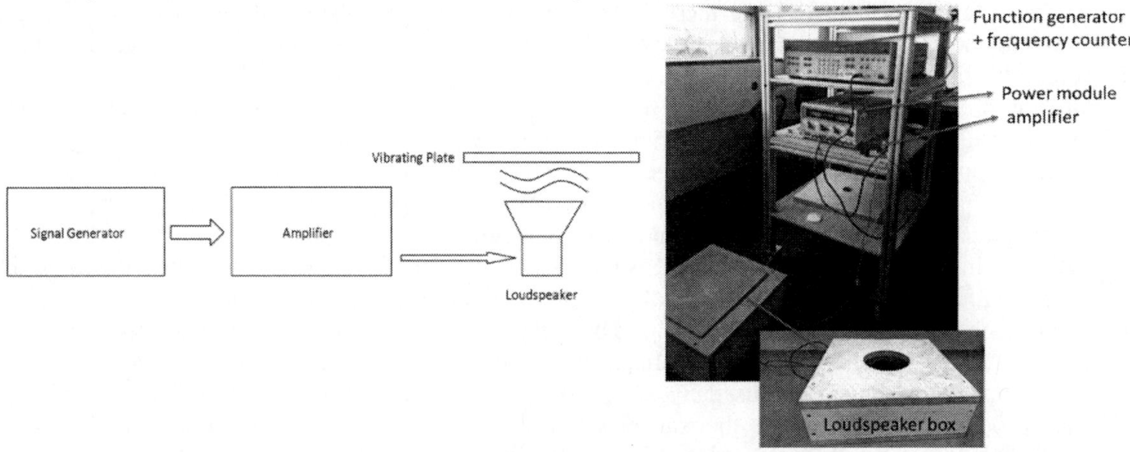

Figure 1. Chladni setup: schematic description (left figure) and portable setup used (right).

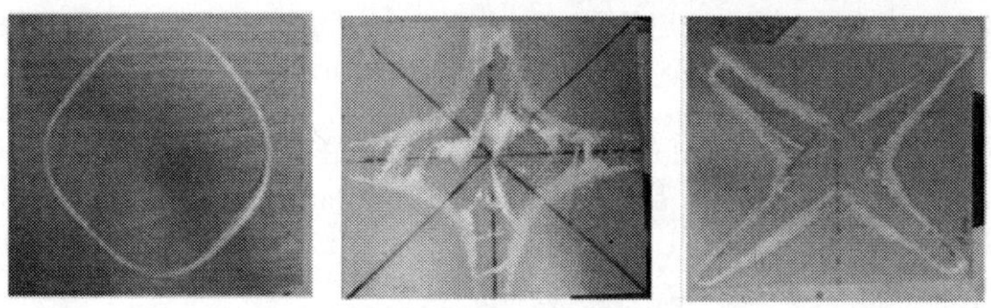

Figure 2. Chladni patterns example: plywood plate O-mode (left), FR4 plate +-mode (centre), FR4 plate X-mode (right).

3. Estimation of the Elastic Constants

Let us consider a thin flat orthotropic plate with free boundary conditions and dimensions $a \times b \times h$ and density ρ (schematically shown in figure 3) characterised by the following elastic properties: Young's moduli in x and y directions (E_x, E_y), in-plane shear modulus G_{xy} and Poisson's ratios of a thin orthotropic plate ν_{xy} and ν_{yx}.

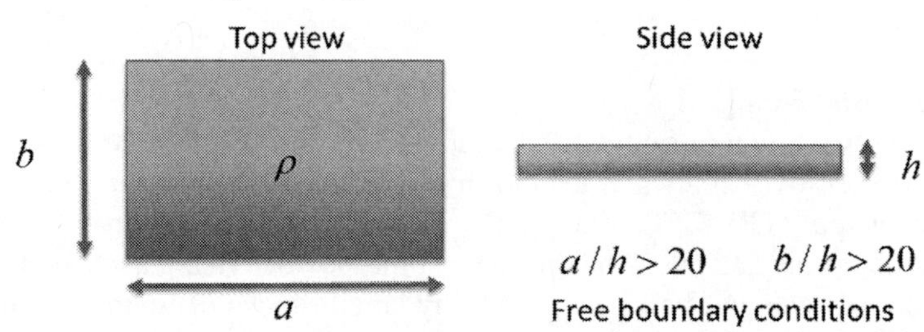

Figure 3. Schematic representation of a thin orthotropic plate with free boundary conditions

The equation describing the out-of-plane harmonic response $w(x, y)$ of the plate at frequency ω is:

$$\rho\omega^2 w = h^2\left[D_1\frac{d^4w}{dx^4} + \left(D_2 + D_4\right)\frac{d^4w}{dx^2dy^2} + D_3\frac{d^4w}{dy^4}\right] \quad (1)$$

Where the Ds are elastic constants associated with [8]:

$$D_1 = \frac{E_x}{12\left(1-\nu_{xy}\nu_{yx}\right)}; \quad D_3 = \frac{E_y}{12\left(1-\nu_{xy}\nu_{yx}\right)}; \quad D_4 = \frac{G_{xy}}{3}; \quad D_2 = \frac{\nu_{xy}E_y}{6\left(1-\nu_{xy}\nu_{yx}\right)} = \frac{\nu_{yx}E_x}{6\left(1-\nu_{xy}\nu_{yx}\right)} \tag{2}$$

Only 4 independent constants need to be established because of the general reciprocal theorem [3]:

$$\nu_{yx}/E_y = \nu_{xy}/E_x \tag{3}$$

Considering free edges and that the plate is cut parallel to the symmetry axes of the material, the boundary conditions of the plate can be expressed as Torque and Shear force at the edges equal to zero, and also the condition that no external torques are applied at the corners of the plate must be enforced [3].

In order to derive the elastic constant expressions, we can use the Rayleigh's principle to derive an approximate estimate (always higher than the actual one) of the natural frequencies as the ratio of the potential energy functional V and the kinetic energy functional T:

$$\omega^2 \approx \frac{\frac{1}{2}\iint h^3\left[D_1\frac{d^2w}{dx^2} + D_2\frac{d^2w}{dx^2}\frac{d^2w}{dy^2} + D_3\frac{d^2w}{dy^2} + D_4\frac{d^2w}{dxdy}\right]\mathrm{d}\,xdy}{\frac{1}{2}\iint \rho h w^2\,\mathrm{d}\,xdy} \tag{4}$$

By using a trial function for the out-of-place function at a specific mode, it is possible to derive a closed-form expression for the natural frequencies as a function of the elastic constants.

3.1. Estimation of the Shear Modulus

A pure twisting mode is characterised by two nodal lines that run parallel to the x and y axes in the middle of the plate (the so-called cross mode), as shown in figure 4.

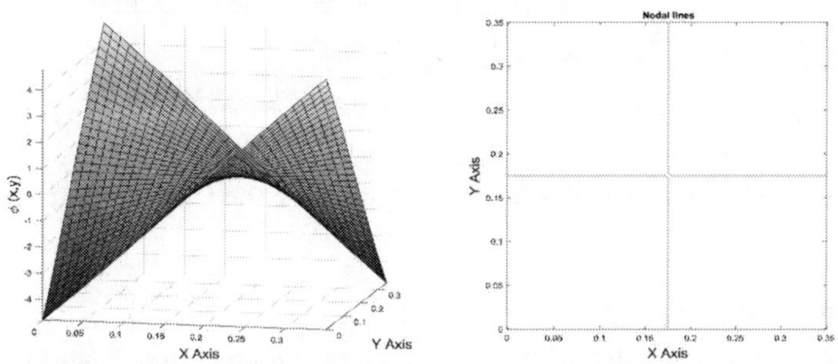

Figure 4. Cross (+) mode: mode shape and nodal lines.

This mode shape involves only one energy term G_{xy}, and the trial function $w = xy$ can be used to approximate this mode shape. Substituting it in equation (4), and integrating from –b/2 to b/2 and –a/2 to a/2, the following relationship is obtained [3]:

$$D_4 \approx 0.274\frac{a^2b^2\rho}{h^2}f_+^2; \qquad f_+ = \frac{\omega_+}{2\pi} \tag{5}$$

Therefore, by estimating with the Chladni the frequency f_+ at which the cross mode occurs, it is possible to derive G_{xy} (using equation (2)) as:

$$G_{xy} \approx 0.822\frac{a^2b^2\rho}{h^2}f_+^2 \tag{6}$$

3.2. Estimation of the Young's moduli in x and y directions

In order to evaluate the Young's moduli, we have to consider the two bending modes of the plates, which are shown in figure 5. These can be referred to as lower (L) and higher (H) bending modes, so that $D_1 \geq D_3$ [3], and such that for the higher mode the nodal lines run parallel to the y axis.

IOP Conf. Series: Journal of Physics: Conf. Series **1106** (2018) 012031 doi:10.1088/1742-6596/1106/1/012031

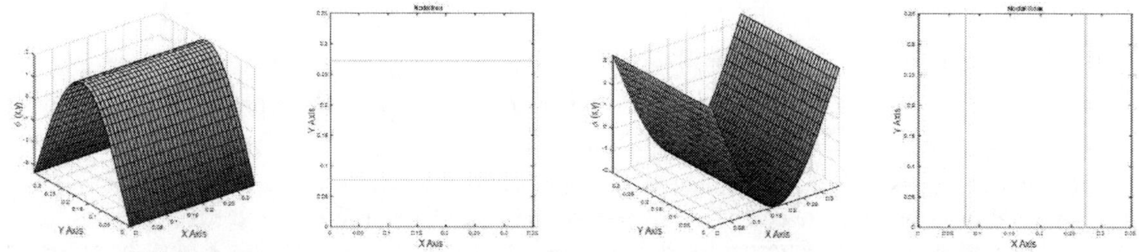

Figure 5. Modal shapes and nodal lines of lower (left) and higher (right) bending modes.

Each plate bending mode shape can be approximated with the first bending mode of an equivalent free-free isotropic Euler-Bernoulli beam of length L and flexural wavenumber $k_n = (\rho A / EI)^{1/4} \omega^{1/2}$ (being A the cross-sectional area and I the second moment of area), which can be found in closed-form (see [6]) as:

$$\cos(k_n L)\cosh(k_n L) = 1 \quad (7)$$

For the first mode n=1 and the first non-zero solution of equation (7) is $k_1 L = 4.73004$. Since the equivalent beam of length $L=a$ as a cross-sectional area of $A=bh$ and $I = bh^3/12$, the Young modulus is obtained as:

$$E_x = \frac{12\rho a^4}{4.73004^4 h^2} f_H{}^2 (2\pi)^2 = 0.0789 \frac{12\rho a^4}{h^2} f_H{}^2 \approx 12 D_1 \quad (8)$$

Being f_H the frequency (Hz) at which it is possible to observe the Chladni pattern of the higher bending mode of the plate. In a similar fashion, it can be found that:

$$E_y = \frac{12\rho b^4}{4.73004^4 h^2} f_L{}^2 (2\pi)^2 = 0.0789 \frac{12\rho b^4}{h^2} f_L{}^2 \approx 12 D_3 \quad (9)$$

Being f_L the frequency (Hz) at which it is possible to observe the Chladni pattern of the lower bending mode of the plate.

3.3. Estimation of the Poisson's ratio

The Poisson's ratio can be derived from the difference of the so-called X and O mode of a thin plate which are shown in figure 6.

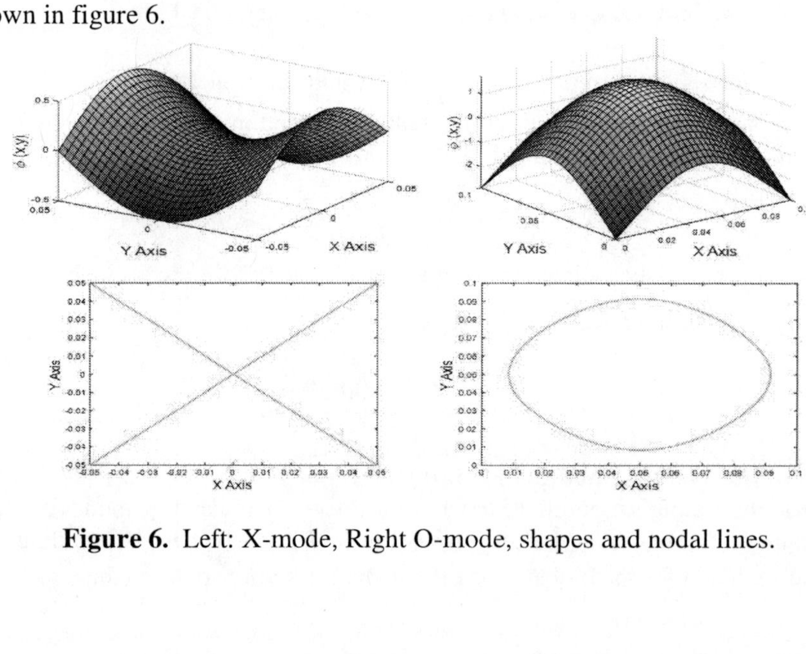

Figure 6. Left: X-mode, Right O-mode, shapes and nodal lines.

IOP Conf. Series: Journal of Physics: Conf. Series **1106** (2018) 012031 doi:10.1088/1742-6596/1106/1/012031

These modes are obtained from the coupling of the lower and higher bending modes and occur for specific aspect ratios of the plate. In particular, they will always occur in an isotropic square plate (*a=b*). For orthotropic plates instead, the size of the plate needs to be adjusted to display those modes. The correct aspect ratio can be found by using equations (8) and (9) and setting $f_H = f_L$ [3]:

$$\frac{a}{b} = \left(\frac{D_1}{D_3}\right)^{1/4} \quad (10)$$

To find a closed-form expression relating the frequency at which the X and O modes occur (f_X, f_O respectively) to the Poisson's ratio, the Rayleigh's principle (equation (4)) is applied again. The trial function to be used for the X mode shape can be expressed as [10]:

$$w = \sin\left[\frac{\pi}{2}\left(\frac{y}{b} - \frac{x}{a}\right)\right]\sin\left[\frac{\pi}{2}\left(\frac{y}{b} + \frac{x}{a}\right)\right] \quad (11)$$

The trial function for the O mode shape can be approximated by a linear combination of the free-free beam functions in the *x* and *y* directions [3]. This can be expressed as [11]:

$$\begin{aligned} w = \sin\left[4.73\left(\frac{x}{a} - 0.5\right)\right] - \frac{\sin(4.73/2)}{\sinh(4.73/2)}\cosh\left[4.73\left(\frac{x}{a} - 0.5\right)\right] + \\ + \cos\left[4.73\left(\frac{y}{b} - 0.5\right)\right] - \frac{\sin(4.73/2)}{\sinh(4.73/2)}\cosh\left[4.73\left(\frac{y}{b} - 0.5\right)\right] \end{aligned} \quad (12)$$

By using equation (4) in combination with equations (12) and (13), and using also equations (8) and (9), the difference between f_X and f_O can be expressed as:

$$\left(f_O^2 - f_X^2\right) \approx \frac{D_2}{0.104\dfrac{a^2b^2\rho}{h^2}} \quad (13)$$

Sato and Woodhouse [4] proposed a very similar expression, in which one coefficient has been corrected from comparison with numerical data and experimental data:

$$\left(f_O^2 - f_X^2\right) \approx \frac{D_2}{0.114\dfrac{a^2b^2\rho}{h^2}} \quad (14)$$

Using the relationships in equation (2), it is therefore possible to evaluate the Poisson's ratio.

4. Refined resonant frequencies using EMA, Chladni and a mathematical model – combined approach

The approximate expressions described in the previous section yield initial values of the elastic constants. These values can be refined by using a Rayleigh-Ritz model or a Finite Element Model of the plate by minimizing the difference between the predicted resonant frequencies and the measured ones. The Chladni procedure can be improved by integrating it with EMA and with mathematical models. The EMA can be used at the very beginning to establish only two average FRFs, and identify the resonances. This measurements can be performed with a relative cheap accelerometer. The Chladni setup can then be used to acoustically excite the plate only at these resonant frequencies and to identify which mode corresponds to which frequency, speeding up the process. A Rayleigh-Ritz *MATLAB* code is then used to refine the initially obtained elastic constants.

5. Comparison of Experimental Modal Analysis, Chladni setup results

Experimental Modal Analysis (EMA) and the Chladni setup are used to evaluate the resonant frequencies, and consequently the material properties of the plywood plate using the approximate expressions derived in the previous section. The plywood plate has dimensions: $a = 350mm, b = 350\ mm, h = 5.50\ mm,$ and density of 568 Kg/m^3. Initially the Chladni test was

IOP Conf. Series: Journal of Physics: Conf. Series **1106** (2018) 012031 doi:10.1088/1742-6596/1106/1/012031

performed to evaluate the 5 mode shapes described in the previous section and the corresponding resonance frequencies. The mode shapes so obtained are shown in figure 7.

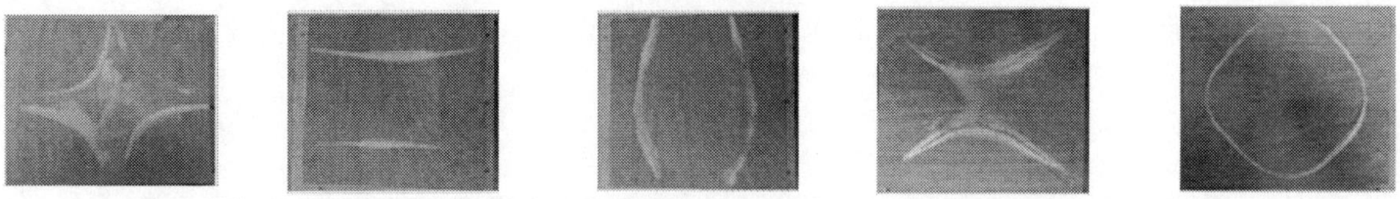

Figure 7. Modal shapes and nodal lines obtained with the Chladni setup

For obtained accurate results with EMA, the plates were vertically suspended using nylon threads to approximate the free boundary conditions that require no energy dissipation. The plates were excited using an instrumented hammer (PCB 086C03 with a rubber tip) and the velocity response was measured at the same point with a single pointer Laser Doppler Vibrometer (Polytec NLV-2500). The sampling rate was 10 kHz and the signal was recorded for 30 seconds. Three measurements were taken to evaluate an average FRF using the standard EMA procedures [6-7]. The mode shapes were not reconstructed by computing the FRFs over a grid of points [6-7], since they were already evaluated with the Chladni setup. The resonant frequencies obtained with the two approaches and with the refining approach described briefly in section 4 (referred to as "combined") are compared in table 1.

Table 1. Resonant frequencies obtained with Chladni and EMA

Method	f_+ (Hz)	f_L (Hz)	f_H (Hz)	f_X (Hz)	f_O (Hz)
Chladni	54.4	123.6	173.6	171.0	181.2
EMA	54.8	128.3	175.8	172.4	182.0
Combined	54.8	128.3	175.8	172.8	183.8

It can be observed that there are small differences in the resonant frequencies estimated between the Chladni and EMA results. This is because there might be small mistakes from the operator in tuning the driving frequency of the Chladni setup to obtain precisely the peak response, and therefore the resonant frequency. Moreover, the Chladni results can be affected by the position of the foam blocks – care must be taken to move them under the nodal lines of each specific mode. It is possible to observe also that the combined approach will improve the prediction of f_X and f_O, which will affect the prediction of the Poisson's ratio.

The corresponding elastic constants (in terms of *D*s) and the % differences obtained with the three approaches are shown in table 2, showing the improvements obtained with the combined approach.

Table 2. Elastic constants based on Chladni and EMA results

	D_1 (MPa)	D_2 (MPa)	D_3 (MPa)	D_4 (MPa)
Chladni	729.9	89.2	369.6	248.9
EMA	748.2	84.5	398.4	252.1
Combined	747.0	84.5	401.6	267.1
% difference Chladni vs EMA	2.44	5.5	7.23	1.27
% difference EMA vs Combined	0.16	0	3.2	5.95
% difference Chladni vs Combined	2.32	5.5	8.65	7.31

IOP Conf. Series: Journal of Physics: Conf. Series **1106** (2018) 012031 doi:10.1088/1742-6596/1106/1/012031

6. Manufacturing Variability Analysis

In manufacturing most of the quality problems are caused by variability in the production process [12]. As a consequence, most of the quality improvement initiatives focus on reducing the causes of variation. Overall when performing a test, variability of the results can be divided into two categories: the variability due to the part-to-part differences and the variability due to the measurement system variation, as schematically shown in figure 8.

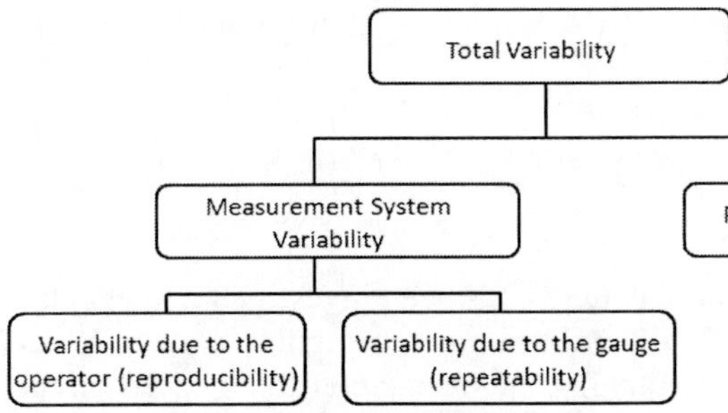

Figure 8. Sources of Variability.

For the measurement system to be acceptable, the measurement system variability must be within a certain limit, so that the part-to-part variability results can be trusted. In the next subsections, the acceptability of the proposed experimental setup as a quality control approach is assessed.

6.1. Measurement system variability

The adequacy of this Chladni setup as a measurement system for the elastic properties was tested. Measurement errors can be classified as bias and precision errors. The bias errors do not affect variability as they consistently shift the measurements in a fixed quantity. Precision errors add variability to the data, as they do not occur in all the readings and they may have different magnitude for different measurements [12,13]. There are two types of precision errors: Repeatability and Reproducibility (R&R) errors. Repeatability or gauge errors are found when the same person measures the same magnitude in several occasions. Reproducibility errors are referred to the variability that may be observed when the same magnitudes are measured by different appraisers.

The Gauge Repeatability and Reproducibility (GR&R) [12] quantifies the precision errors and assesses the ability of the measuring system to discriminate changes in the measured magnitude. To evaluate the GR&R, one plate was measured by three different appraisers. Each appraiser measured the plate five times. An ANOVA (Analysis of Variance) study was performed. To assess the acceptability of the GR&R" result obtained, two ratios commonly used in industrial practises were considered [12]: (i) "GR&R" divided by the total variation found from experiments performed on an ensemble of plates; (ii) "GR&R" divided by the tolerance of the specification applicable (higher limit minus lower limit of the specification). The acceptability criteria are shown in table 3.

Table 3. Acceptability criteria for "GR&R" results commonly used during industrial practice [12].

Acceptability	Excellent	Good	Marginal
GR&R/Total Variation (%)	< 10 %	< 20 %	< 30 %
GR&R/Tolerance (%)	< 10 %	< 20 %	< 30 %

6.2 Assessment of the measurement system

Part-to-part variability of 30 plates was measured with the Chladni setup. The mean, standard deviation and total variation (6 sigma), and histograms were computed. The relative sizes of the GR&R to the sizes of the total variation of the distributions were first calculated. The results are shown in table 4. Acceptability of the measuring system for D_2 are not shown since this would require to cut the 30 plates to satisfy the aspect ratio that would lead to the appearance of the X and O mode. While accurate estimates of D_1 and D_3 (which are related to the Young's moduli) can be obtained with the proposed setup, D_4 measurements (which is related to the shear modulus) are unacceptable. It was found that acceptability of the measurement system can be improved if operators get appropriate training and rigorously follow operating procedures.

Table 4. Acceptability of the measurement system according to the 'GR&R' / total variation (%).

	GR&R f (Hz)	GR&R D's (MPa)	Total Variation D's (MPa)	'GR&R' / total variation (%)	Acceptability
D_1 - Mode h	1.6743	14.1	288.6	4.88%	Excellent
D_3 - Mode l	1.5036	9.0	106.2	8.47%	Excellent
D_4 - Mode +	1.5445	14.4	47.0	30.6%	Unacceptable

7. Conclusion

An inverse procedure based on the Chladni patterns has been investigated to evaluate the manufacturing variability of the elastic properties of thin orthotropic rectangular plates. A detailed mathematical derivation of the equations needed for estimating the elastic constants based on the knowledge of five natural frequencies has been presented, and a combined approach using the Chladni setup with EMA and a mathematical model has been described and applied to a plywood plate, showing very good results. The manufacturing variability of an ensemble of 30 "nominally identical" plywood plates was investigated, showing that the variability of Young's moduli can be estimated accurately, while care must be taken by the operator when assessing the shear modulus.

Acknowledgments

The experiments were performed in the Dynamics, Vibration and Uncertainty Lab (University of Oxford), which has been established thanks to the John Fell Fund (163/029). FI is grateful to Pembroke College for providing funding to participate to MPSVA 2018. AC gratefully acknowledges the financial support provided by Balliol College for her Career Development Fellowship.

8. References

[1] Carlsson L A and Adam D F and Pipes R B 2014 Experimental characterization of advanced composite materials, fourth edition (Prentice-Hall)
[2] Menard HP 2008 Dynamic Mechanical Analysis: a practical introduction 2nd Edition *CRC Press*
[3] McIntyre ME and Woodhouse J 1988 On Measuring the Elastic and Damping Constants of Orthotropic Sheet Material *Acta metallica* 36 1397-1416
[4] Sato J, Hutchings I M and Woodhouse J. 2007 Determination of the dynamic elastic properties of paper and paperboard from the low-frequency vibration modes of rectangular plates. *In: 61st Appita Annual Conference and Exhibition, Gold Coast, Australia.*
[5] Talbot JP and Woodhouse J 1997 The Vibration Damping of Laminated Plates *Composites Part A* **28**A pp 1007-1012
[6] Avitabile P 2001 Modal Testing: A Practitioner's Guide. (Wiley).
[7] Brandt A, 2011 Noise and Vibration Analysis: Signal Analysis and Experimental Procedures, (John Wiley and Sons)
[8] Hearmon R F S 1961 Introduction to applied anisotropic elasticity. Oxford University Press.
[9] Gonçalves P J P, Brennan M J and Elliot S J. 2007 Numerical evaluation of high-order modes of vibration in uniform Euler-Bernoulli beams *Journal of Sound and Vibrations* 301:1035-1039.
[10] K.E. Fallstrom 1991 Determining Material Properties in Anisotropic Plates Using Rayleigh's Method. Division of Experimental Mechanics Lulea University of Technology Lulea Sweden
[11] A.W. Leissa 1969 *Vibration of Plates* Ohio State University Columbus Ohio
[12] T.T. Allen 2006 *Introduction to Engineering Statistics and Six-Sigma* Springer
[13] Q. Brook 2010 *Lean Six Sigma & Minitab* 3rd edition The complete Toolbox Guide for all Lean Six Sigma Practitioners Open Resources Ltd

IOP Conf. Series: Journal of Physics: Conf. Series **1106** (2018) 012032 doi:10.1088/1742-6596/1106/1/012032

Operational vibration shape measurement of piezoceramic disc actuator using digital image correlation vibrometry with a single reference signal

I A Jones[1,2], M Jabbal[1] and T M Buss[1]

[1] Faculty of Engineering, University of Nottingham, University Park, Nottingham NG7 2RD, UK

Abstract. This paper describes the enhancement and application of a digital image correlation (DIC) vibrometry method which avoids the need for high speed cameras. The underlying technique, described earlier by members of the present team (Warburton et al, Exp. Mech. 56(7) 1219–1230), was used to measure the operational vibration shapes of a piezoceramic synthetic jet actuator disc. Following the method previously reported, pairs of images were captured using the Dantec Q-400 hardware at a frequency of 0.25 Hz, with synchronised flash illumination but not synchronised with the excitation signal, which was captured simultaneously with each image pair. The images were processed using the Dantec Istra4D software and were post-processed largely as described by Warburton in order to obtain amplitude and phase maps for each frequency, but instead of using an externally-generated quadrature signal as a reference, an alternative reference was taken from the median displacement over an area of the specimen, in order to resolve the phase position of each image pair. Although the raw measured deflected shapes were generally dominated by the main dishing mode of the actuator, alternative higher order shapes such as asymmetric and trefoil shapes occurred in quadrature with these dishing-mode deflections.

1. Introduction

Earlier work by members of the present team has described an approach to the use of low-speed digital image correlation (DIC) equipment for measurement of operational deflected shapes as close approximations to mode shapes. The approach involves exciting a specimen at resonant frequency and using an asynchronous, free-running DIC system to image the vibration shape. The approach was described in detail by Warburton *et al.* [1], and has some similarities with (but also major differences from) an approach described by Fruehmann *et al.* [2] As described in reference [1], the data acquisition unit in the DIC system is used to capture the excitation signal, and in order to resolve the phase ambiguity of that signal a second signal, integrated using an integrating amplifier so that it is approximately in quadrature with the first signal, is used as an additional reference.

The aims of the present paper are twofold: to describe a variant of the approach described in reference [1] which avoids the need for the integrating amplifier, and to apply both the original method and the variant to the practical case study of a piezoelectric actuator disc used within a synthetic jet actuator mechanism for fluid mechanics applications [3], in order to determine the shapes with which it vibrates operationally. Specifically, the vibration modes of a typical commercial (PZT-5A) polycrystalline actuator disc will be explored, and a comparative study will be undertaken regarding a

[2] To whom correspondence may be addressed.

PZT disc and a custom-made single-crystal (PMN-PT) disc of the same dimensions, comparing the results of the present study against those in [3].

2. Summary of existing approach

The approach described this paper involves a modification to the apparatus and the processing procedure described by Warburton *et al.* [1]. It is not the present intention to duplicate the background or description given in that paper but a brief summary will be given before outlining the modification of the approach which is the subject of the present paper.

2.1. Existing apparatus and procedure before incorporation of novel referencing approach

In its original form requires the following apparatus including the underlying commercial system:

- A 3D digital image correlation system, in this case the Dantec Q-400 system equipped with two 5 MP cameras. This incorporated a data acquisition and interfacing unit capable of inputting several analogue signals. It also outputs a TTL (5V logic) digital trigger signal to the cameras, which is also used for triggering the flash.
- A signal generator with (if required) a power amplifier to drive the excitation system. The output from the signal generator is also fed into one of the analogue inputs of the Q-400 system to provide part of the phase position identification for signal capture.
- A means of providing intense illumination, specifically a conventional studio flashlamp system (Bowens ProLite 60)
- A special circuit to trigger the flashlamp system from the TTL camera trigger signal using a MOSFET to emulate the closing of electrical contacts
- Some means of mechanically exciting the vibrations e.g. a shaker coupled to the specimen via a stinger, or a loudspeaker unit acoustically coupled to the specimen; in the present case the actuator itself is the subject of the experiment.
- An integrating amplifier circuit based on a 741 operational amplifier, used to generate a second reference signal approximately in quadrature with the excitation signal and hence provide a means of identifying the phase position within a given cycle.

The specimen, in this case one side of the actuator disc, is prepared with a speckle pattern using (for example) carefully-applied aerosol spray. The 3D DIC system is calibrated in the usual way using a chequerboard target following the automated procedures provided within the Q-400 system. The procedure for each resonant frequency involves manually tuning the frequency of the signal generator until maximum amplitude of vibration is observed (either by ear or by observing a "dancing bead" on the specimen). The Q-400 system is then set to capture a large number of image pairs (typically 60) in free-running mode i.e. not synchronised to the excitation signal, at a frequency e.g. 0.25 Hz which allows time for the flash system to re-charge between images. Indeed, it is important that excitation signal is not an exact multiple of the capture frequency, though no special care is normally needed to achieve this. Each capture event involves the recording of two images at very short exposure (1/10000 s) illuminated by the flash system, along with the two analogue signals representing the directly measured and integrated excitation signals. A diagram of the apparatus, including the items omitted when using the variant of the method to be described, is given in figure 1.

The images are processed in the standard manner using the ISTRA4D software to obtain measured displacement fields in the local coordinate system of the specimen for each capture event. These are then post-processed, again as described by Warburton [1], firstly to find the phase relationship of the directly-captured and integrated excitation signals (by fitting an elliptical Lissajous figure to the points obtained by plotting the integrated signal against the directly captured one), and hence to find the phase position or circular time (ωt) within each cycle of excitation. The displacement vector at any point on the specimen is fitted to a sinusoidal function of ωt to recover the amplitude and phase angle of the vibration response at each point. Finally, the displaced shapes can be recovered at an arbitrary number of phase positions within each cycle (e.g. every 10°) can be reconstructed.

While the above procedure is effective, a practical issue is that the gain of the integrating amplifier needs to be adjusted to match both the frequency and amplitude of the input signal to avoid either insufficient amplitude or clipping, noting that the gain of an integrating amplifier is inversely proportional to frequency.

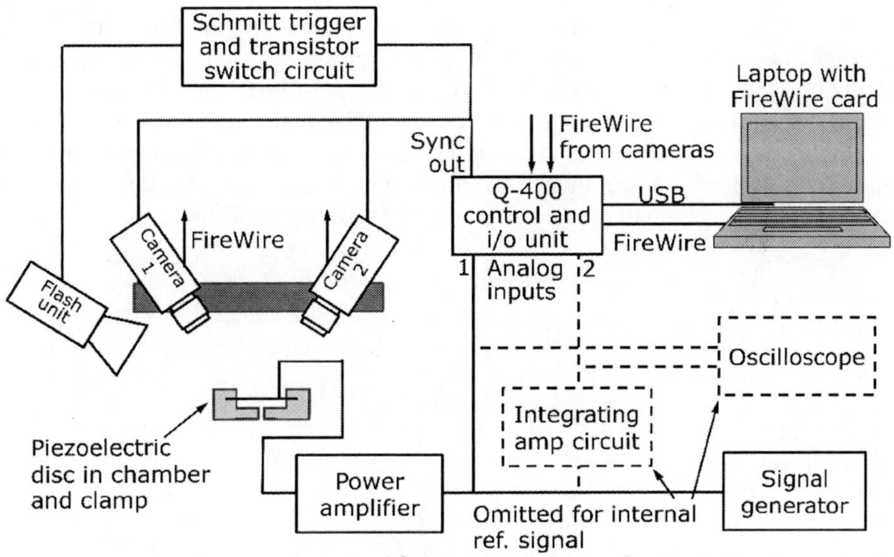

Figure 1: Apparatus for DIC vibrometry on synthetic disc actuator, including items omitted when using internal reference signal (modified from reference [1]).

2.2. Modified approach avoiding need for second reference signal

The present variant of the method arose because of the non-availability of the integrating amplifier used in the original implementation of the process at a time when indicative measurements were needed on a sample of the piezoelectric actuator, and relies upon the fact that not every displacement on the specimen is in phase with the excitation signal. The principle is that an area of the specimen is chosen and the median displacement over this area is calculated for each of the time points. This value is then used in place of the integrated signal in plotting the Lissajous figure, and hence used for disambiguating the phase information from the excitation signal. Clearly this approach will work best if there is significant phase shift between the normal mode vibrations of a structure and the excitation signal, or if there is significant phase difference between the vibrations at different points on the structure i.e. if the structure is exhibiting complex modes and there is a region with significant phase difference. Conversely, if all the points on the structure are vibrating approximately in phase with (or in phase opposition to) the excitation signal the Lissajous figure will be close to a straight line, and the inevitable noise on the results will make it very difficult to fit an ellipse to the results in a way which successfully resolves the phase ambiguity of a given point. As well as avoiding the need for an integrating amplifier, this new variant of the process avoids the issues caused by the frequency-dependent gain of the integrating amplifier.

3. Apparatus and procedure for present experiments.

Two sets of experiments were undertaken. The first was an initial trial used to explore the problem and understand the typical vibration modes, for which the above method was used to obtain results in the absence of an available integrating amplifier. The specimen was a commercial (APC International) 50 mm diameter PZT disc of total thickness 0.23 mm, clamped within the actuator cavity, and it was prepared in the usual manner with a speckle spray pattern. The apparatus was set up as shown in figure 1 with the integrating amplifier and oscilloscope omitted, and the frequency tuned manually to detect seven different resonant frequencies.

The experiment was re-run at a later date with a replacement integrating amplifier. Two piezoelectric actuator discs were tested: a commercial PZT disc smaller than that used in the familiarisation study (27 mm diameter, 0.45 mm thickness), and a single-crystal PMN-PT disc of the type intended for driving the air jet manufactured to the same dimensions. For each disc, two frequencies were excited, each at two different voltage levels, which before amplification were approximately 70 and 140 mV (rms) respectively, in order to give peak amplitudes after amplification of 10 V and 20 V respectively to match reference [3]. Two different frequencies were used: one (1.8 kHz) corresponds to to the Helmholtz frequency of the cavity into which the disc is inserted, and the other (2.8 kHz) is close to the mechanical/structural resonant frequency of the PMN-PT disc itself. Graphs of air velocity and disc displacement *vs.* frequency for both kinds of actuators are presented in reference [3]. The voltage levels used here were somewhat lower than in the preliminary experiment, to avoid any risk of fracturing the single crystal actuator which is far more costly than the polycrystalline ones.

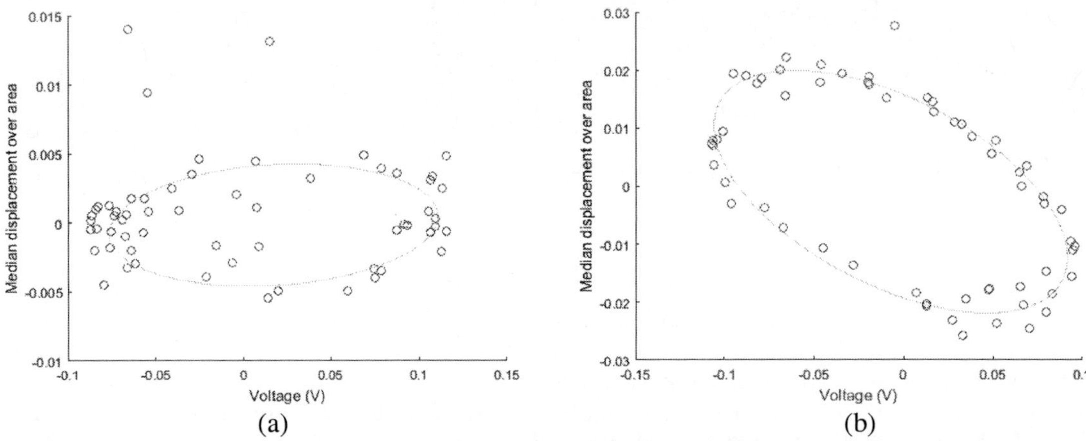

Figure 2: Lissajous figures showing relationship between median displacement over an area, and excitation voltage for (a) 347 Hz and (b) 1940 Hz. The outliers were manually excluded from the fitting of the ellipses shown.

4. Results

The results of the familiarisation experiment were processed using the modified approach described in Section 2.2. The use of an averaged displacement as the second reference signal is seen (for example, in figure 2) to give a rather scattered Lissajous figure at each of the frequencies though the elliptical trend is very clear in most cases; in some cases the trend was less clear with the figure being more of an elliptical cloud. Some clear outliers are visible in that figure; this is believed to be due to physical disturbance of the apparatus (due to floor deflections) at the start of each experiment as the operator starts the capture process before leaving the area. These outliers were omitted manually from the ellipse-fitting process; outliers are automatically eliminated from the sinusoidal fitting of displacement to ωt as described in [1]. Despite this relatively poor fit of the ellipse in some cases, the use of the resulting values of phase position ωt (termed here "circular time") resulted in clear sinusoidal trends in the displacement *vs.* circular time, and the clearly-defined operational deflected shapes in figure 3 give confidence in the results.

The results of the second set of experiments were processed as described by Warburton *et al.* without the modification described in section 2.2. Unlike in the preliminary experiment, the range of frequencies investigated was limited to two frequencies of interest rather than covering a wider range of possible resonances. No obvious complex modes are exhibited and the main mode of deformation appears to be the dishing mode in each case. The results for the polycrystalline (PZT) disc are given in figure 4 and those for the single crystal (PMN-PT) disc are given in figure 5.

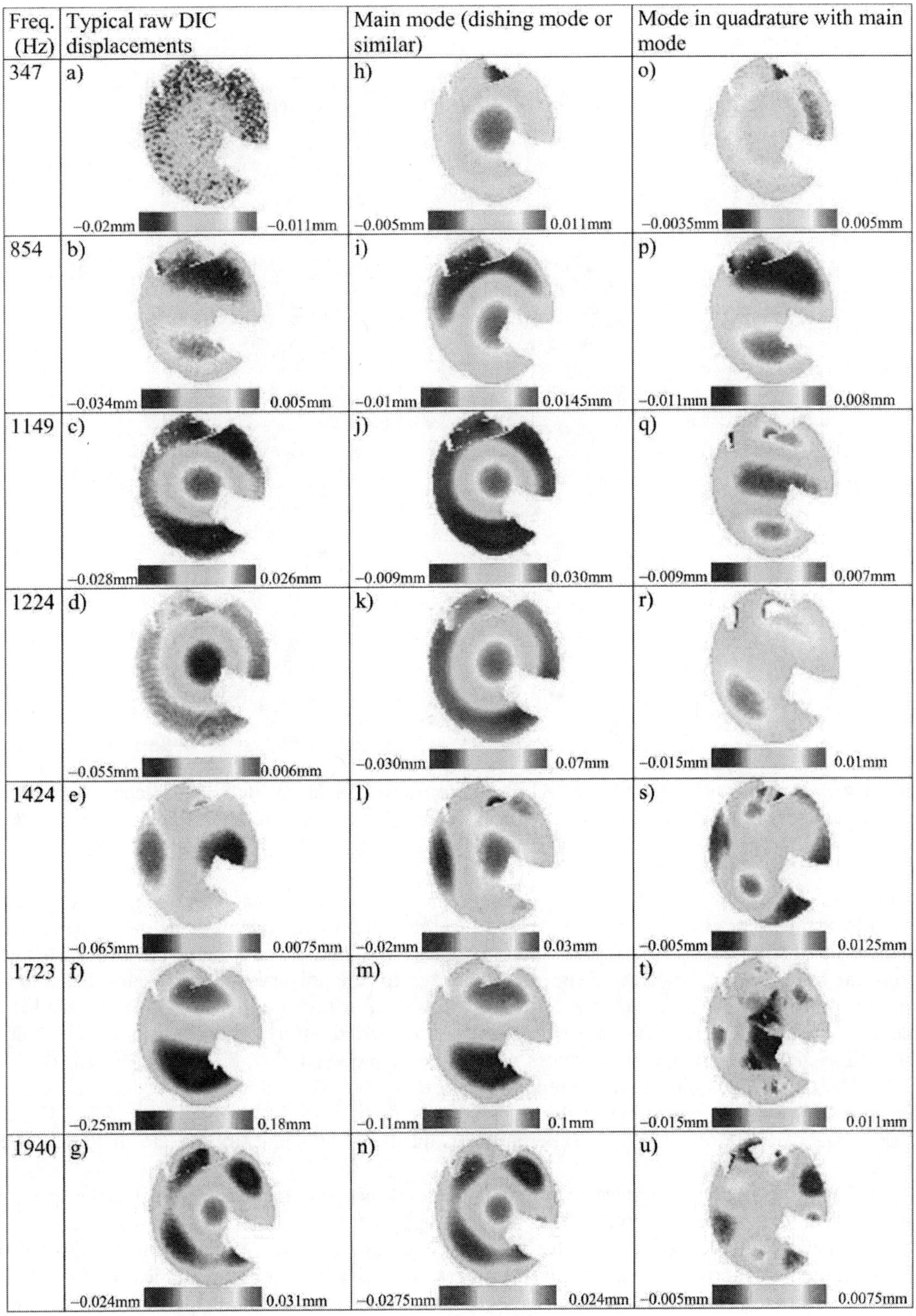

Figure 3: (a-g) Typical raw maps of deflection from DIC system; (h-n) main deflected shapes measured using DIC vibrometry with internal reference signal; (o-u) higher order deflected shapes in quadrature with main modes.

Frequency & voltage (rms)	Typical raw DIC displacements	Main mode (dishing mode or similar)
1.8 kHz 71 mV	a) 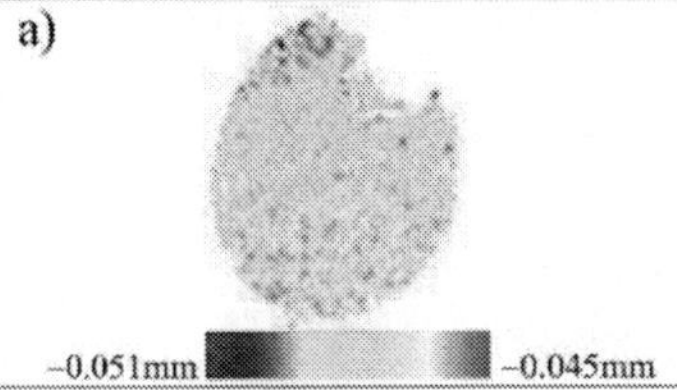	e) 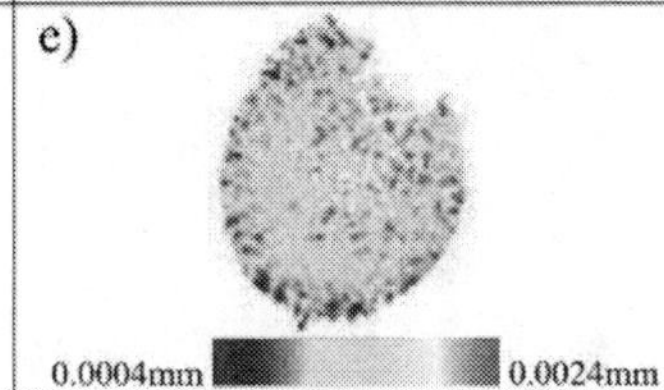
1.8 kHz 141 mV	b) 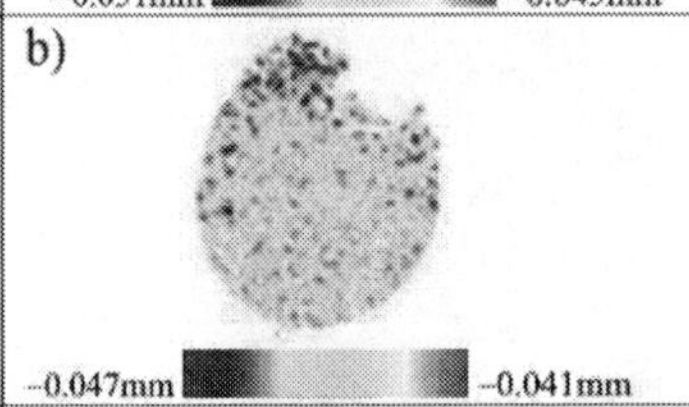	f) 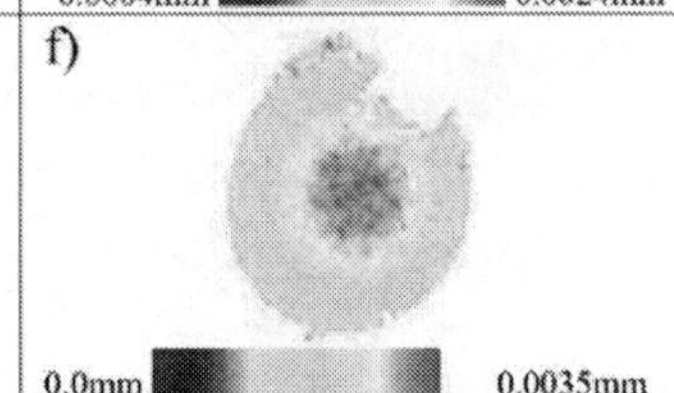
2.8 kHz 71 mV	c) 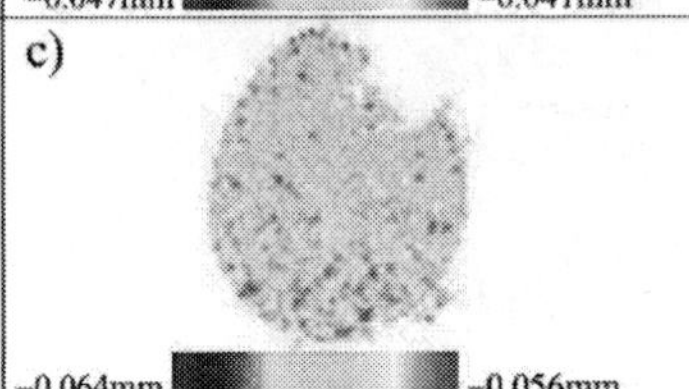	g) 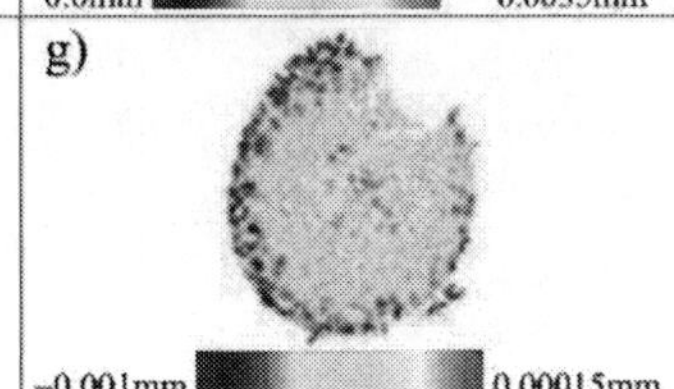
2.8 kHz 143 mV	d) 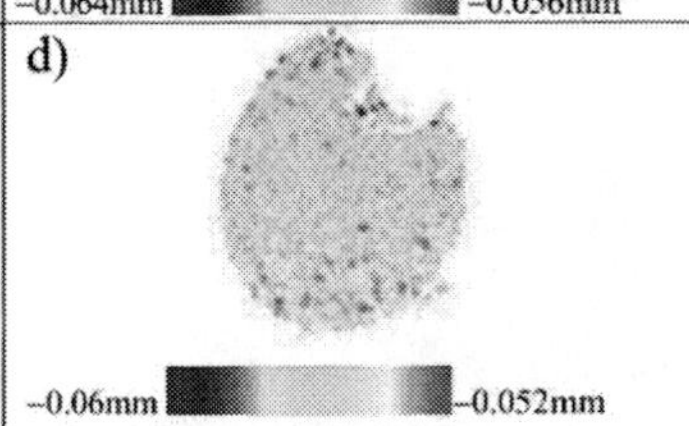	h) 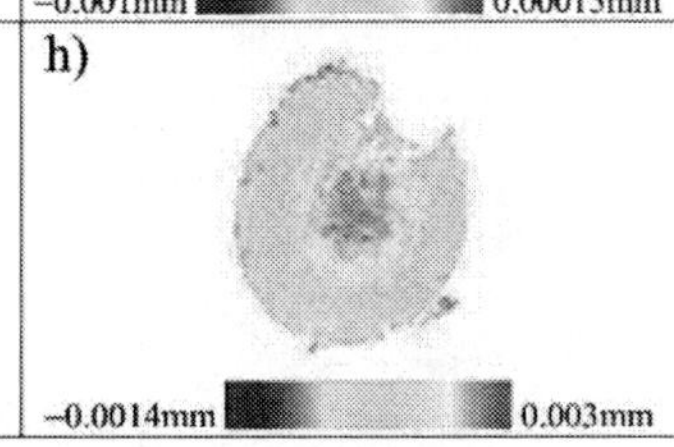

Figure 4: Results for polycrystalline (PZT) actuator: (a-d) typical raw deflected shapes from DIC system; (e-h) deflected shapes from DIC vibrometry using external reference signal. Note that no higher-order shapes were detected in quadrature with these deflections.

5. Discussion

The results of the preliminary experiment, relating to the 50 mm polycrystalline actuator, show seven resonances up to around the lower frequency investigated in the main experiments. Minor resonances covering the same kind of frequency range were reported for the polycrystalline actuator tested in [3], though it should be noted that the disc used in this test was of different dimensions from that in [3]. It should be noted that the detection of these resonances was carried out by ear and was not dependent up on the DIC system. For example, resonances noted here included 347 Hz, 854 Hz, 1149 Hz, 1224 Hz and 1723 Hz; COMSOL finite element predictions for this 50 mm actuator gave frequencies for mode shapes (not shown here) corresponding to the present quadrature displacement shapes at 312 Hz, 651 Hz, 1068 Hz, 1218 Hz and 1563 Hz (alternatively, the last could plausibly relate to the 1424 Hz resonance).

The results from the preliminary experiment are not directly comparable with [3] as the specimen dimensions and voltages are different. However, they appear to provide a rational explanation for the numerous small amplitude peaks in [3] for the polycrystalline actuator, in the form of higher-order deformation modes superimposed upon (and in quadrature with) the dishing mode desired for actuator operation. Specifically, in most of the cases here (the exception being 1723 Hz) there is a dishing mode, sometimes axisymmetric and sometimes less so, which is the desired deformation mode for the intended diaphragm-like air actuation. In quadrature with each of these is a higher order mode, typically with some form of rotational symmetry, for example a having two, four six or eight antinodes, though these are not always well defined.

IOP Conf. Series: Journal of Physics: Conf. Series **1106** (2018) 012032 doi:10.1088/1742-6596/1106/1/012032

Frequency & voltage (rms)	Typical raw DIC displacements	Main mode (dishing mode or similar)
1.8 kHz 74 mV	a) 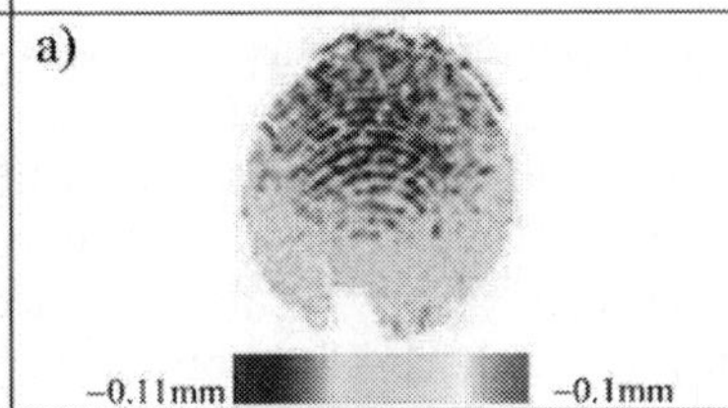	e) 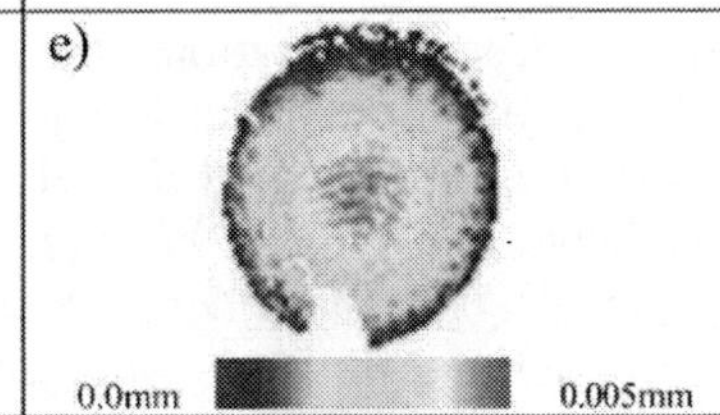
1.8 kHz 142 mV	b) 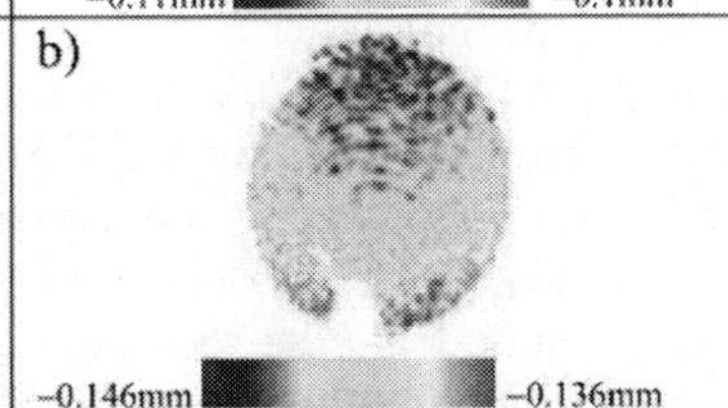	f) 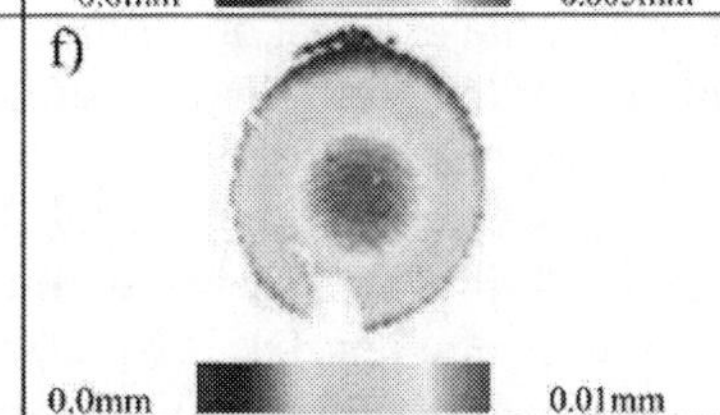
2.8 kHz 71 mV	c) 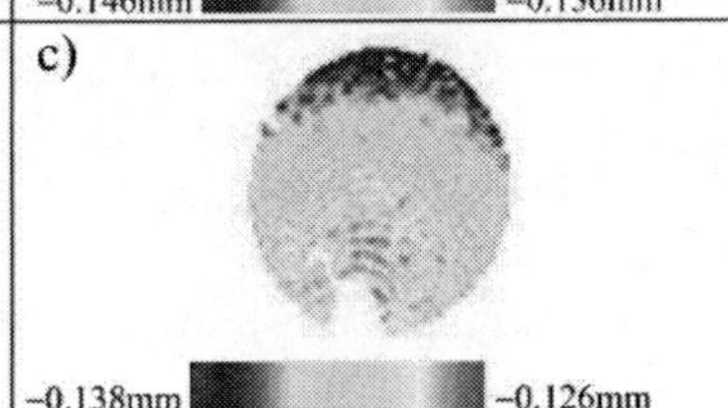	g) 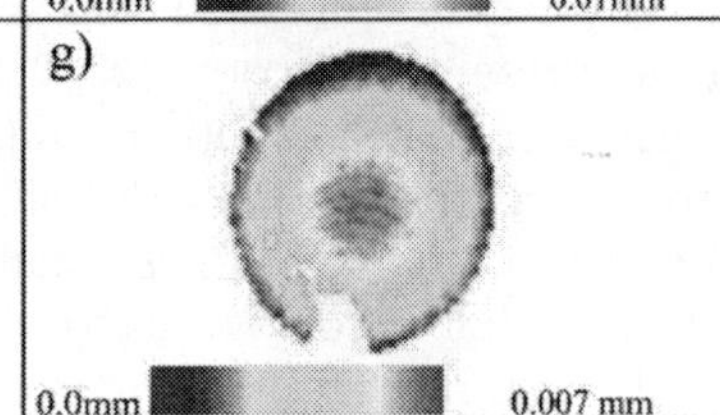
2.8 kHz 144 mV	d) 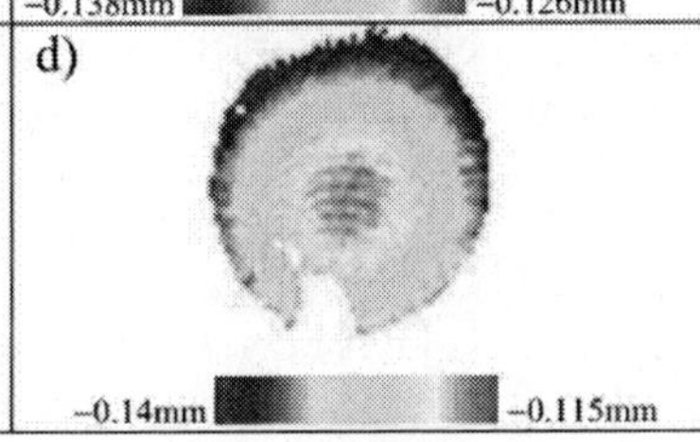	g) 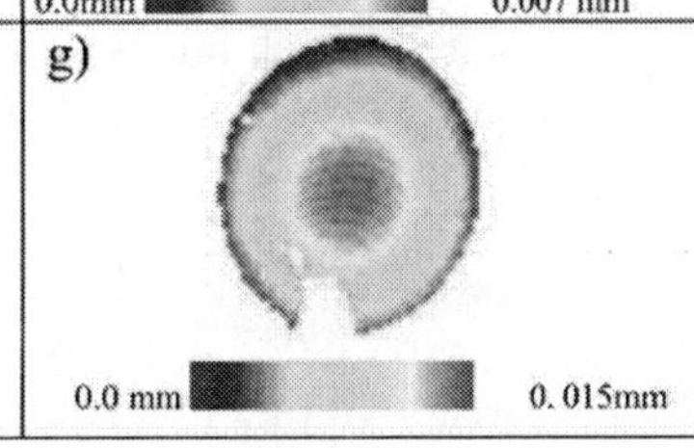

Figure 5: Results for single crystal (PMN-PT) actuator: (a-d) typical raw deflected shapes from DIC system; (e-h) deflected shapes from DIC vibrometry using external reference signal. Again, no higher-order shapes were detected in quadrature with these deflections.

The displacement results from the second set of experiments are, however, more directly comparable with those from reference [3] and show some agreement and some variation. Those for the polycrystalline (PZT) actuator are generally in good agreement, for example the peak-peak displacement at 1.8 kHz is measured here as 0.007 mm (0.0035 mm × 2) which is very close to that given in [3]. At 2.8 kHz the peak-peak displacement of the actuator centre relative to the edge appears to be around 0.009 mm, compared with 0.018 mm from [3], perhaps suggesting that the 2.8 kHz resonance is not so pronounced in the present experiment or that sensitivity of the system is reduced at higher frequencies. A similar though more exaggerated comparison is observed for the single crystal (PMN-PT) actuator: at 1.8 kHz a peak-peak displacement of 0.02 mm is measured here, compared with 0.012 mm in [3]. However, the very large peak-peak displacement of 0.087 mm in [3] is not encountered here, with the value measured here being only 0.030 mm. It is noted that the resonance shown in [3] is quite sharp so any departure from resonant frequency, or additional damping, would greatly affect the measured amplitude. It should also be noted that repeated measurements were undertaken on the second set of experiments; for space reasons only one set is given here though the results were generally repeatable except in cases of failed experiments; for example the peak-peak displacement at 2.8 kHz on the single crystal model was alternatively measured at around 0.028 mm.

One reason why the amplitudes at 2.8 kHz are lower than expected is likely to be because the electronic shutter time, set to 1/10000 s, covers over a quarter of the vibration cycle period, leading to smearing of the measured displacements and consequent loss of apparent amplitude (this may be shown to give around 13% reduction). Further work will therefore need to explore the use of shorter exposure times to ensure that this effect does not occur, provided that such shorter times can be used without loss of image quality.

In terms of measured operational deflected shapes, it is reassuring that all those measured at 1.8 kHz and 2.8 kHz are of the desired dishing mode required for operation as a synthetic jet actuator.

More generally, the results of the DIC vibrometry show that, even when a quadrature reference signal is not available, useful data can be recovered, indeed with little or no apparent loss of fidelity. A more thorough investigation will need be undertaken on the results obtained from the second set of experiments, to determine the loss of accuracy which results from using the internal referencing approach rather than the external quadrature reference signal.

Examination of the preliminary results might indicate that, for the main vibration modes at least, little is gained by performing the sinusoidal fitting etc. described by Warburton [1] rather than merely observing the displacement field from the DIC data. However, recovery of the higher-order mode shapes is not possible just by observation but becomes very clear once the DIC displacement are post-processed. Moreover, the raw DIC displacement results from the second set of experiments generally show very little by way of vibration mode shape information due to static deflections and noise, and it is only when these results are post-processed that the vibration mode shapes become clear. It is very encouraging that operational vibration mode shapes with amplitudes as little as a few microns are successfully recovered; only the polycrystalline results for 1.8 kHz and 0.71 mV, with an amplitude of around 0.001 mm, were too noisy to be meaningful.

Close examination of some of the displacement maps shows the presence of fringe-like features which arise within the raw DIC displacement fields rather than as a result of the post-processing of vibration data; as explored in [1], this is understood to be an artefact arising from the presence of very small speckles, and later versions of the DIC system are understood to overcome this problem using smoothing of the raw images.

6. Conclusions

The DIC vibrometry approach described by Warburton [1] has been successfully applied to a case study consisting of the forced vibration analysis of a synthetic jet actuator, even though (for initial experiments) the equipment needed to generate a quadrature reference signal was not available. Operational vibration mode shapes, including high-order modes superimposed upon, and in quadrature with, the main dishing modes of the actuator, have successfully been recovered in all cases except those where the vibration amplitude is very small (less than 0.002 mm). Displacement amplitudes are broadly in agreement with those obtained from laser vibrometry on different instances of the same types of discs though there were some differences in values relating to resonances. It may be concluded that, with or without an external quadrature reference signal, the low speed DIC vibrometry approach provides a useful alternative to scanning laser vibrometry as a means of measuring operational vibration shapes.

Acknowledgements

The authors are grateful to Mr J. Jeyalingam for performing the COMSOL simulations of the 50 mm PZT actuator, used here for providing frequency comparisons.

References

[1] Warburton JR, Lu G, Buss TM, Docx H, Matveev MY and Jones IA 2015 Digital image correlation vibrometry with low speed equipment *Exp. Mech.* **56**(7) 1219–30.

[2] R.K. Fruehmann RK, Dulieu-Barton JM, Quinn S and Tyler JP 2015 The use of a lock-in amplifier to apply digital image correlation to cyclically loaded components *Opt. Laser Eng.* **68** 149–59.

[3] Jabbal M, Buss T and Jones A 2018 Efficiency enhancement of synthetic jet actuators using single crystal piezoceramics, accepted for presentation at RAeS conference "The Future of Aerodynamics" Bristol, 24 - 26 July.

IOP Conf. Series: Journal of Physics: Conf. Series **1106** (2018) 012033 doi:10.1088/1742-6596/1106/1/012033

Flight envelope expansion based on active mitigation of flutter via a V-stack piezoelectric actuator

Ioan Ursu[1], Dragos Daniel Ion Guta[1], Daniela Enciu[1*], George Tecuceanu[1], Aurelian Andrei Radu[2]

[1]INCAS – National Institute for Aerospace Research "Elie Carafoli"
Bd. Iuliu Maniu 220, 061126 Bucharest, Romania

[2]ISS - Institute of Space Science, Str. Atomistilor 409
077125 Magurele, Ilfov, Romania

*Corresponding author: enciu.daniela@incas.ro

Abstract. The instability of the aircraft's flexible control surfaces caused by uncontrolled vibrations represents a challenging issue of the aviation security. Flutter is a violent vibration whose amplitude grows strongly in a short time. Once the flutter is reached, the plane is destabilized and it can no longer be controlled. This paper proposes a specially designed demonstrator intended to extend the flight envelope by raising the speed limit at which the flutter occurs. The anti-flutter demonstrator is an intelligent airplane wing model made from a longeron covered by an aerodynamic layer. The wing has an aileron as a primary flight control surface, at one end, and a flange conceived to fix the wing in the aerodynamic tunnel, at the other end. The actuator consists in two V-shaped piezo stacks. The main advantage of the piezo actuator, the bandwidth (about 30 Hz), is exploited. The aero-elastic control is efficient if the deflection of the control surface is of a few degrees while the frequency is of at least 25-30 Hz. The control law is obtained through the receptance method of eigenvalues assignment using the on line input/output measured transfer function rather than the knowledge of system matrices.

1. Introduction

The primary flight surfaces of the airplane are flexible surfaces, subject to aeroelastic forces. The interaction between inertial, elastic and aerodynamic forces can trigger a disastrous aeroelastic dynamic phenomenon called flutter. The flutter is a self-sustaining unstable oscillation that increases quickly in intensity. It is a complex and difficult process to study. In the case of planes, as speed increases, there is a threshold beyond which structural vibrations can no longer be damped, and they begin to increase in amplitude by accumulating energy in the structure.

At the dawn of aviation, there were several flight disasters caused by unstable vibrations [1], [2]. Gradually, flight flutter tests were introduced [1], [3]. At present, all aircraft get their approval to fly after passing a series of tests, including a test for establishing the flight envelope, in which a safety margin is considered. However, no flight regime is really immune to flutter [1]. To counteract this phenomenon, passive methods such as mass balancing, increased rigidity of the structure and change of geometry were initially used. Unfortunately, these methods led to an increase in the total weight and lowered the aircraft's performance. Recently, active flutter control methods have been implemented using the actuators of the aircraft primary flight controls [4]-[9].

IOP Conf. Series: Journal of Physics: Conf. Series **1106** (2018) 012033 doi:10.1088/1742-6596/1106/1/012033

As flight flutter tests are expensive and time consuming, often leading to irreversible destruction of the aircraft, complementary methods are considered: mathematical modelling accompanied by numerical simulations, optimization of control laws implemented on primary flight controls, reconsideration and design of better aerodynamic structures, ground tests and wind tunnel tests. A comparison between theoretical methods and numerical simulations during flight is presented in [10].

The present paper rolls out a wind tunnel tested demonstrator for the active control of the structural vibrations of a wing model with aileron and piezoelectric actuator.

2. Physical model description

The anti-flutter demonstrator is an intelligent airplane wing model made from a longeron covered by an aerodynamic layer (profile NACA 0012). The wing has an aileron as a primary flight control surface, at one end, and a flange conceived to fix the wing in the subsonic tunnel, at the other end. The spar is a 1 mm thick rectangular tube (1200x120x25 mm) provided with notches to control its stiffness. The elements defining the aerodynamic surface are made from wood or ROHACELL 71S resin. The wing structure is depicted in Fig. 1.

Fig. 1. CATIA model of the longeron (left) and wing (right)

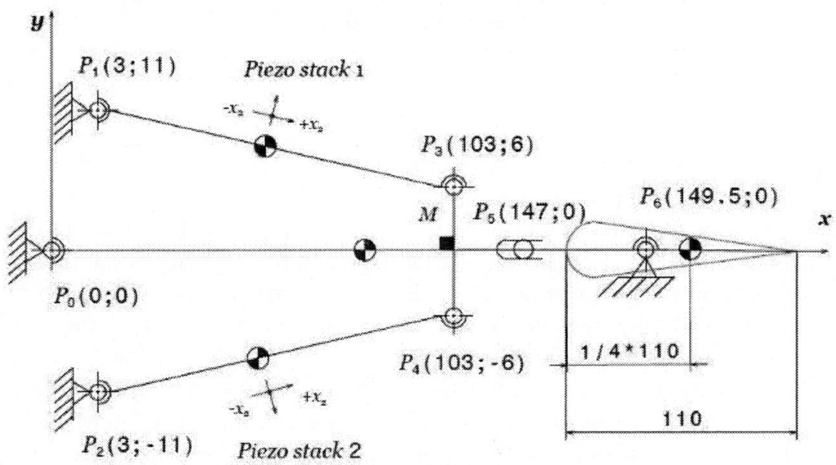

Fig. 2. Kinematic scheme of aileron actuation with 2 V-shaped piezo stacks; coordinates in mm

The actuator is made up of two V-shaped piezo stacks. The advantages of the piezo actuators are the reduced size, the wide bandwidth and, last but not least, the high energy density. There is a disadvantage related to their reduced stroke, but the kinematic scheme of Fig. 2 shows how to achieve a multiplication of the output stroke, at the point P5, with the price of a force demultiplication. The piezo stacks are arranged along the segments P_1P_3 and P_2P_4. When the stack P_1P_3 is activated by increasing the supply

IOP Conf. Series: Journal of Physics: Conf. Series **1106** (2018) 012033 doi:10.1088/1742-6596/1106/1/012033

voltage V, $V_{01}+\Delta V(t-\tau)$, which determines its extension to move to the right and slightly below the articulated point P_3, the stack P_2P_4 is supplied with the voltage $V_{02}-\Delta V(t)$

$$\Delta V(t-\tau)=\begin{cases}0, t\le\tau\\ \Delta V, t>\tau\end{cases}$$

which causes the withdrawal to the left and slightly downward of the articulated point P_4 therefore not to withstand to downward movement of the articulated point P_5 in the slider crank mechanism. This provides the movement with positive angle (up) δ of the aileron. The two stacks are successively active versus passive, in the sense of the presented description. The constant τ defines a delay in the application of voltage to "active" stack, to ensure the withdrawal of the passive stack. The left side of the scheme is actually in a different parallel plane to the right side, the one in which the aileron and the slider crank mechanism, sketched in the point P_6, are represented. The mathematical model of the actuator and the numerical simulations are found in paper [11]. The deflection of the control surface must be at least 5-6 degrees, with the frequency band of at least 25-30 Hz, as specified in work [12]. A particular attention is paid to the bandwidth and this clause is clearly in favour of piezo actuators. Obviously, the classical actuators do not have resources for this high bandwidth value; but even for piezoelectric actuators, stroke requirements may be problematic. The piezo stacks NAC2022-H98-AO1 were bought from NOLIAC and have the following basic properties: height 98 mm, stroke 148.8 μm, capacitance 19010 nF, maximal force developed 4200 N, maximum operating temperature 150°C, material NCE51F.

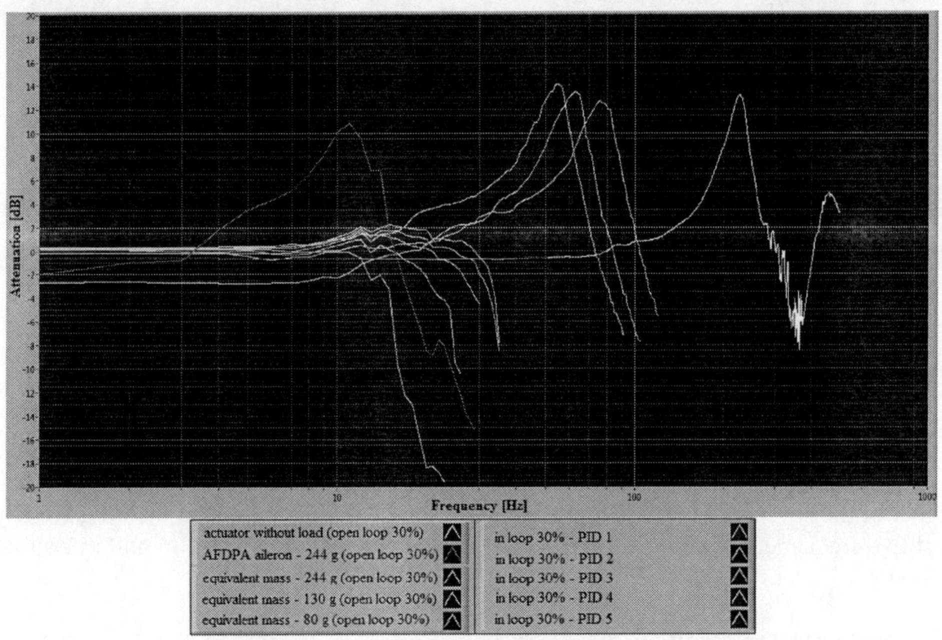

Fig. 3: Frequency characteristics of the piezoelectric actuator

3. Preliminary results of benchmark tests in the Mechatronic Lab

Various tests were carried out to determine the frequency characteristics of the piezoelectric actuator without load (white line), with aileron as inertial load (red line) and with equivalent mass (blue line), see Fig. 3. The actuator without inertial load has strong resonances at frequencies of several hundred Hz. In the case of the actuator with mounted aileron, the bandwidth decreases to 15 Hz with an unacceptable resonance peak at about 12 Hz. Green curves are made with PID control with various coefficients experimentally determined. On the right side, the system is more dynamic (e.g. with increased bandwidth), and has an acceptable resonance peak of 2 dB. With a suitable choice of parameters for the PID controller, a suitable bandwidth of at least 30 Hz can be provided.

IOP Conf. Series: Journal of Physics: Conf. Series **1106** (2018) 012033 doi:10.1088/1742-6596/1106/1/012033

The model was designed to create a realistic, elastic wing, unlike most rigid models with external springs simulating elasticity as found in literature [13]. A second criterion was to ensure the occurrence of flutter for the uncontrolled system under the conditions of the INCAS subsonic wind tunnel. Measurements of vibration modes were performed in the Mechatronic Lab using the experimental assembly shown in Fig. 4. Table 1 compares the calculated and the measured values.

Table 1. Natural frequencies, determined by CATIA, and measured, respectively

mode	CATIA frequency [Hz]	measured frequency [Hz]
1	6.23	5.93 bending
2	10.21	11.70 torsion
3	20.83	22.73 bending
4	26.32	
5	29.83	

Fig. 4. Vibration measurements on wing model, experimental mountings

The wing with aileron and piezoelectric actuator was mounted in a vertical position in the subsonic wind tunnel to determine the air velocity at the time of the flutter. The air velocity gradually increased to verify the moment when the flutter occurs.The experiment led to the following results: flutter velocity - 41 m/s, flutter frequency - 5.8 Hz. As it was expected, the wing structure, specifically the longeron, suffered major damages during the flutter (Fig. 5), therefore a new specimen had to be manufactured. The designer of the new specimen had to provide a wing with a lighter aileron meeting the requirements of the first two frequencies estimated by CATIA package (6.23 and 10.21 Hz) and measured (5.93 Hz and 11.7 Hz).

4. Wind tunnel tests and results

Based on the experimentally obtained frequency response, the transfer functions for the pole allocation were determined. This operation was required by the receptance method for the synthesis of the control law [14]. The bandwidth is consistent in the context of a 5.8 Hz measured flutter frequency.

The new wing has been tested to increase the flight envelope based on active vibration control. Initially, a roots locus in the open loop was made at successive air velocity of 0 m/s, 5 m/s, 10 m/s, ..., 25 m/s, based on the identification of the transfer functions from the actuator to the two accelerometers. With the uncontrolled system, the evolution towards the instability of a mode was followed by noting the velocity of the air flow which is responsible for the occurrence of the unstable poles (Fig. 6). The next step was to repeat the tests, this time in a closed loop observing, as it was expected, the widening of the aeroservoelastic stability range, namely increasing the aeroelastic stability limit relative to the airflow velocity, Figs 7-9.

Fig. 5. Left and middle: sequences of flutter evolution of the wing at 41 m/s air speed; right: wing system with actuator during complex tests to validate the method of increasing the speed of the flutter; at the top of the wing is the piezolectric actuator. At the top right are the two accelerometers

5. Discussion and conclusions

The experimental results of the tests performed on the wing model and the piezoelectric actuator indicate 1) the piezo actuator's compatibility with the flutter control objectives and 2) the efficiency of the receptance method in the synthesis of the control law. Such an objective would have been more difficult to achieve with a hydraulic or electric actuator, given their low cut-off frequencies. The receptance method has proven to be consistent: the input-output transfer functions have been identified with good accuracy, allowing to determine the amplification factors for position and speed reactions and, consequently, a consistent attenuation of approx. 8 dB of the resonance vibration (Fig. 9).

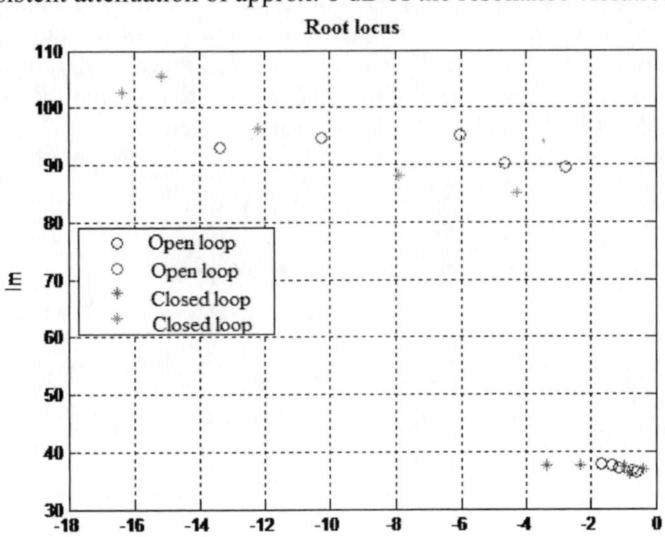

Fig. 6 The location of basic bending and torsional poles in open and closed circuits, depending on the air velocity. Only poles with positively imaginary side are represented

IOP Conf. Series: Journal of Physics: Conf. Series **1106** (2018) 012033 doi:10.1088/1742-6596/1106/1/012033

Fig. 6 is instructive in terms of the evolution of poles of open and closed loop transfer functions, as well as the air velocity. One may notice that poles move away from the real axis (based on a more pronounced damping) as air velocity increases, an effect that is blunted in the absence of control, but amplified in its presence. The speeds represented in figure (10, 15, 20, 25, 30 m/s) are outside the speed of the flutter. Near the speed of the flutter, the data from the literature show that for the torsion mode there is a precipitated turn of the poles towards the imaginary axis [15].

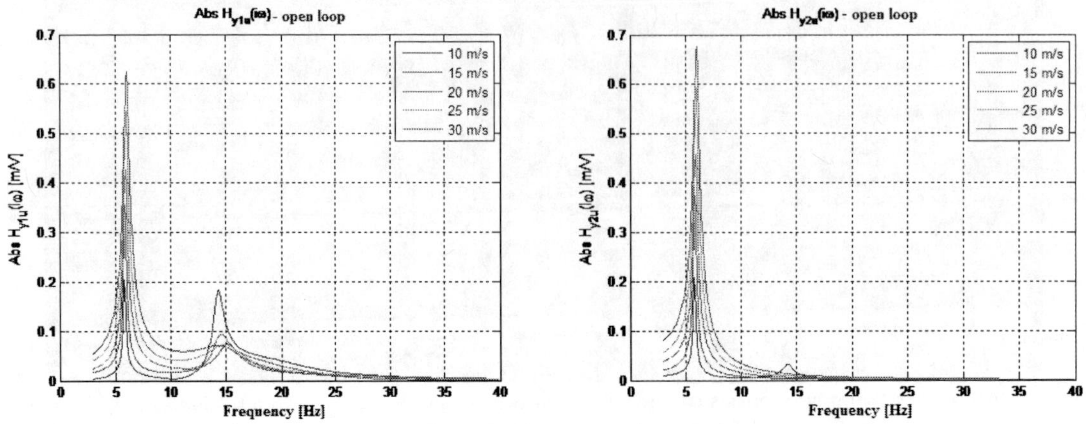

Fig. 7. Frequency amplitude characteristics of open loop transfer functions $H_{y1u}(s)$ and $H_{y2u}(s)$ corresponding to air velocities V=10, 15, 20, 25, 30 m/s (notation: u – control; $y1$ – displacement recorded from accelerometer 1; $y2$ – displacement recorded from accelerometer 2)

The pole allocation strategy was the following: a) the transfer functions are identified at a certain air velocity; b) the control law based on the allocation of poles for one mode or two basic modes is determined; c) the poles prescribed for allocation to the closed loop system are chosen to double the bending mode damping factor and increase the torsion mode's own frequency by 2 Hz, in the idea of increasing the spread of the two modes, since, in flutter, the two modes will overlap. The idea of increasing the spread is translatable in terms of "robustness" of the control law.

Here are a few words about active control technology based on the receptance method. It actually has the data of an emerging technology. A comprehensive analysis of the concept could be found in [16]: the attributes would be (i) *radical novelty*, (ii) *relatively fast growth*, (iii) *coherence*, (iv) *prominent impact*, and (v) *uncertainty and ambiguity*! From the perspective of the attribute (v) an emerging technology is still regarded with scepticism by the control engineers.

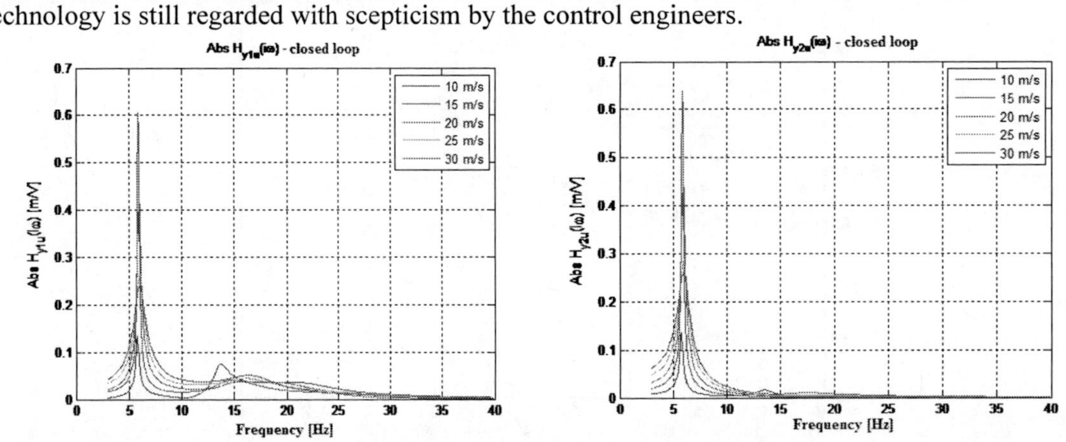

Fig. 8. Frequency amplitude characteristics of closed loop transfer functions $H_{y1u}(s)$ and $H_{y2u}(s)$ corresponding to speeds V=10, 15, 20, 25, 30 m/s

Basically, the receptance method only requires on-line measurements of the frequency response of the structure, eluding the need to know the ***M***, ***C***, ***K*** (mass, damping and rigidity) and ***B*** (control influence) matrices. In the receptance method, neither the model order reduction nor the synthesis of an estimator for unmeasured state is required. In principle, this controller can be continuously corrected on the basis of the measurements, with beneficial consequences in the performance of feedback. These attributes provide the method with a rapid growth in applications, once gained the confidence of those interested. The method is derived from a matrix analysis result, the Sherman-Morrison formula [13], which gives the inverse of a matrix in case of a change in rank, depending on the inverse of the original matrix.

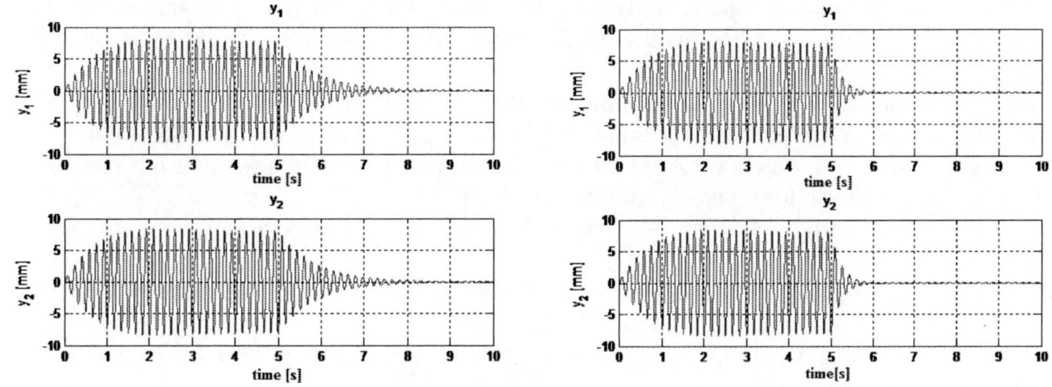

Fig. 9. The response of the system to resonant excitation applied for 5 seconds: amplitude of 2 degrees and frequency equal to its own bending mode frequency of 5.8 Hz. Left: open loop; right: closed loop, the control is introduced at time $t = 5$ sec

In conclusion, the most significant result of the paper is the advanced testing in the wind tunnel of a powerful method of control synthesis, the method of receptance, having a pronounced character of novelty (the latest data on the net refer to a PhD thesis from 2015 [17], from Louisiana State University, a theoretical and not experimental work).

Acknowledgments
This work was supported by a grant of the Romanian Ministry of Research and Innovation, CCCDI – UEFISCDI, project number 87PCCDI/2018, project registration code PN-III-P1-1.2-PCCDI-2017-0868, within PNCDI III. Also, many thanks are addressed to colleagues from INCAS and STRAERO for the success of this project.

References
[1] M. W. Kehoe, "A historical overview of flight flutter testing", Dryden Flight Research Center, NASA -TM-4720, 1995.
[2] J. E. Cooper, "Towards faster and safer flight flutter testing", in RTO AVT Symposium on "Reduction of Military Vehicle Acquisition Time and Cost through Advanced Modelling and Virtual Simulation", France, 2002.
[3] I. E. Garrick and W. H. Reed, "Historical development of aircraft flutter, AIAA 81-0491", Journal of Aircraft, vol. 18, no. 11, pp. 897-912, 1981.
[4] R. J. Newsom, "Active control of aeroelastic response", NASA TM_83179, 1981.
[5] I. Ursu and F. Ursu, Active and semiactive control (in Romanian), Bucharest: Publishing House of the Romanian Academy, 2002.
[6] H. Ashley, S. M. Rock, R. Digumarthi and K. Chaney, "Active control for fin buffet alleviation", WL-TR-93-3099, 1994.

[7] J. Ko, A. J. Kurdilla and T. W. Strganac, “Nonlinear dynamics and control for a structurally nonlinear aeroelastic system, TX 77843-3141”, Department of Aerospace Engineering, Texas A&M University, 1997.
[8] I. Ursu, M. Stoia-Djeska and F. Ursu, “Active control laws for flutter suppression”, Annales of Universit of Craiova, Electrical Engineering Series, vol. 27, no. 27, pp. 62-70, 2004.
[9] J. L. Pinkerton, A.-M. R. McGowan, R. W. Moses, R. C. Scott and J. Heeg, “Controlled aeroelastic response and airfoil shaping using adaptive materials and integrated systems”, in SPIE's 1996 Symposium on Smart Structures and Integrated Systems, San Diego, CA, 1996.
[10] R. Lind, “Flight-test evaluation of flutter prediction methods”, Journal of Aircraft, vol. 40, no. 5, pp. 964-970, 2003.
[11] I. Ursu, D. D. Ion Guta, D. Enciu, A. Toader, M. Dan, C. Donciu and V. Turcan, “Mathematical modelling of a V-stack piezoelectric aileron actuation”, INCAS Bulletin, vol. 8, no. 4, pp. 141-155, 2016.
[12] E. V. Ardelean, M. A. McEver, D. G. Cole and R. L. Clark, “Active Flutter Control with a V-Stack Piezoelectric Flap Actuator”, Journal of Aircraft, vol. 43, no. 2, pp. 482-486, 2006.
[13] J. J. Block, and T. W. Strganac, “Applied active control for a nonlinear aeroelastic structure”, Journal of Guidance, Control, and Dynamics, vol. 21, No. 6, 1998
[14] J. E. Mottershead and Y. M. Ram, “Receptance method in active vibration control”, AIAA Journal, vol. 45, no. 3, pp. 562-567, 2007.
[15] R. Huang, Y. Zhao, H. Hu, “Wind tunnel tests for active flutter control and closed-loop flutter identification”, AIAA Journal, vol. 54, no. 7, July 2017
[16] D. Rotolo et al., What Is an Emerging Technology, Research Policy, SPRU Working Paper Series, Univ of Sussex
[17] A. Maha, Study of Active Vibration Control using the Method of Receptances, Thesis, 2015, Louisiana State University.

IOP Conf. Series: Journal of Physics: Conf. Series **1106** (2018) 012034 doi:10.1088/1742-6596/1106/1/012034

On the vibration performance assessment of a plate with damaged constrained layer damping patches

B Di Dato, A Cicirello, M Chatzis

Department of Engineering Science, University of Oxford, Oxford, UK

E-mail: alice.cicirello@eng.ox.ac.uk

Abstract. The purpose of this study is to investigate the metrics to be used for the vibration performance assessment of a plate with damaged CLD patches in operating conditions. In particular, Experimental Modal Analysis is used to determine the effects of the CLD patches on the Frequency Response Function (FRF). EMA can be performed during maintenance operations or in a laboratory. However, during operating conditions the excitation cannot be measured. Alternative output-only strategies, e.g. Frequency Domain Decomposition (FDD), can be implemented to extract information about some of the structural properties of the system. Three damaging conditions of the CLD are investigated. It is shown that the deterioration of the CLD can actually improve the vibrational performance of the plate in the desired frequency range. Two metrics are therefore suggested to assess the vibration performance: acceleration peak and frequency shift.

1. Introduction

Constrained-Layer Damping (CLD) is a damping treatment used for passively reducing structural vibrations and noise of lightweight structural components. It consists of a viscoelastic polymer layer sandwiched between a structural element, whose vibration need to be reduced, and a relative stiffer layer [1]. When the out-of-plane vibrations of the structural element occur, the vibrational energy is dissipated through shear deformations of the viscoelastic layer [1]. These treatments, due to their inherent advantages (low weight, no structural modifications) and high efficiency, are widely used in the aeronautical and aerospace industry. CLD can be applied uniformly to the surface of a structural element or in patches. While the uniform configuration can have an effect on a broad frequency range, patches can be used to target specific modes or smaller frequency bands. The performance of a CLD can be assessed by quantifying the drop in amplitude of the Frequency Response Function (FRF) and/or by considering the shift in resonant peaks [1-3].

Experimental Modal Analysis (EMA) [4-5] is a popular approach for the evaluation of the FRF. This approach requires measuring both excitation and response of the structure, and the implementation of signal processing strategies [4-5]. EMA can be performed during maintenance operations, or in a laboratory. However, measurement of the excitation is often impractical. Alternative output-only strategies, e.g. Frequency Domain Decomposition (FDD) [6-7], can be implemented to extract some modal information, such as modal damping, natural frequencies and mode shapes, but not the FRF peak amplitude. The application of this two approaches for the the vibration performance assessment of a plate with damaged CLD patches is investigated in this paper.

IOP Conf. Series: Journal of Physics: Conf. Series **1106** (2018) 012034 doi:10.1088/1742-6596/1106/1/012034

An aluminium plate with and without CLD patches is investigated at room temperature. EMA is used to determine the effects of the CLD patches on the FRF. The FDD approach is then applied to yield modal damping and natural frequencies showing a very good agreement with the results yielded by the FRF. Subsequently, the viscoelastic layer and the aluminium layer of the CLD patches are damaged. In particular, three different damage configurations are investigated. For each configuration, EMA and FDD are applied, showing again a very good agreement.

Interestingly the results of this experimental campaign shows that the deterioration of the CLD can actually improve the vibrational performance of the plate in the desired frequency range. Because of these results, it is suggested to assess the vibration performance of a plate with damaged constrained layer damping patches by monitoring the acceleration peak (in the time domain) and the frequency shift. The paper is structured as follow: the test setup is described in Section 2; the CLD is described in section 3 and the FRF obtained for the bare plate and for the plate with CLD are compared; in section 4 the vibration performance of damaged CLD is experimentally investigated. FDD is introduced in section 5 and applied to output only measurements to yield the modal parameters.

2. Test setup

The test specimen is an aluminium plate with dimensions 200 mm x 300 mm x 2 mm, density 2800 kg/m^3, Young modulus 72 GPa, Poisson ratio 0.3. The plate was suspended with thin lightweight nylon wires attached to a frame to approximate free boundary conditions, as shown in Figure 1. The plate was excited with an instrumented hammer (PCB 086C03 with a nylon tip) and the velocity response was measured with a single pointer Laser Doppler Vibrometer (Polytec NLV-2500). The sampling rate was 10 kHz and the signal was recorded for 30 seconds. Three averages were taken for each measurement point to evaluate an average FRF.

Figure 1. Test setup

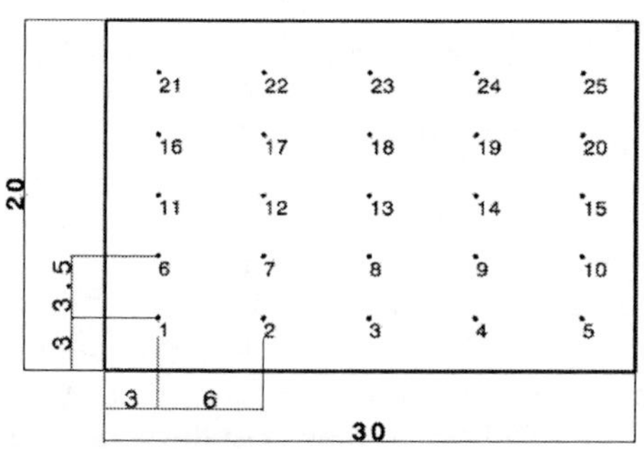

Figure 2. Grid of points

A grid of 25 points was drawn on the plate, as schematically shown in Figure 2. The Laser Doppler Vibrometer was used to measure the velocity response at point 7, while the hammer was used to excite the plate in the 25 grid points, to yield 25 averaged FRFs. These FRFs were used to characterise the mode shapes of the plate, and evaluating the damping ratio (using the circle fitting method) at each resonance. In what follows, only the FRF measured at point 7 is considered.

3. Constrained Layer Damping

The damping layer used in this test campaign is the 3M Vibration Damping Tape 435 [8], shown in Figure 3. This damping layer is characterised by a top aluminium layer of thickness 0. 20 mm, and a damping polymer with thickness of 0.14 mm [8], and the tape has a width of 50 mm. The dynamic mechanical properties of the damping polymer vary with frequency and temperature, and will also vary with time. 3M provides the shear storage modulus and the loss factor of a viscoelastic adhesive in the form of a nomograph to account for the frequency and temperature dependence [8]. The room temperature was monitored with a temperature probe, and it was varying between 18 °C and 22.5 °C. In this temperature range there was no substantial variation of the CLD properties. The CLD was used to

IOP Conf. Series: Journal of Physics: Conf. Series **1106** (2018) 012034 doi:10.1088/1742-6596/1106/1/012034

partially cover the plate along the edges, as shown in Figure 4, with the aim of affecting mostly the first two modes (at 106 Hz and 115 Hz), which are a twisting and a bending modes.

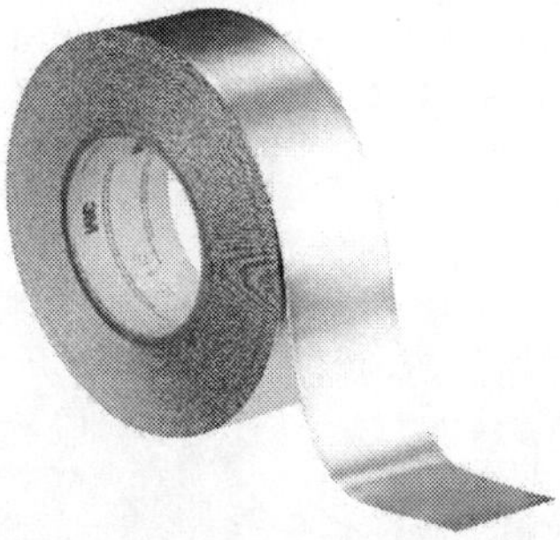

Figure 3. The 3M Vibration Damping Tape 435 [8]

Figure 4. Plate with CLD

The Frequency Response Functions (in terms of acceleration divided by force) obtained for the bare plate and for the plate with the CLD partial treatment are shown in Figure 5. It can be observed that the peaks at the first two resonance frequencies have dropped of more than 10 dB. The shifts in resonant frequencies and the changes in damping ratios are detailed in Tables 1 and 2.

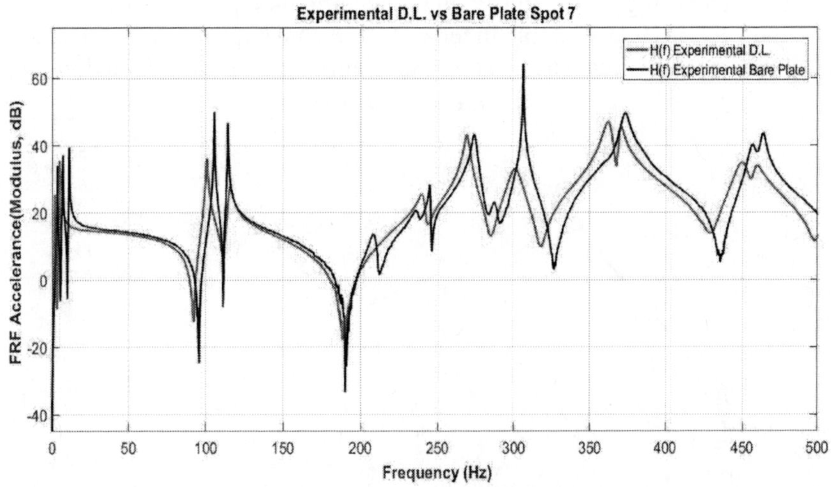

Figure 5. Comparison between the FRF obtained for the Bare Plate (black curve) and for the Plate with CLD treatment (red curve)

Table 1. Resonances obtained for the Bare Plate and for the Plate with CLD treatment.

	f_1 (Hz)	f_2 (Hz)	f_3 (Hz)	f_4 (Hz)	f_5 (Hz)
Bare Plate	105.58	114.51	244.93	274.13	306.2
Treated plate	100.89	116.11	239.84	268.28	300.7

Table 2. Damping ratio obtained for the Bare Plate and for the Plate with CLD treatment.

	ζ_1	ζ_2	ζ_3	ζ_4	ζ_5
Bare Plate	0.0005	0.00077	0.0023	0.005	0.003
Treated plate	0.004	0.012	0.0107	0.0054	0.0114

4. Vibration performance of damaged Constrained Layer Damping

In this section the effect of a partially damaged CLD on the average FRF is investigated. In particular, three damage cases were considered, indicated A, B and C, as shown in Figure 6. While A is a light damage condition mostly at the edges of the CLD, case B is characterised by a more diffused damage of the viscoelastic and aluminium layer. Case C represents the most severe type of damage: part of the damping layer is removed, and it can be observed that in many areas the CLD is not attached to the aluminium plate.

Figure 6. Damaged CLD: case A (left), case B (centre) and case C (right)

The averaged FRFs obtained at point 7 (where both the input and output are measured) in the frequency range 85 Hz – 130 Hz (focusing on the twisting and bending mode) are shown in Figure 7. The effect of the damage on the CLD is clearly visible in terms FRF resonant peaks drop and resonant frequencies shift. Variations in the resonant frequencies and in the damping ratio are detailed in Tables 3 and 4.

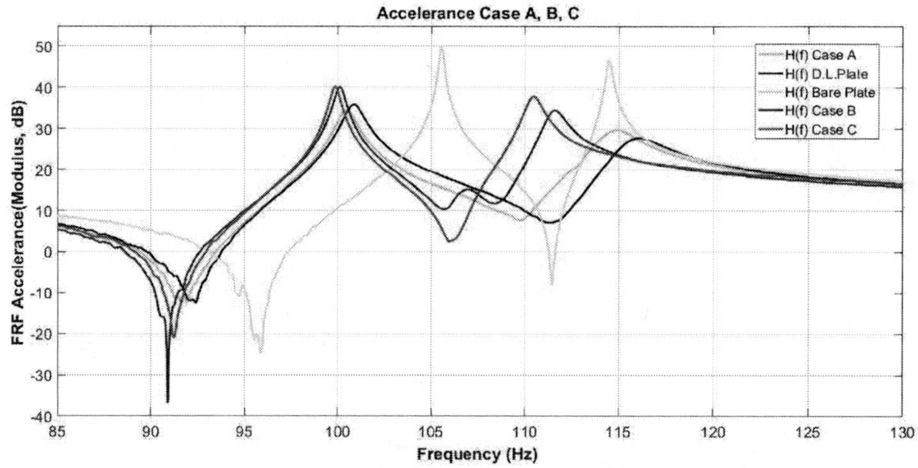

Figure 7. Comparison between the FRF obtained for the Bare Plate (yellow curve) and for the Plate with CLD treatment (black curve, indicated as DL Plate), and the Damaged CLD cases A (green), B (blue) and C (red).

Table 3. First two resonances obtained for the Bare plate, the plate with CLD treatment (CLD), and the plate with damaged CLD (cases A, B and C).

	CLD	A	B	C	Bare
f_1 (Hz)	100.89	100.49	100.13	99.88	105.58
f_2 (Hz)	116.11	114.93	111.62	110.48	114.51

IOP Conf. Series: Journal of Physics: Conf. Series **1106** (2018) 012034 doi:10.1088/1742-6596/1106/1/012034

Table 4. Damping ratio for the first two resonances obtained for the Bare plate, the plate with CLD treatment (CLD), and the plate with damaged CLD (cases A, B and C).

	CLD	A	B	C	Bare
ζ_1	0.0045	0.0043	0.0022	0.002	0.0005
ζ_2	0.0120	0.0102	0.0047	0.0037	0.0008

From these results, it can be observed that the damping ratio decreases as damage increases. Damaging the CLD leads to less rigidity of the CLD assembly, which in turn will produce less shear deformations and therefore less energy dissipation. However, other energy dissipation mechanisms can occur, such as friction. An additional effect of damage to the CLD is that of a small reduction of the resonance frequencies. This can be attributed to the CLD performing as an added mass to the system due to loss of rigidity.

A second test campaign was performed on the same plate. Damage of the CLD was produced by detaching and reattaching the CLD on the plate. The three damage conditions (A1, B1, and C1) are shown in Figure 8, and the corresponding averaged FRFs are shown in Figure 9.

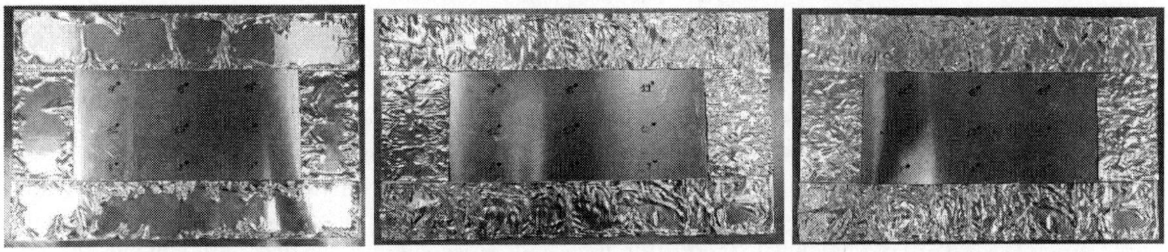

Figure 8. Damaged CLD study 2: case A1 (left), case B1 (centre) and case C1 (right)

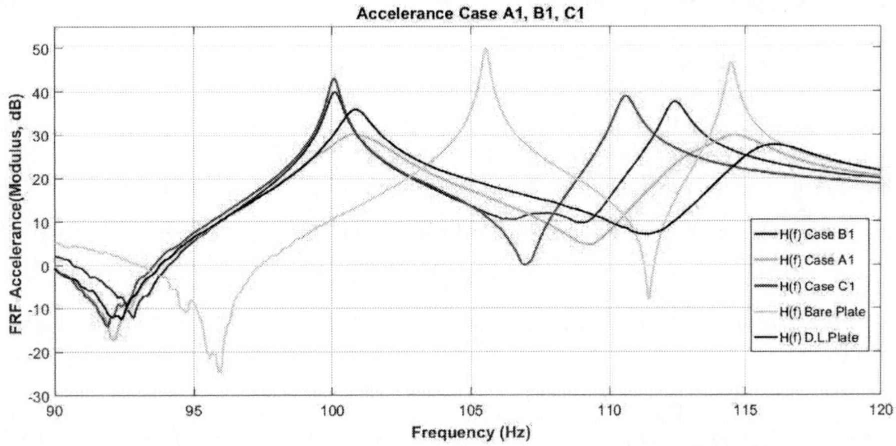

Figure 9. Comparison between the FRF obtained for the Bare Plate (yellow curve) and for the Plate with CLD treatment (black curve), and the Damaged CLD cases A1 (green), B1 (blue) and C1 (red). Variations in the resonant frequencies and in the damping ratio are detailed in Tables 5 and 6.

These results show that the damping ratio does not decreases for case A1, although they do for the other damage cases. A reduction of the resonance frequencies is observed also in this case as damage is increased. It can be also observed that the first resonance peak is not increasing as the damage increases.

In particular, Case A1 shows a bigger peak drop that the undamaged CLD case. It might be concluded that specific forms of low-level damage may be beneficial.

Table 5. First two resonances obtained for the Bare plate, the plate with CLD treatment (CLD), and the plate with damaged CLD (cases A1, B1 and C1).

	CLD	A1	B1	C1	Bare
f_1 (Hz)	100.89	100.86	100.13	100.11	105.58
f_2 (Hz)	116.11	114.64	112.42	110.62	114.51

Table 6. Damping ratio for the first two resonances obtained for the Bare plate, the plate with CLD treatment (CLD), and the plate with damaged CLD (cases A1, B1 and C1).

	CLD	A1	B1	C1	Bare
ζ_1	0.0045	0.0091	0.002	0.0013	0.0005
ζ_2	0.0120	0.0098	0.0031	0.0025	0.0008

Overall it can be concluded that: (i) modal damping may not be an optimal metric to determine the performance of the plate with CLD treatment; (ii) damaging the CLD patches may in some case of low level damage produce a bigger peak drop compared to the undamaged CLD; (iii) damaging the CLD patches can induce a reduction of the resonance frequencies. Therefore, the deterioration of the CLD could potentially improve the vibrational performance of the plate in the desired frequency range. Because of these results, it is suggested to assess the vibration performance of a plate with damaged constrained layer damping patches by acquiring data with an accelerometer to monitor the acceleration peak in the time domain, and using output-only strategies to assess the frequency content, and therefore monitoring resonance shifts. In the next section, the use of the FDD to this type of problems is investigated. The objective of the proposed analysis is to improve the structural performance, not to identify or quantify the damping layer damage.

5. Frequency Domain Decomposition

Frequency Domain Decomposition (FDD) [6-7] is a technique for output-only modal identification. The spectral density matrix is decomposed into a set of single degree of freedom system, and the individual SDOF auto-spectral density functions are transformed back to time domain to identify damping and frequency [6-7]. Given a standard input ($x(t)$)-output ($y(t)$) relationship in the frequency domain:

$$Y(\omega) = H(\omega)X(\omega) \tag{1}$$

Being $H(\omega)$ the Frequency Response Function. This can be rearranged in terms of Power Spectral Densities (PSD):

$$G_{yy}(\omega) = H^*(\omega)G_{xx}(\omega)H^T(\omega) \tag{2}$$

Where the apex * indicates the complex, and T indicates the transpose operator. The FDD approach assumes that the output PSD $G_{xx}(\omega)$ is constant. This assumption is valid for broadband random inputs. The first step of the FDD is to estimate the power spectral density matrix $\mathbf{G}_{yy}(j\omega)$ at discrete frequencies $\omega=\omega_i$. This can be done using the Matlab function cspd(x,x). $\mathbf{G}_{yy}(j\omega)$ is then decomposed by using the Single Value Decomposition (SVD):

$$\mathbf{G}_{yy}(j\omega) = \mathbf{U}_i\mathbf{S}_i\mathbf{U}_i^{\mathrm{H}} \tag{3}$$

IOP Conf. Series: Journal of Physics: Conf. Series **1106** (2018) 012034 doi:10.1088/1742-6596/1106/1/012034

where the matrix $\mathbf{U}_i = [u_{i1}, u_{i2}, \ldots, u_{im}]$ is a unitary matrix with entries corresponding to the singular vectors u_{ij}, , $\mathbf{S}_i$ is a diagonal matrix with entries corresponding to the scalar singular values s_{ij}, and the apex H indicates the Hermitian. At certain frequencies, only a limited number of modes will contribute significantly (assuming low modal overlap). Thus, according to the FDD theory [6-7], the first singular vector u_{i1} is an estimate of that mode shape:

$$\hat{\phi} = u_{i1} \tag{4}$$

and the corresponding singular value defines the auto-PSD of the corresponding SDOF system.
This PSD function is identified around a peak by comparing the mode shape estimate $\hat{\phi}$ with singular vectors for the frequency lines around the peak [6-7], and using the modal assurance criterion (MAC). The resonant frequency and damping ratios are obtained from the SVDs around the isolated peak singular value [6-7].
FDD makes the assumption of a white noise input excitation or a free vibration response. Hence in hammer tests, the part of the response of the plate after the impact is over can be used, where indeed the data correspond to free vibration response.
The modal parameters identified from EMA (using input and output data) and FDD (using output only data) are compared in Tables 7 and 8.

Table 7. Modal parameters for the first resonance peak – second and first test campaign

1st mode	Frequency (Hz)	Frequency FDD (Hz)	Damping Ratio	Damping Ratio FDD
Damped Plate	100.88	100.85	0.0045	0.0048
Case A1	100.86	100.85	0.0091	0.0093
Case B1	100.13	100.09	0.002	0.0022
Case C1	100.11	100.09	0.0013	0.0016
Bare Plate	105.57	105.58	0.0005	0.0007

1st mode	Frequency (Hz)	Frequency FDD (Hz)	Damping Ratio	Damping Ratio FDD
Damped Plate	100.88	100.85	0.0045	0.0048
Case A	100.48	100.40	0.0043	0.0044
Case B	100.13	100.09	0.0022	0.0029
Case C	99.88	99.79	0.002	0.0027
Bare Plate	105.57	105.58	0.0005	0.0007

Table 8. Modal parameters for the second resonance peak – second and first test campaign

2nd mode	Frequency (Hz)	Frequency FDD (Hz)	Damping Ratio	Damping Ratio FDD
Damped Plate	116.11	116.11	0.0119	0.0122
Case A1	114.64	114.59	0.0098	0.0113
Case B1	112.42	112.45	0.0031	0.0029
Case C1	110.62	110.62	0.0025	0.0011
Bare Plate	114.51	114.43	0.0008	0.0007

2nd mode	Frequency (Hz)	Frequency FDD (Hz)	Damping Ratio	Damping Ratio FDD
Damped Plate	116.11	116.11	0.0119	0.0122
Case A	114.93	114.89	0.0102	0.0108
Case B	111.62	111.53	0.0047	0.0034
Case C	110.48	110.47	0.0037	0.0021
Bare Plate	114.51	114.43	0.0008	0.0007

It can be observed that in both investigation cases, the FDD method provides natural frequencies and damping ratios values very close to that obtained by EMA. The FDD can be used as a strategy for real-time vibration performance assessment of a plate with damaged CLD patches.

6. Conclusion

This paper has addressed the vibration performance assessment of an aluminum plate with damaged constrained layer damping patches. In particular, two test campaigns with three different damage configurations were investigated. For each configuration, EMA and FDD were applied, showing a very good agreement. The results of this experimental campaign have shown that some level of the deterioration of the CLD can improve the vibrational performance of the plate in a desired frequency range. Because of these results, it is suggested to assess the vibration performance of a plate with damaged constrained layer damping patches by acquiring data with an accelerometer to monitor the acceleration peak in the time domain, and using FDD to assess the frequency content, and therefore monitoring resonance shifts.

Acknowledgments

The experiments were performed in the Dynamics, Vibration and Uncertainty Lab (University of Oxford), which has been established thanks to the John Fell Fund (163/029). AC gratefully acknowledges the financial support provided by Balliol College for her Career Development Fellowship.

References

[1] Hambric S A, Sung S H, and Nefske D J 2016 Engineering Vibroacoustic Analysis. (*John Wiley and Sons, UK*).

[2] Moreira R A and Rodrigues J D 2002 The modellisation of Constrained Damping Layer treatments using the Finite Element Method: spatial model and viscoelastic behaviour. *Proc. International Conference on Structural Dynamics Modelling, Test, Analysis, Correlation and Validation (Portugal).*

[3] Moreira R A and Rodrigues J D 2004 Constrained Damping Layer Treatments: Finite Element Modelling. *Journal of Vibration and Control* **10** 575-95.

[4] Avitabile P 2001 Modal Testing: A Practitioner's Guide. (*Wiley*).

[5] Brandt A, 2011 Noise and Vibration Analysis: Signal Analysis and Experimental Procedures, John Wiley and Sons, New York

[6] Brincker R, Ventura C and Andersen P, 2011 Damping Estimation by Frequency Domain Decomposition. *Conference: IMAC2001: A Conference on Structural Dynamics, At Kissimmee, Florida, USA*

[7] Brincker R. Zhang L and Andersen P 2010, Modal identification of output-only system using frequency domain decomposition. *Smart Materials and Structures* **10** 441-445.

[8] 3M Vibration Damping Tapes 434-435-436 datasheet 2013

IOP Conf. Series: Journal of Physics: Conf. Series **1106** (2018) 012035 doi:10.1088/1742-6596/1106/1/012035

Computation of a well–conditioned dynamic stiffness matrix for elastic layers overlying a half–space

Andrew T. Peplow

Department of Life and Environmental Sciences, College of Natural & Health Sciences, Zayed University, Abu Dhabi, UAE.

E-mail: andrew.peplow@zu.ac.ae

Abstract. In the context of range–independent solid media, we propose a well–conditioned dynamic stiffness matrix for an elastic layer sitting over an elastic half–space. This formulation overcomes the well–known problem of numerical ill-conditioning when solving the system of equations for deep-layered strata. The methodology involves the exact solutions of transformed ordinary differential equations in the wavenumber domain, a projection method based on the transformed equations with respect to the depth coordinate. By re–arranging the transformed equations the solutions remain numerically well-conditioned for all layer depths. The inverse transforms are achieved with a numerical quadrature method and the results presented include actual displacement fields in the near-field of the load.

1. Introduction

A two–dimensional model is considered to demonstrate the effect of a harmonic finite strip–load over layered strata, [1]. The results derived by Fourier transform are valid for any frequency and more importantly any depth of layer. In principle, following the well-known traditional methods, [2], we could use displacement and stress–continuity boundary conditions at the bottom of the layer with equations at the ground surface to generate equations for four subsequent unknowns of stress and displacement. However, this direct approach leads to formidable numerical problems. The reason for this work is that if traditional expressions for the characteristic wave functions, such as cosh or sinh, are employed these can have a dramatic effect on the numerical evaluation of solutions. Problems arise due to the cancellation or division of either very small or very large numbers. To overcome this Karasalo [3] derived a well–conditioned propagator matrix for radially symmetric problems. In this work, though, we construct a single stiffness matrix for the physical layer for plane–strain problems which conveniently avoids these difficulties. We therefore deduce a new global dynamic stiffness matrix for functions that do not cause numerical problems. A *scaled* dynamic stiffness matrix for the layer is derived. To do this the vibration components in the wavenumber domain for layer depths and a half–space are considered and arranged into a single matrix formulation. For the forced response, the load is modelled as an infinite strip, so that the problem is plane. The purpose of the present study is to present a computational method which does not suffer numerical evaluation difficulties when predicting vibration transmission, in particular its attenuation on the surface of a deep layer. The usefulness of the method is illustrated by presenting numerical results from two potentially computationally intensive application examples.

IOP Conf. Series: Journal of Physics: Conf. Series **1106** (2018) 012035 doi:10.1088/1742-6596/1106/1/012035

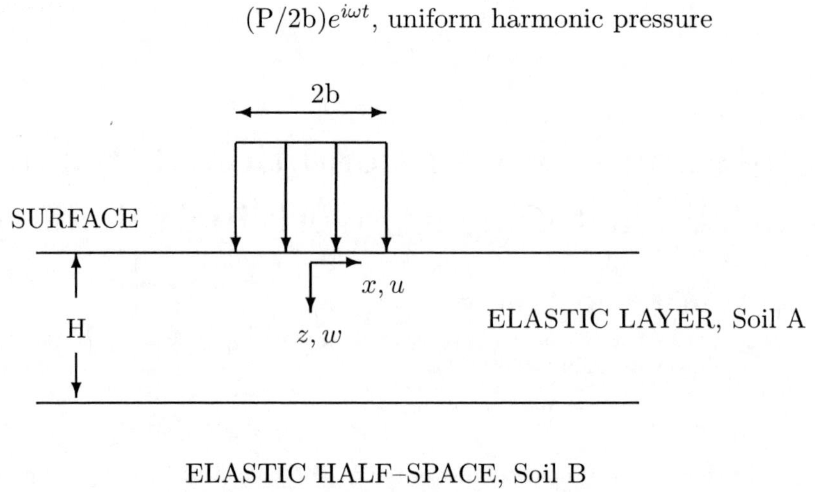

Figure 1. Diagram of the model

Although not developed in this short–communication, solutions to characteristic equations which establish wave propagation parameters can now be determined more efficiently for dynamic [4] or moving–load problems [5], for example, due to the reduced number of equations.

2. Vibration transmission

The model considered is shown in Figure 1. The strip load has a length $2b$, and is aligned with respect to the z-axis. It rests on an homogeneous, isotropic, elastic layer, with material properties E (Young's modulus), ρ (density) and ν (Poisson's ratio). An harmonic vertical load acts uniformly over the strip. The elastic layer of finite depth, H, of homogeneous and isotropic material, overlies a half–space of flexible material. The model is two–dimensional, and the co–ordinate system and parameters are shown in Figure 1.

Much of the analysis necessary for the derivation of the dynamic stiffness matrix has been presented in references [1], [2] and [6], so this will only be briefly summarized. For plane strains the behaviour of the elastic material is described by Navier's elastodynamic equations, [1]. In the absence of body forces, these can be written as:

$$(\lambda+\mu)\frac{\partial\Delta}{\partial x} + \mu\nabla^2 u = \rho\frac{\partial^2 u}{\partial t^2}, \; (\lambda+\mu)\frac{\partial\Delta}{\partial z} + \mu\nabla^2 w = \rho\frac{\partial^2 w}{\partial t^2}, \tag{1}$$

where u, w are the components of the displacement in the x and z directions and Δ is the dilatation, ρ is the density of the material and λ, μ are complex Lamé constants. The stress–strain relations are also included here :

$$\sigma_{zx} = \mu\left(\frac{\partial u}{\partial z}+\frac{\partial w}{\partial x}\right), \sigma_{zz} = \lambda\left(\frac{\partial u}{\partial x}+\frac{\partial w}{\partial z}\right) + 2\mu\frac{\partial w}{\partial z}, \tag{2}$$

The quantities c_1 and c_2 are respectively the P and S wave speeds, given by:

$$c_1^2 = \frac{\lambda+2\mu}{\rho} = \frac{E(1-\nu)}{\rho(1+\nu)(1-2\nu)}, \quad c_2^2 = \frac{\mu}{\rho} = \frac{E}{2\rho(1+\nu)}. \tag{3}$$

Table 1. Material properties

Soil	A	B
$E\ (Pa)$	2.69×10^8	204×10^8
$\rho\ (kgm^{-3})$	1550	2450
ν	0.257	0.179
$c_R\ (ms^{-1})$	242	1706
$c_2\ (ms^{-1})$	263	1879
$c_1\ (ms^{-1})$	459	3005

the boundary conditions for this problem are a uniform strip–load if width $2b$ acting over the surface and at $z = H$ continuity of displacement and traction between the upper and lower–media. Equation (1) may be solved by introducing potentials ϕ and H such that

$$u = \partial\phi/\partial x - \partial H/\partial z \text{ and } w = \partial\phi/\partial z + \partial H/\partial x \tag{4}$$

where, given wavenumbers $k_1 = \omega/c_1$ and $k_2 = \omega/c_2$

$$\alpha_1^2 = \zeta^2 - k_1^2 \text{ and } \alpha_2^2 = \zeta^2 - k_2^2. \tag{5}$$

It is common practice to write solutions to the ensuing ordinary differential equations in terms of cosh and sinh functions. For computational purposes this choice of characteristic functions is not convenient for problems involving spatial domains chosen to be extremely deep. Hence, we propose the general solutions may be written as a *scaled formulation*:

$$\begin{aligned} \overline{\phi} &= A_1\, e^{-\alpha_1 z} + A_2\, e^{\alpha_1 (z-H)}, 0 < z < H \\ \overline{H} &= B_1\, e^{-\alpha_2 z} + B_2\, e^{\alpha_2 (z-H)},\ \ 0 < z < H. \end{aligned} \tag{6}$$

The reasons for choosing the scaled exponential characteristic functions over the hyperbolic functions is clear. Essentially, this choice ensures the characteristic functions do not grow unbounded with depth, H including other expressions involving variables, α_1 or α_2, Eq. (5)
.

Substituting the values $z = 0$ and $z = H$ into the equations (6) yields the first matrix equation

$$\{\overline{u}\} = [C]\,\{A\}\,. \tag{7}$$

where $\overline{u} = [i\overline{w}_0,\, \overline{u}_0,\, i\overline{w}_H,\, \overline{u}_H]^T$ and the 4x4 complex-valued matrix $[C]$ is given by

$$[C] = \begin{bmatrix} -i\alpha_1 & i\alpha_1 e^{-\alpha_1 H} & -\zeta & -\zeta e^{-\alpha_2 H} \\ i\zeta & i\zeta e^{-\alpha_1 H} & \alpha_2 & -\alpha_2 e^{-\alpha_2 H} \\ -i\alpha_1 e^{-\alpha_1 H} & i\alpha_1 & -\zeta e^{-\alpha_2 H} & -\zeta \\ i\zeta e^{-\alpha_1 H} & i\zeta & \alpha_2 e^{-\alpha_2 H} & -\alpha_2 \end{bmatrix}. \tag{8}$$

Now, further developing the system of equations from the stress equations (2),

$$\{\overline{\sigma}\} = [S]\,\{A\}\,, \tag{9}$$

$$[S] = \begin{bmatrix} -i\alpha_1^2(\lambda+2\mu) + i\lambda\zeta^2 & \left(-i\alpha_1^2(\lambda+2\mu) + i\lambda\zeta^2\right) g_1 & -2\mu\zeta\alpha_2 & 2\mu\zeta\alpha_2\zeta\, g_2 \\ 2i\mu\zeta\alpha_1 & -2i\mu\zeta\alpha_1\zeta\, g_1 & \mu\left(\alpha_2^2+\zeta^2\right) & \mu\left(\alpha_2^2+\zeta^2\right) g_2 \\ i\left(\alpha_1^2(\lambda+2\mu) - \lambda\zeta^2\right) g_1 & i\left(\alpha_1^2(\lambda+2\mu) - \lambda\zeta^2\right) & 2\mu\zeta\alpha_2\, g_2 & -2\mu\zeta\alpha_2 \\ -2i\mu\zeta\alpha_1\, g_1 & 2i\mu\zeta\alpha_1 & -\mu\left(\alpha_2^2+\zeta^2\right) g_2 & -\mu\left(\alpha_2^2+\zeta^2\right) \end{bmatrix} \tag{10}$$

IOP Conf. Series: Journal of Physics: Conf. Series **1106** (2018) 012035 doi:10.1088/1742-6596/1106/1/012035

and $g_i = e^{-\alpha_i H}, i = 1, 2$ and $\overline{\sigma} = [-i\overline{\sigma}_0,\ -\overline{\tau}_0,\ i\overline{\sigma}_H,\ \overline{\tau}_H]^T$. We now combine equations (7) and (9), to arrive at a single matrix expression which expresses the displacements and stresses at the surface and the interface in the wavenumber–domain:

$$[T]\,\{\overline{u}\} = \{\overline{\sigma}\}\,. \tag{11}$$

where $[T] = [S][C]^{-1}$. The algebraically complicated unstructured matrix $[T]$ is given in the Appendix. Specifically $[T]$ is the dynamic stiffness matrix for a single elastic layer valid for any depth $H > 0$. To include the half–space, we utilize the matrix equations presented in [6] which leads to a matrix system

$$[P] = \frac{1}{D}\begin{bmatrix} (\lambda+2\mu)\alpha_2 k_1^2 & 2\mu\zeta(\alpha_1\alpha_2-\zeta^2)+(\lambda+2\mu)\zeta k_1^2 \\ 2\mu\zeta(\alpha_1\alpha_2-\zeta^2)+\mu\zeta k_2^2 & \mu\alpha_1 k_2^2 \end{bmatrix}\begin{bmatrix} i\overline{w}_H \\ \overline{u}_H \end{bmatrix} = \begin{bmatrix} i\overline{\sigma}_H \\ \overline{\tau}_H \end{bmatrix} \tag{12}$$

where $D = 1/(\alpha_1\alpha_2 - \zeta^2)$ and it is understood the soil parameters are related to the half–space below the upper–layer. Equations can now be combined to give a single matrix equation for an elastic layer over an elastic halfspace, involving the *scaled* stiffness matrix for the elastic layer $[T_{ij}]$ and the matrix for the halfspace $[P_{ij}]$. The general matrix form for any global domain becomes a 4x4 complex valued matrix

$$[T]_G = \left[\begin{array}{cc:cc} T & T & T & T \\ T & T & T & T \\ \hdashline T & T & T\text{-}P & T\text{-}P \\ T & T & T\text{-}P & T\text{-}P \end{array}\right] \tag{13}$$

It is straightforward to generalise this technique to n elastic layers over a half–space where the size of the dynamic stiffness matrix will become a single complex–valued $2(n+1)$ x $2(n+1)$ matrix.

3. Problems in numerical evaluation of stiffness matrix

Generally, for non–dimensional wavenumbers $k_s h \geq 13$, where k_s is the shear wavenumber, the conventional approach "breaks down". That is, for depths greater than around two shear wavelengths, $h \geq 2\lambda_s$ a numerical bottleneck problem arises when solving the linear system of algebraic equations. Equally for high frequency computations can become ill–conditioned. Note that we cannot show results where "bottleneck" occurs or is about to occur as matrix elements can become unbounded so solutions are not presentable.

The choice of projected method permits a stable numerical evaluation for entries in the stiffness matrix for all soil types, frequencies and layer–depths. This avoids numerical round–off errors especially division of large numbers by small numbers and the subtraction of very large numbers. For example, due to round–off errors, it can be shown that for large $zeta_1, \zeta_2 > 0$ the relative error corresponding to the evaluation of the subtraction of two functions, such as $|\cosh(\zeta_1) - \cosh(\zeta_2)|$ can grow like $\exp(|\zeta|)$, $\zeta_1 < \zeta < \zeta_2$ where the difference $|\zeta_1 - \zeta_2|$ is small. In themselves the numerical evaluation of hyperbolic functions cosh and sinh can also be problematic for large values of their arguments.

4. Numerical results

Soil characteristics values used are presented in Table 1, that is two sets of soil characteristics which represent different conditions for a single frequency, 64 Hz on a strip of width $2b = 1.5$ m subjected to a uniform unit load $P = 1$ N.

Not shown here but functions in wavenumber domain which determine the vibration response for elastic layers over half–space with the same material have been computed. The absolute errors between various cases from shallow to very deep layers were all negligible. Nonetheless the results presented, calculated using the Clenshaw–Curtis numerical quadrature method, are related to the ground parameters shown in Table 1.

5. Free vibration

In Figure 2 is shown the variation of wavenumber with frequency for the first six natural modes. Included in the figure shows the variation of wavenumber for a thick layer of material, soil B, which overlies the half–space of softer material, soil A. Figures 3(a) and (b) compare results

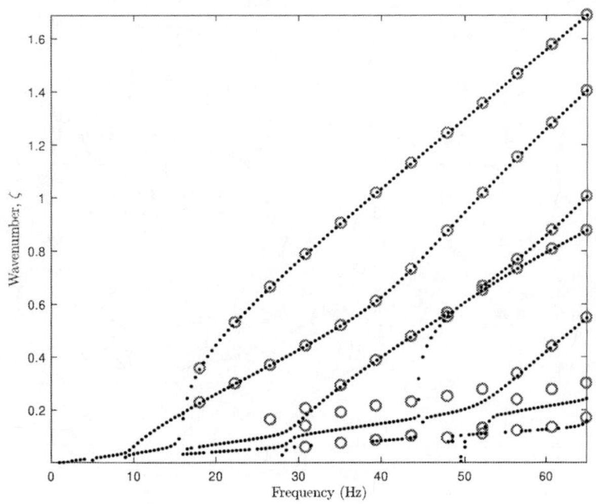

Figure 2. Variation of the wavenumber with frequency of the first six propagating modes for the 7.0m layer over a half–space (dots) and the same layer over thick 9.0m hard strata.

from a simple half–space model [6], to the scaled layered models for 2.0m, 10.0m and 100.0m layers over half–spaces. Especially comparing results between the elastic layer of depth 100.0m compared to the half–space model it is clear that the scaled model approach allows analysis for very deep layers.

6. Conclusions

A two–dimensional model has been developed for investigating the propagation of surface vibration over arbitrary depth elastic–layers. The model consists of an elastic, isotropic and homogeneous layer which overlies a half–space. A well-conditioned dynamic stiffness matrix has been developed for this model, which is derived by projecting the characteristic functions onto the end–points in the depth dimension.

Given the general validity of this formulation for dynamic stiffness matrices many new problems may be modelled, where plane–strain conditions apply, are within easy reach. This method may also be easily developed to take into account sub–layers of different material within the strata and extended to three-dimensional problems..

6.1. Acknowledgements

This work has been carried out under Zayed University RIF Grant (R16054) during a research stay at Aalborg University, Denmark during the summer of 2017.

7. References

[1] Ewing W M, Jardetzky W S and Press F 1957 *Elastic waves in layered media* vol 104 (McGraw–Hill, New York)
[2] Jones D V and Petyt M 1993 *Journal of Sound and Vibration* **161** 1–18
[3] Karasalo I 1994 *Journal of Sound and Vibration* **172** 671–688

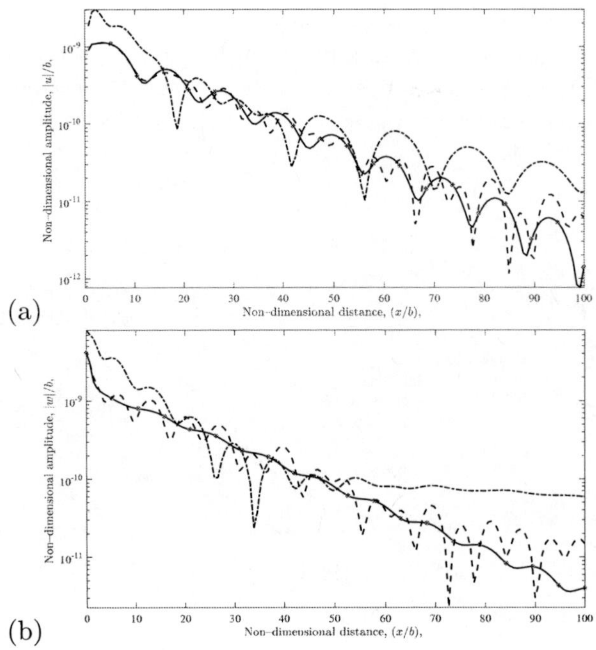

Figure 3. Absolute values of transformed components horizontal and vertical displacement, plotted against non–dimensional distance, elastic layer (soil A_1) overlying a half–space model (soil B_1), $H = 2.0m$ (dash–dot) and $H = 10.0m$ (dashed) $H = 100.0m$ (solid) at frequency $64Hz$. Red-dots represent displacements calculated using the half–space model, [6].

[4] Liu C L and Ai Z Y 2017 *International Journal for Numerical and Analytical Methods in Geomechanics* **41** 1711–1723
[5] Sun H, Cai Y and Xu C 2009 *International Journal for Numerical and Analytical Methods in Geomechanics* **33** 529–550
[6] Jones D V and Petyt M 1991 *Journal of Sound and Vibration* **147** 155–166

IOP Conf. Series: Journal of Physics: Conf. Series **1106** (2018) 012035 doi:10.1088/1742-6596/1106/1/012035

Appendix

We follow the notation that the element in the ith row and jth column is denoted T_{ij}. The matrix $[T]$ is however not symmetric but we have $T_{31} = T_{13}$, $T_{32} = -T_{14}$, $T_{33} = T_{11}$,$T_{34} = -T_{12}$,$T_{41} = -T_{23}$,$T_{42} = T_{24}$, $T_{43} = -T_{21}$ and $T_{44} = T_{22}$. The remaining elements of the matrices are as follows:

$$\begin{aligned}
T_{11} &= -\alpha_2 (\lambda + 2\mu) k_1^2 \left(\left(\zeta^2 - \alpha_1\alpha_2\right)^2 \left(e^{-2(\alpha_1+\alpha_2)H} - 1\right) + \right. \\
&\quad \left. \left(\zeta^2 + \alpha_1\alpha_2\right)^2 \left(e^{-2\alpha_1 H} + e^{-2\alpha_2 H}\right)\right) /D, \qquad (14)\\
T_{12} &= \zeta \left\{ \left(\zeta^2 - \alpha_1\alpha_2\right) \left((\lambda + 2\mu)\alpha_1^2 - 2\mu\alpha_1\alpha_2 - \lambda\zeta^2\right) \left(e^{-(\alpha_1+\alpha_2)2H} + 1\right) + \right. \\
&\quad 4\alpha_1\alpha_2 \left((\lambda + 2\mu)\alpha_1^2 - (\lambda - 2\mu)\zeta^2\right) e^{-(\alpha_1+\alpha_2)H} - \\
&\quad \left. \left(\zeta^2 + \alpha_1\alpha_2\right) \left((\lambda + 2\mu)\alpha_1^2 + 2\alpha_1\alpha_2 - \lambda\zeta^2\right) \left(e^{-2\alpha_1 2H} + e^{-2\alpha_2 H}\right)\right\} /D, \qquad (15)\\
T_{21} &= -\mu\zeta \left\{ \left(\zeta^2 - \alpha_1\alpha_2\right) \left(\zeta^2 + \alpha_2^2 - 2\alpha_1\alpha_2\right) \left(e^{-2(\alpha_1+\alpha_2)H} + 1\right) + \right. \\
&\quad 4\alpha_1\alpha_2 \left(\alpha_2^2 + 3\zeta^2\right) e^{-(\alpha_1+\alpha_2)H} + \\
&\quad \left. \left(\zeta^2 + \alpha_1\alpha_2\right) \left(\zeta^2 + 2\alpha_1\alpha_2 + \alpha_2^2\right) \left(e^{-2\alpha_1 H} + e^{-2\alpha_2 H}\right)\right\} /D, \qquad (16)\\
T_{22} &= \mu\alpha_1 k_2^2 \left\{ \left(\zeta^2 + \alpha_1\alpha_2\right) \left(e^{-2\alpha_1 H} - e^{-2\alpha_2 H}\right) - \left(\zeta^2 - \alpha_1\alpha_2\right) \left(e^{-2(\alpha_1+\alpha_2)H} - 1\right)\right\} /D, \qquad (17)\\
T_{13} &= 2\alpha_2 k_1^2 (\lambda + 2\mu) \left\{\zeta^2 e^{-\alpha_2 H} \left(e^{-2\alpha_1 H} - 1\right) + \alpha_1\alpha_2 e^{-\alpha_1 H} \left(1 - e^{-2\alpha_2 H}\right)\right\} /D, \qquad (18)\\
T_{14} &= -2\alpha_1\alpha_2\zeta k_1^2 (\lambda + 2\mu) \left\{e^{-\alpha_1 H} \left(e^{-2\alpha_2 H} + 1\right) - e^{-\alpha_2 H} \left(e^{-2\alpha_1 H} + 1\right)\right\} /D, \qquad (19)\\
T_{23} &= 2\mu\zeta\alpha_1\alpha_2 k_2^2 \left\{e^{-\alpha_1 H} \left(e^{-2\alpha_2 H} + 1\right) - e^{-\alpha_2 H} \left(e^{-2\alpha_1 H} + 1\right)\right\} /D, \qquad (20)\\
T_{24} &= -2\mu\alpha_1 k_2^2 \left\{\zeta^2 e^{-\alpha_1 H} \left(1 - e^{-2\alpha_2 H}\right) - \alpha_1\alpha_2 e^{-\alpha_2 H} \left(1 - e^{-2\alpha_1 H}\right)\right\} /D. \qquad (21)
\end{aligned}$$

Here

$$\begin{aligned}
D &= \left(\zeta^2 - \alpha_1\alpha_2\right)^2 (\exp(-2(\alpha_1 + \alpha_2)H) + 1) - \\
&\quad \left(\zeta^2 + \alpha_1\alpha_2\right)^2 (\exp(-2\alpha_1 H) + \exp(-2\alpha_2 H)) + \\
&\quad 8\zeta^2\alpha_1\alpha_2 \exp(-(\alpha_1 + \alpha_2)H) \qquad (22)
\end{aligned}$$